NEUROSCIENCE RESEARCH PROGRESS

ENCYCLOPEDIA OF NEUROSCIENCE RESEARCH

VOLUME 3

NEUROSCIENCE RESEARCH PROGRESS

Additional books in this series can be found on Nova's website
under the Series tab.

Additional e-books in this series can be found on Nova's website
under the e-books tab.

NEUROSCIENCE RESEARCH PROGRESS

ENCYCLOPEDIA OF NEUROSCIENCE RESEARCH

VOLUME 3

EILEEN J. SAMPSON
AND
DONALD R. GLEVINS
EDITORS

Nova Science Publishers, Inc.
New York

Copyright © 2012 by Nova Science Publishers, Inc.

All rights reserved. No part of this book may be reproduced, stored in a retrieval system or transmitted in any form or by any means: electronic, electrostatic, magnetic, tape, mechanical photocopying, recording or otherwise without the written permission of the Publisher.

For permission to use material from this book please contact us:
Telephone 631-231-7269; Fax 631-231-8175
Web Site: http://www.novapublishers.com

NOTICE TO THE READER

The Publisher has taken reasonable care in the preparation of this book, but makes no expressed or implied warranty of any kind and assumes no responsibility for any errors or omissions. No liability is assumed for incidental or consequential damages in connection with or arising out of information contained in this book. The Publisher shall not be liable for any special, consequential, or exemplary damages resulting, in whole or in part, from the readers' use of, or reliance upon, this material. Any parts of this book based on government reports are so indicated and copyright is claimed for those parts to the extent applicable to compilations of such works.

Independent verification should be sought for any data, advice or recommendations contained in this book. In addition, no responsibility is assumed by the publisher for any injury and/or damage to persons or property arising from any methods, products, instructions, ideas or otherwise contained in this publication.

This publication is designed to provide accurate and authoritative information with regard to the subject matter covered herein. It is sold with the clear understanding that the Publisher is not engaged in rendering legal or any other professional services. If legal or any other expert assistance is required, the services of a competent person should be sought. FROM A DECLARATION OF PARTICIPANTS JOINTLY ADOPTED BY A COMMITTEE OF THE AMERICAN BAR ASSOCIATION AND A COMMITTEE OF PUBLISHERS.

Additional color graphics may be available in the e-book version of this book.

Library of Congress Cataloging-in-Publication Data

Encyclopedia of neuroscience research / editors, Eileen J. Sampson and Donald R. Glevins.
  p. cm.
Includes index.
ISBN 978-1-61324-861-4 (hardcover)
1. Neurosciences--Research. I. Sampson, Eileen J. II. Glevins, Donald R.
RC337.E53 2011
616.80072--dc23
                            2011017613

Published by Nova Science Publishers, Inc. † New York

Contents

VOLUME 1

Preface		ix
Chapter I	Prefrontal Morphology, Neurobiology and Clinical Manifestations of Schizophrenia *Tomáš Kašpárek*	1
Chapter II	Prefrontal Cholinergic Receptors: Their Role in the Pathology of Schizophrenia *M. Udawela, E. Scarr and B. Dean*	47
Chapter III	Participation of the Prefrontal Cortex in the Processing of Sexual and Maternal Incentives *Marisela Hernández González and Miguel Angel Guevara*	73
Chapter IV	From Conflict to Problem Solution: The Role of the Medial Prefrontal Cortex in the Learning of Memory-Guided and Context-Adequate Behavioral Strategies for Problem-Solving in Gerbils *Holger Stark*	115
Chapter V	The Orbitofrontal Cortex and Emotional Decision-Making: The Neglected Role of Anxiety *Sabine Windmann and Martina Kirsch*	141
Chapter VI	PEEF: Premotor Ear-Eye Field. A New Vista of Area 8B *Bon Leopoldo, Marco Lanzilotto and Cristina Lucchetti*	153
Chapter VII	Prefrontal Cortex: Its Roles in Cognitive Impairment in Parkinson's Disease Revealed by PET *Qing Wangab, Kelly A. Newella, Peter T. H. Wongd and Ying Luc*	171

Chapter VIII	Noradrenergic Actions in Prefrontal Cortex: Relevance to AD/HD Amy F. T. Arnsten	177
Chapter IX	Prefrontal Cortex: Brodmann and Cajal Revisited Guy N. Elston and Laurence J. Garey	199
Chapter X	Developmental Characteristics in Category Generation Reflects Differential Prefrontal Cortex Maturation Julio Cesar Flores Lázaro and Feggy Ostrosky-Solís1	215
Chapter XI	Common Questions and Answers to Deep Brain Stimulation Surgery Fernando Seijo, Marco Alvarez-Vega1, Beatriz Lozano, Fernando Fernández-González, Elena Santamarta and Antonio Saíz	227
Chapter XII	Deep Brain Stimulation and Cortical Stimulation Methods: A Commentary on Established Applications and Expected Developments Damianos E. Sakas and Ioannis G. Panourias	255
Chapter XIII	Cortical Stimulation versus Deep Brain Stimulation in Neurological and Psychiatric Disorders: Current State and Future Prospects Damianos E. Sakas and Ioannis G. Panourias	271
Chapter XIV	Invasive Cortical Stimulation for Parkinson's Disease and Movement Disorders B. Cioni, A. R. Bentivoglio, C. De Simone, A. Fasano, C. Piano, D. Policicchio, V. Perotti and M. Meglio	303
Chapter XV	Deep Brain Stimulation in Epilepsy: Experimental and Clinical Data M. Langlois, S. Saillet, B. Feddersen, L. Minotti, L. Vercueil, S Chabardès, O. David, A. Depaulis P. Kahane and C. Deransart	319
Chapter XVI	Psychosurgery of Obsessive-Compulsive Disorder: A New Indication for Deep Brain Stimulation? Dominique Guehl, Abdelhamid Benazzouz, Bernard Bioulac and Pierre Burbaud,Emmanuel Cuny and Alain Rougier, Jean Tignol and Bruno Aouizerate	339
Chapter XVII	Deep Brain Stimulation in Adult and Pediatric Dystonia Laura Cif, Simone Hemm, Nathalie Vayssiere and Philippe Coubes	353

VOLUME 2

Chapter XVIII	Deep Brain Stimulation of the Subthalamus: Neuropsychological Effects *Rita Moretti, Paola Torre, Rodolfo M. Antonello and Antonio Baval*	365
Chapter XIX	Subthalamic High-Frequency Deep Brain Stimulation Evaluated by Positron Emission Tomography in a Porcine Parkinson Model *Mette S. Nielsen, Flemming Andersen, Paul Cumming, Arne Møller, Albert Gjedde, Jens. C. Sørensen and Carsten R. Bjarkam1*	381
Chapter XX	Current and Future Perspectives on Vagus Nerve Stimulation in Treatment-Resistant Depression *Bernardo Dell'Osso, Giulia Camuri, Lucio Oldani and A. Carlo Altamura*	395
Chapter XXI	Cognitive Aspects in Idiopathic Epilepsy *Sherifa A. Hamed*	407
Chapter XXII	Cognitive Impairment in Children with ADHD: Developing a Novel Standardised Single Case Design Approach to Assessing Stimulant Medication Response *Catherine Mollica, Paul Maruff and Alasdair Vance*	443
Chapter XXIII	Novel Therapies for Alzheimer's Disease: Potentially Disease Modifying Drugs *Daniela Galimberti, Chiara Fenoglio and Elio Scarpini*	475
Chapter XXIV	Cognitive Interventions to Improve Prefrontal Functions *Yoshiyuki Tachibana, Yuko Akitsuki and Ryuta Kawashima*	499
Chapter XXV	Insights from Proteomics into Mild Cognitive Impairment, Likely the Earliest Stage of Alzheimer's Disease *Renã A. Sowell and D. Allan Butterfield*	521
Chapter XXVI	Animal Models for Cerebrovascular Impairment and its Relevance in Vascular Dementia *Veronica Lifshitz and Dan Frenkel*	541
Chapter XXVII	The Critical Role of Cognitive Function in the Effective Self-administration of Inhaler Therapy *S. C. Allen*	561
Chapter XXVIII	Foetal Alcohol Spectrum Disorders: The 21st Century Intellectual Disability *Teresa Whitehurst*	573

Chapter XXIX	Where There are no Tests: A Systematic Approach to Test Adaptation *Penny Holding, Amina Abubakar and Patricia Kitsao Wekulo*	**587**
Chapter XXX	Paid Personal Assistance for Older Adults with Cognitive Impairment Living at Home: Current Concerns and Challenges for the Future *Claudio Bilotta, Luigi Bergamaschini, Paola Nicolini and Carlo Vergani*	**599**
Chapter XXXI	Neurotoxicity, Autism, and Cognitive Impairment *Rebecca Cicha, Brett Holfeld and F. R. Ferraro*	**609**

VOLUME 3

Chapter XXXII	Molecular Mechanisms Involved in the Pathogenesis of Huntington Disease *Claudia Perandones, Martín Radrizzani and Federico Eduardo Micheli*	**617**
Chapter XXXIII	Huntingtin Interacting Proteins: Involvement in Diverse Molecular Functions, Biological Processes and Pathways *Nitai P. Bhattacharyya, Moumita Datta, Manisha Banerjee, Srijit Das and Saikat Mukhopadhyay*	**655**
Chapter XXXIV	DNA Repair and Huntington's Disease *Fabio Coppedè*	**677**
Chapter XXXV	Putting Together Evidence and Experience: Best Care in Huntington's Disease *Zinzi Paola1, Jacopini Gioia1, Frontali Marina and Anna Rita Bentivoglio*	**691**
Chapter XXXVI	Oral Health Care for the Individual with Huntington's Disease *Robert Rada*	**705**
Chapter XXXVII	Suicidal Ideation and Behaviour in Huntington's Disease *Tarja-Brita Robins Wahlin*	**717**
Chapter XXXVIII	Making Reproductive Decisions in the Face of a Late-Onset Genetic Disorder, Huntington's Disease: An Evaluation of Naturalistic Decision-Making Initiatives *Claudia Downing*	**745**
Chapter XXXIX	The Control of Adult Neurogenesis by the Microenvironment and How This May be Altered in Huntington's Disease *Wendy Phillips and Roger A. Barker*	**799**
Index		**857**

In: Encyclopedia of Neuroscience Research
Editors: Eileen J. Sampson and Donald R. Glevins
ISBN 978-1-61324-861-4
© 2012 Nova Science Publishers, Inc.

Chapter XXXII

Molecular Mechanisms Involved in the Pathogenesis of Huntington Disease

Claudia Perandones[1], Martín Radrizzani[2] and Federico Eduardo Micheli[3]

[1]Movement Disorders and Parkinson´s Disease Program, Hospital de Clínicas, José de San Martín, University of Buenos Aires, National Administration of Laboratories and Institutes of Health, ANLIS, Dr. Carlos G. Malbrán, Buenos Aires, Argentina

[2]Neuro and Molecular Cytogenetics Laboratory, School of Science and Technology, National University of San Martín. (CONICET)

[3]Department of Neurology, Movement Disorders and Parkinson´s Disease Program, Hospital de Clínicas, José de San Martín, University of Buenos Aires

Abstract

Huntington's disease (HD) is an autosomal-dominant, progressive neurodegenerative disorder with a distinct phenotype, including chorea, incoordination, cognitive decline, and behavioral difficulties. The underlying genetic defect responsible for the disease is the expansion of a CAG repeat in the gene coding for the HD protein, huntingtin (htt). This CAG repeat is an unstable triplet repeat DNA sequence, and its length inversely correlates with the age at onset of the disease. Expanded CAG repeats have been found in 8 other inherited neurodegenerative diseases. Despite its widespread distribution, mutant htt causes selective neurodegeneration, which occurs mainly and most prominently in the striatum and deeper layers of the cortex.

Remarkable progress has been made since the discovery of HD gene in 1993. Animal models to study the disease process, unraveling the expression and function of wild-type and mutant huntingtin (htt) proteins in the central and peripheral nervous systems, and understanding expanded CAG repeat containing mutant htt protein interactions with CNS proteins in the disease process have been developed . Consequently, it has been concluded that htt may cause toxicity via a wide range of different mechanisms. The primary consequence of the mutation is to confer a toxic gain of function on the mutant htt protein, and this may be modified by certain normal activities that are impaired by the mutation. It is likely that the toxicity of mutant htt is only revealed after a series of cleavage events leading to the production of N-terminal huntingtin fragment(s) containing the expanded polyglutamine tract. Although aggregation of the mutant protein is the hallmark of the disease, the role of aggregation is complex and the arguments for protective roles of inclusions are discussed. HD progression has been found to involve several pathomechanisms, including expanded CAG repeat protein interaction with other CNS proteins, transcriptional dysregulation, calcium dyshomeostasis, defective mitochondrial bioenergetics and abnormal vesicle trafficking. Notably, not all the effects of mutant htt are cell-autonomous, and it is possible for abnormalities in neighboring neurons and glia to also have an impact on connected cells.

The present review focuses on HD, outlining the effects of mutant htt in the nucleus and cytoplasm as well as the role of cell-cell interactions in the HD pathology. The widespread expression and localization of mutant htt and its interactions with a variety of proteins suggest that mutant htt is engaged in multiple pathogenic pathways. A better understanding of these mechanisms will lead to the development of more effective therapeutic targets.

1. Huntington's Disease

1.1. Introduction: Unstable Expanding Repeats as a Novel Cause of Disease

Prior to this decade, the adage that goes "like begets like" fitted nicely with the dogma of human genetics where the genetic material was considered stable upon transmission, and in the rare instance of mutation, the new variant itself was stably inherited. Although there had been clear instances when this did not hold true, the human genetics community did not embrace the notion of "dynamic mutations".

Clinical researchers had long suspected something was amiss when carefully examining families with dominant Myotonic Dystrophy, where they found that off-springs of affected individuals often had a more severe form of the disease.

The term "anticipation" was used to define this progressive increase in expressivity of the identical mutation over a number of generations.

Since anticipation did not easily fit with the biological dogma of the genetics of that era, the concept of anticipation was summarily dismissed as an ascertainment bias. Many years later, the recognition of increasing penetrance through subsequent generations in Fragile X syndrome, later known as the Sherman Paradox, resembled the genetic anticipation of Myotonic Dystrophy.

While the observation in Fragile X syndrome -the Sherman paradox- was more easily accepted by the scientific community, anticipation could not be readily explained at a molecular level.

Even when proven true, the phenomenon represents one of the darkest black holes in molecular genetics.

In 1991, the genes responsible for the Fragile X syndrome and Spinobulbar Muscular Atrophy were found to contain unstable, expanded trinucleotide repeats. The following year,

Myotonic Dystrophy was also found to be the result of an expanded trinucleotide repeat. These findings were soon followed by a remarkable number of neurological disorders, each sharing the mutational mechanism of the unstable expansion of a repeat, most often a triplet in form.

Soon, the behavior of these repeats in affected families clearly revealed a pattern where the increasing penetrance (Sherman Paradox) or increasing expressivity (Anticipation), through subsequent generations, correlated with increasing lengths of the triplet repeat. Thus, a biological basis of anticipation was defined -the dynamic mutation.

1.2. Common Features of Diseases due to Unstable Expanding Repeats

Trinucleotide repeat expansions now account for at least 16 neurological disorders ranging from childhood developmental disorders such as X-linked mental retardation syndromes to late onset neurodegenerative disorders such as Huntington disease and inherited ataxias. The variability in repeat size underlies the broad spectrum of phenotypes seen in each of these disorders. The repeats show somatic and germline instability. Successive generations of families affected by such dynamic mutations experience anticipation or earlier onset age and more rapid disease progression owing to intergenerational repeat instability. For example, the onset of Myotonic Dystrophy ranges from adulthood in parents and grandparents to birth in children and grandchildren, depending on the size of the repeat.

Although unstable trinucleotides are the most common repeats to cause neurological disorders, others such as tetranucleotides and pentanucleotides expand to cause type 2 Myotonic Dystrophy and Spinocerebellar Ataxia type 10, respectively.

Several developmental and neuromuscular disorders are caused by either an insertion or a duplication of a small trinucleotide repeat (GCG)n typically encoding alanine. Examples of such disorders include hand-foot-genital syndrome, synpolydactyly, oculopharyngeal muscular dystrophy, and the X-linked mental retardation caused by mutation in the Aristaless-related homeobox gene (Brais et al., 1998, Stromme et al., 2002, Utsch et al., 2002).

These disorders differ from the classic trinucleotide expansion disorders because the expansions are small and are not as dynamic.

Tables 1, 2 and 3 provide a concise description of these disorders, the mutational basis, gene product, and key clinical features. The disorders have been divided according to pathogenic mechanism, i.e., whether mutations cause loss of function of the protein or gain of function of the RNA or protein, or cause a yet-to-be-determined mechanism.

The expanded triplet repeats can be found in transcribed RNA destined to be untranslated (either 5´ or 3´ such as in Fragile X Syndrome or Myotonic Dystrophy, respectively), spliced out intronic sequence (such as in Friedreich´s Ataxia), or coding exonic sequence (such as the dominant ataxias) (Figure 1). In general, the noncoding repeats are able to undergo massive expansions from a normal number of 6-40 triplets to an abnormal range of many hundreds or thousands of repeats. This leads either to transcriptional suppression, as in the case of Fragile X Syndrome, or to abnormal RNA processing, limiting the number of mature cytoplasmic messages, as in the case of Myotonic Dystrophy. In contrast, the coding expansions undergo much more modest expansions from a normal range of approximately 10-35 repeats to an abnormal range of approximately 40 to 90 triplets. Since these are CAG repeats coding for polyglutamine tracts, constraints of the individual protein structures significantly modify this range. This is most apparent in SCA 6 where the largest normal allele contains 18 repeats, whereas the smallest abnormal allele contains 21 repeats. In all coding expansions, the mechanism(s), while still poorly understood, appear to reflect a gain or change of function of the abnormal protein, eventually leading to neurodegeneration.

Common features of the diseases caused by expansion of an unstable CAG repeat within a gene include:

- They are all late-onset neurodegenerative diseases, and except for Kennedy disease, they are all dominantly inherited.
- The expanded allele is transcribed and translated.
- The trinucleotide repeat encodes a polyglutamine tract in the protein.
- There is a critical threshold repeat size, below which the repeat is nonpathogenic and above which it causes disease.
- The larger the repeat above the threshold, the earlier the onset age (on average; predictions cannot be made for individual patients, but there is a clear statistical correlation).
- The CAG repeat disease genes identified so far are widely expressed and encode proteins of unknown function. When the polyglutamine tract exceeds the threshold length, there is protein aggregation, forming an inclusion body that apparently kills the cell.
- The different clinical features of each disease reflect killing of different cells, presumably because of interactions with other cell-specific proteins.
- Neuronal cell death caused by protein aggregates is a common thread in the pathology of CAG repeat diseases, Alzheimer disease, Parkinson disease and prion diseases; the mechanism and their general significance remain to be discovered.

1.3. Clinical and Genetic Aspects of HD

In 1872, Dr. George Huntington published a seminal work titled "On Chorea". This manuscript described the features of an illness observed among the members of three families residing in East Hampton, New York. Dr. Huntington called the disease "hereditary chorea" and his observation on the mode of inheritance is remarkable because the basic tenants of genetics, as defined by Gregor Mendel, had not been recognized at this time. Dr. Huntington described the transmission of the disease in the following terms: "When either or both of the parents have shown manifestations of the disease... one or more of the offspring almost invariably suffer from the disease, if they live to adult age. But if by any chance these children go through life without it, the thread is broken and the grandchildren and great-grand children of the original shakers may rest assured that they are free from the disease". This description provides a functional description for the autosomal dominant mode of transmission with complete penetrance characteristic of HD.

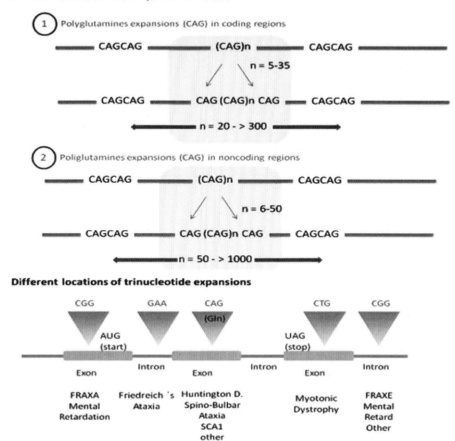
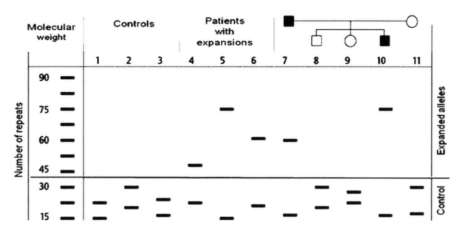

Panel 1. Polyglutamine (CAG) expansions in coding regions
Panel 2. Polyglutamine (CAG) expansions in noncoding regions
Panel 3. Different locations of trinucleotide expansions
Panel 4. Laboratory diagnosis of trinucleotide repeat disorders

Figure 1. Location of trinucleotide expansions in humans.

Table 1. Unstable repeat disorders caused by loss-of-function, RNA mediated, or unknown mechanisms

A. Loss of function mechanism

Disease	MIM number	Repeat Unit	Gene Product	Normal Repeat	Expanded Repeat	Main clinical features
FRAXA	309550	(CGC)n	FMRP	6-60	>200 (full mutation)	Mental retardation, macroorchidism, connective tissue defects, behavioral abnormalities
FRAXE	309548	(CCG)n	FMR2	4-39	200-900	Mental retardartion
FRDA	229300	(GAA)n	Frataxin	6-32	200-1700	Sensory ataxia, cardiomyopathy, diabetes.

B. RNA-mediated pathogenesis

Disease	MIM number	Repeat Unit	Gene Product	Normal Repeat	Expanded Repeat	Main clinical features
DM1	160900	(CTG)n	DMPK	5-37	50-10.000	Myotonia, weakness, cardiac conduction defects, insulin resistance, cataracts, testicular atrophy, and mental retardation in congenital form
DM2	602668	(CCTG)n	ZNF9	10-26	75-11.000	Similar to a DM1 but no congenital form
FXTAS	309550	(CGG)n	FMR1 RNA	6-60	60-200	Ataxia, tremor, Parkinsonism and dementia.

A. Disorders produced by a loss of function mechanism
B. Disorders produced by RNA-mediated pathogenesis

Dr. Huntington´s treatise displayed such clarity that it was reprinted in prominent neurology texts and termed "Huntington´s Chorea". More recently, the observation that some cases, and particularly those with a young onset, may present without chorea, has led to the widespread adoption of the designation Huntington´s Disease (HD) to refer to this affliction.

HD is a midlife-onset disease which strikes at a mean age of 40 years, but onset may vary from 4 to 80 years of age (Harper, 1992).

It is a progressive neurodegenerative disorder with primary neuropathological involvement in the basal ganglia. HD is invariably fatal, without periods of remission, and the course from onset to death averages 15 to 17 years (Myers, Sax et al. 1991). The disease is characterized mainly by involuntary choreiform movements, cognitive impairment, and personality disorder featuring depression, anger and temper outbursts, as well as cognitive disorders.

Table 2. Unstable repeat disorders caused by unknown pathogenic mechanisms

Disease	MIM number	Repeat Unit	Gene Product	Normal Repeat	Expanded Repeat	Main clinical features
SCA 8	608768	(CTG)n	SCA 8 RNA	16-34	> 74	Ataxia, slurred speech, nistagmus.
SCA 10	603516	(ATTCT)n		10-20	500-4500	Ataxia, tremor, dementia.
SCA 12	604326	(CAG)n	PPP2R2B	7-45	55-78	Ataxia and seizures
EHL2	606438	(CTG)n	Junctophilin	7-28	66-78	Similar to HD.

Table 3. Polyglutamine disorders caused by gain-of-function mechanisms

Disease	MIM number	Gene Product	Normal Repeat	Expanded Repeat	Main clinical features
Huntington disease	143100	Huntingtin	6-34	36-121	Chorea, dystonia, cognitive deficits, psychiatric disorders
SCA 1	164400	Ataxin 1	6-44	39-82	Ataxia, slurred speech, spasticity, cognitive impairments.
SCA 2	183090	Ataxin 2	15-24	32-200	Ataxia, polyneuropathy, decreased reflexes, infantile variant with retinopathy.
SCA 3	109150	Ataxin 3	13-36	61-84	Ataxia, Parkinsonism, spasticity.
SCA 6	183086	CACNA1a	4-19	10-33	Ataxia, disarthria, nistagmus, tremors.
SCA7	164500	Ataxin 7	4-35	37-306	Ataxia, blindness, cardiac failure in infantile form.
SCA 17	607136	TBP	25-42	47-63	Ataxia, cognitive decline, seizures, and psychiatric disorders.
SBMA	313200	Androgen Receptor	9-36	38-62	Motor weakness, swallowing difficulties, gynecomastia, decreased fertility.
DRPLA	125370	Atrophin	7-34	49-88	Ataxia, seizures, choreoathetosis, dementia.

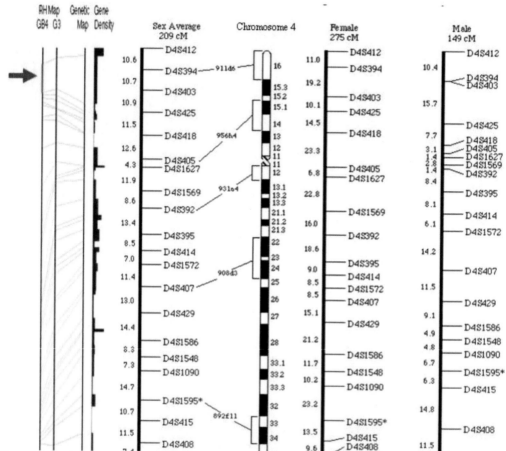

The cytogenetic bands of the short arm of chromosome 4 (4p) are depicted with the location of the HD defect in 4p16.3 from the genetic linkage and physical mapping according to what studies indicate. The map of 4p16.3 provides an expanded view of this telomeric cytogenetic band to illustrate the location of the defect relative to polymorphic DNA markers used to sequentially narrow the minimal HD genetic region by recombination analysis, linkage disequilibrium, and haplotype analysis.

Figure 2. Huntingtin gene localization in Chromosome 4.

The HD gene was genetically linked to an anonymous marker located in 4p16.3 (Figure 2) (Gusella, Wexler et al. 1983) and, following a 10-year long molecular genetic search, was finally discovered to be an unstable CAG trinucleotide repeat (The Huntington's Disease Collaborative Research Group, 1993). The mutation extends a polymorphic stretch of CAG codons in the first exon of the HD gene, lengthening a segment of polyglutamine near the amino terminus of the HD protein which has been named huntingtin. The mutation which leads to the expression of HD is an approximate doubling in the number of the triplet repeats from a normal number of about 18 to an expanded number of 40 or more.

1.4. Epidemiology of HD

The prevalence of HD is estimated at 5 to 10 affected persons per 100,000 among individuals of European descent but is less common among other ethnic groups (Harper, 1992; Al-Jader, Harper et al., 2001; Bertram and Tanzi, 2005).

Indeed, the prevalence of HD among the native African population is so low as to make an accurate estimate difficult, and its prevalence among Japanese and Asian populations is approximately one-tenth of that observed in Caucasians. Since for each affected individual, there is an estimate of two living carriers who are still too young to manifest symptoms, the prevalence of HD gene carriers in the general Caucasian population may be estimated at 15 to 30 per 100,000, and there is an equal number of siblings at risk, who are the ones who have not inherited the HD defect. Thus, approximately 1 in every 2,500 births is an individual born at risk for the disease, making HD the most common of the CAG repeat expansion diseases.

Therefore, genotype-phenotype correlations in HD will most likely enable to describe uncommon events such as new mutations or alleles with reduced penetrance, which may be too rare to be observed in other triplet repeat diseases.

1.5. Clinical Correlates in HD

A. Chorea and motor impairment

The term chorea comes from a Greek word that means "dance". Chorea refers to a prominent feature of the gait disturbance characteristic of HD and the "dance-like" quality of movement. Choreic movements are involuntary, slow, and random and may involve any voluntary muscle group. Thus the involuntary aspect of the movement disorder in HD features random twitching of both upper and lower extremities, either on the left and/or right sides.

These movements may begin as mild myoclonic-like movements involving the fingers, toes, or facial muscles early in the course of the disease, and evolve to large movements of the limbs and trunk as the illness advances. Progressively, movements may interfere with gait, speech, chewing, and swallowing, and other aspects of motor function associated with daily life activities. However, the extent of the movement impairment may vary substantially from person to person, and in some cases there may be little evidence of involuntary movement. In the late stages of HD, chorea may lessen and may be replaced by rigidity and dystonia.

In addition to the involuntary choreic movement, there is a slowing of voluntary movement. This bradykinesia may be seen in tasks involving rapid alternating movements, eye movements, and finger and tongue tapping. Some studies have proposed that the progression of involuntary movements may be more uniform across different stages of HD and may therefore represent a more consistently progressing feature for assessment of disease course than the severity of chorea.

The relation of motor impairment to CAG repeat size has not been widely studied. Andrew et al. (Andrew, Goldberg et al.1993) found no association between the repeat length and the form of clinical presentation (e.g., motor, mood, or cognitive disturbance as the initial feature).

B. Age at onset

The age at onset of HD is highly variable. The earliest cases have been reported with onset at 2 to 3 years of age, while the latest onsets approach 80 years of age or even older. The average is at age 40 and is not gender related. Approximately 7% of all cases present before the age of 20 and this group of individuals has traditionally been termed "juvenile" onset HD.

The impact of the HD repeat upon phenotypic expression has been heavily researched. The relationship between age at onset and repeat size is unequivocal and very strong but it is not linear; most investigators have noted that the best fit is a model in which the repeat predicts the log10 of onset age (Ranen, Stine, et al. 1995). For the 220 HD cases depicted in Figure 3, the correlation between repeat size and log of onset age is r = - 0.84 and accounts for about 70% of the variance in onset age.

Kremer et al (Kremer, Squitieri, et al. 1993), noted that for late-onset cases, with onset at age 50 or older, the correlation of onset with repeat size is reduced (r=-0.29).

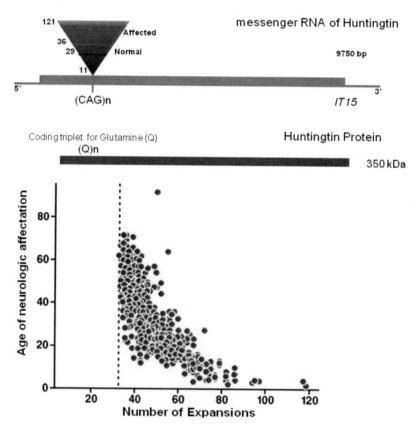

Relationship between the length of the expanded HD CAG repeat and age at neurological onset of disease. The CAG repeat sizes for 220 individuals affected with HD diagnosed through the New England Huntington´s Disease Research Center are presented in relation to the age at onset of motor impairment. Repeat size is strongly related to age at onset. Onset age before age 20 is usually associated with a repeat size of more than 60 CAG units. Among persons with adult onset, the range in onset age for a given repeat is large and may vary by 35 years or more.

Figure 3. Relationship between length of CAG repeat and age at neurologic onset of disease.

It has been noted that there is a +/- 18-year, 95% confidence interval around the estimated onset age for a given repeat size. Although one group recently suggested that the confidence interval may be smaller than this estimate, all published reports confirm that the range in onset age for a given repeat size exceeds 30 years among persons with mid and late-life disease.

Thus, for more than 90% of presymptomatic HD cases, the relationship of repeat size xxx is not strong enough to predict the onset age.

C. Rate of disease progression

The course of disease in HD is remarkably variable, but there have been only a few studies of factors related to progression (Myers, Sax, et al. 1991). There is evidence that age at onset is related to rate of progression, and it has been suggested that a single mechanism influences both the age at onset and the rate of disease progression. Although not strong enough to predict onset pre-symptomatically, the HD repeat size appears to be the primary determinant of onset age and thus it is reasonable to expect that the repeat size is related to the rate of disease progression as mediated either by its effect upon onset age or by a direct effect upon disease progression.

A study of a set of monozygotic twins raised apart suggested that progression is largely determined by genetic factors and that the most likely determinant is the trinucleotide repeat size (Sudarsky, Myers, et al., 1983).

Many authors have reported a significant association between repeat size and rate of disease progression.

Another factor is the gender of the affected parent, with offspring of affected fathers having a more rapid progression than offspring of affected mothers. In addition, lower body mass index early in the disease is related to more rapid disease progression.

While the gender of the affected parent and paternal transmission is now known to be related to the transmission of an expanded repeat, it is not known whether the repeat size influences weight loss or lower body mass index (Marder, Zhao H, et al. 2009).

These studies are relevant when considering trials for therapeutic interventions, which will need to consider the HD repeat in the randomization of study participants.

D. Neuropsychology

1. Cognitive function in HD

While language functioning remains relatively intact, the most striking cognitive deficits involve areas of executive system functioning (e.g., strategies in problem solving and cognitive flexibility), short-term memory, and visuospatial performance. The memory disorder is characterized by inconsistent retrieval of information. In the early stages of the disease, recognition memory for verbal information is robust, while spontaneous recall of information may be more frequently impaired.

Early in the disease process, the cognitive deficits are relatively focal. A global, progressive subcortical dementia does not evolve until the disease advances significantly. Although this dementia is significant and debilitating in the late and final stages of the disease, it is qualitatively different from the cortical dementia observed in Alzheimer's disease (Duff, Beglinger, et al. 2009).

2. CAG relationship to cognition

Andrew et al. (Andrew, Goldberg, et al. 1993) found no association between the CAG repeat length and the form of clinical presentation (e.g. motor, mood, or cognitive disturbance as the initial feature), and others have also failed to find a relationship between the repeat length and either the type of symptom onset or disease progression or the type of psychiatric involvement. However, these observations seem to be valid only for early or mid-life age onset cases of the disease. A retrospective observational study of thirty-four individuals with late onset of Huntington Disease (HD) (onset range 60-79 years) performed by Lipe and Bird (Lipe and Bird, 2009) showed that when CAG trinucleotide expansion size ranged from 38-44 repeats, motor problems were the initial symptoms at onset.

Many papers have tried to establish correlations between cognitive functions and different parameters of progression or severity in HD.

Peinemann et al. (Peinemann, Sabine, et al. 2005) conducted a study whose purpose was to clarify whether cognitive dysfunction in early stages of HD is correlated with loco-regional structural changes in 3D-MRI. HD patients demonstrated robust regional decreases of gray matter volumes in the caudate and the putamen. Executive dysfunction was significantly correlated with the areas of highest significant differences out of voxel-based morphometry (VBM), results that were located bilaterally in the caudate. Moreover, subgroup analyses revealed marked insular atrophy in HD patients who performed worse in the single executive tasks.

Two aspects were most remarkable in this correlational study: (i) striatal atrophy in HD patients in early stages plays an important role not only in impaired motor control but also in executive dysfunction, and (ii) extrastriatal cortical areas, i.e., the insular lobe, seem to be involved in executive dysfunction as assessed by neuropsychological tests requiring planning and problem solving, stimulus response selectivity and concept formation.

E. Clinical correlates of mitochondrial function in HD muscle

Mitochondrial function analyses on HD postmortem brain demonstrated a severe defect of mitochondrial complex II/III activity in the striatum. Studies have also shown abnormal bioenergy in the HD brain in vivo, with elevated lactate and reduced N-acetyl aspartate within the striatum. Htt expression is not restricted to the CNS and has been documented in skeletal muscle and myoblast cultures from the R6/2 transgenic mouse model of HD.

Many studies reported a defect of mitochondrial ATP synthesis capacity in HD skeletal muscle in vivo, which inversely correlated with the length of the CAG repeat --i.e. the longer the repeat, the worse the defect (Lodi R, Schapira AH, Manners D, et al. 2000). Furthermore, gene expression profiles of R6/2 HD transgenic mice and HD skeletal muscle have also demonstrated changes, with a transition from fast to slow-twitch fibers in the mutants, suggesting an adaptive response to muscle that is less dependent on oxidative phosphorylation (Strand, Aragaki, et al. 2005).

The study by Turner et al. (Turner, Cooper, et al. 2007) demonstrated that individual HD skeletal muscle complex II/III activity correlated with clinical disease progression and with the cognitive score on the Unified HD Rating Scale (UHDRS), both measures of functional status, as well as with the age of the patient, disease duration and the repeats/ years score.

The reduction in the complex II/III function showed a strong correlation with disease progression on several clinical scales independently of an age-related phenomenon.

It seems possible that a progressive reduction in complex II/III activity occurs in HD patient muscle, and this could cause a sequence of biochemical events that may include damage, as suggested for HD striatum.

Mitochondrial dysfunction, and specifically complex II/III activity, might be involved in the sequence of HD pathogenesis and this can be correlated with worsening of several clinical parameters. Prolonged treatment with free radical scavengers may ameliorate the progressive loss of complex II/III activity and modify disease progression.

1.6. Isolation of the HD Mutation

The clinical and genetic characteristics of HD, including its distinct symptoms, midlife onset, unambiguous mode of inheritance, high penetrance, and prevalence in the general population, made it an ideal disease to approach using strategies based on genetic linkage. In 1983, the HD defect was mapped in the vicinity of an anonymous polymorphic DNA marker, D4S10, following the inheritance of restriction fragment length polymorphisms in two large HD kindred, of American and Venezuelan descent (Gusella, Wexler, et al. 1983).

Genetic linkage to D4S10 established that virtually all cases of HD were likely to arise from defects in the same gene (Conneally, Haines et al., 1989). These early linkage studies also demonstrated that D4S10 was ~4cM (~4% recombination) from HD, providing the basis for the molecular diagnosis of HD in asymptomatic at-risk individuals able to participate in a genetic linkage test.

Linkage to D4S10 revealed the complete phenotypic dominance of HD. Affected homozygotes, possessing two copies of the HD defect, were discovered to be clinically indistinguishable from their heterozygous siblings with one copy of the mutation (Wexler, Young et al., 1987; Myers, Leavitt et al., 1989).

A consortium of investigators employed a combination of genetic and physical mapping techniques to confine the mutation to a segment of ~ 2 million base pairs of 4p16.3. Simultaneously, they generated genetic and physical maps of the region, developing overlapping clone sets, and refining the defect's location using recombination analysis in disease pedigrees and linkage disequilibrium between the disorder and genetic markers.

The location of the HD gene (Figure 2) was ultimately pinpointed by an analysis of haplotypes formed by multiallele markers from across the region, which revealed that although ~two thirds of disease chromosomes had unrelated haplotypes and were likely to have independent origins, the remaining ~one third shared a small (275 kb) region between D4S95 y D4S180. This provided strong evidence that this subset of disease chromosomes was likely to descend from a common ancestral chromosome (MacDonald, Novelleto et al., 1992).

Analysis of candidate genes from this subregion finally led to the discovery of a polymorphic CAG repeat in the 5' end of interesting transcript 15 (IT15), which was expanded and unstable on disease chromosomes.

1.7. The HD Trinucleotide Repeat Mutation

The pure stretch of CAG trinucleotide repeat that is expanded on disease chromosomes is located near the 5' end of a novel 4p16.3 gene (Figures 1 and 7), immediately adjacent to a broken array of CAG/CCG codons containing a mildly polymorphic (6 a 12 repeats), stably transmitted stretch of CCG triplets.

The results of initial genotype-phenotype correlation studies quickly revealed that most normal chromosomes and the majority ($\geq$ 90 %) of disease chromosomes possess 7 CCG repeats (Rubinztein, Barton, et al., 1993), whereas the adjacent array of pure CAG repeats was found to be highly polymorphic. Normal chromosomes possessed from 6 to 34 CAG repeats that are inherited in a Mendelian fashion, and HD chromosomes from 39 to ~ 86 units that are inherited in a strikingly non-Mendelian manner. Moreover, rare alleles with 36-39 repeats were found in exceptional individuals, the unaffected elderly relatives of sporadic de novo cases of the disease.

These initial findings made it possible to offer a direct DNA test -Polymerase Chain Reaction (PCR)-, which determines the amplification of the HD CAG repeat length, to many individuals at risk in a wide variety of testing situations. The findings of subsequent genotype-phenotype studies performed in research and clinical diagnostic laboratories around the world have provided critical data for the generation of verifiable hypotheses for the action mechanism of the HD mutation. These results provided evidence, for example, that the length of CAG repeat in the expanded range shows a marked correlation with the age at onset of the disease, where age at onset decreases as the CAG repeat size increases (Figure 3: Relationship between length of CAG repeat and age at neurologic onset of disease).

They also pointed out that any given expanded CAG repeat is associated with a broad range of onset ages, highlighting the relevance of other factors in determining age at onset in any single individual.

Currently four CAG repeat size intervals are recognized as associated with varying disease risk in HD (Figure 4: Repeat size and HD risk).

These ranges have been defined by the U.S. H.D. Genetic Testing Group (USHDGTG) and are derived from information gleaned from more than 1000 HD tests. Nevertheless, the disease-related risk corresponding to the 29 to 39 repeats range is often based upon fewer than 10 observations at each repeat size, and thus, these risk estimates can be expected to change with additional information.

a) Normal: repeat sizes up to 26 units

Individuals with repeats in this range do not develop HD, nor has there been any confirmed instances of a child inheriting HD from a parent with a repeat in this range.

b) Nonpenetrant with paternal meiotic instability: repeats of 27 to 35 units.

Repeats in this range are rare and represent approximately 1% of expanded alleles seen in HD testing protocols. There have been no confirmed reports of persons with repeats in this range expressing HD. There are, however, confirmed cases of paternally transmitted meiotic instability such that descendants of fathers with repeats in this range are known to have inherited an expanded allele in the clinical range. There are no reported cases of maternally transmitted meiotic instability in this range, producing offspring with repeats in the clinical range (Semaka, Creighton, et al. 2006).

c) Reduced penetrance with meiotic instability: Repeats of 36 to 39 units

Repeats in this range are rare and represent approximately 1 to 2% of expanded alleles seen in HD testing programs. Some individuals with repeats in this range develop HD and others live into their 90s without evidence of the disease (Myers, MacDonald et al., 1993; McNeil, Novelleto et al., 1997). There is clear evidence that penetrance increases with increasing allele size in this range. Penetrance has been roughly estimated at 25% for 36 repeats, 50% at 37 repeats, 75% at 38 repeats, and 90 % at 39 repeats based upon USHDGTG data.

d) HD: Repeats of 40 units or larger

It is currently believed that all persons with repeats in the range of 40 or more will eventually develop HD.

However, some individuals with repeats at the low end of this range are reported to exhibit initial symptoms at ages older than common life expectancy and, thus, there may be some reduced penetrance among carriers of 40 or 41 repeats (Brinkman, Mezei et al., 1997).

Kenney et al (Kenney, Powel, et al., 2007) reported an autopsy proven HD case with 29 trinucleotide repeats; the patient presented HD classic phenomenology. Many supplementary studies were carried out and a great number of potential HD phenocopies with overlapping features were excluded.

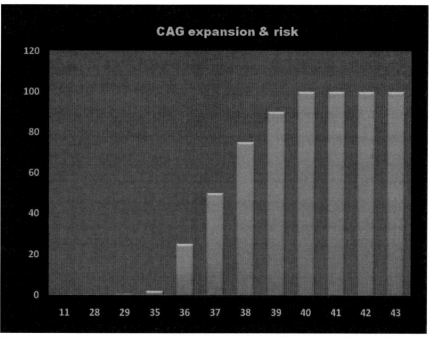

Four CAG repeat size intervals are recognized as associated with varying disease risk in HD. These ranges have been defined by the US HD Genetic Testing Group (USHDGTG) and are derived from information gleaned from more than 1000 HD tests.

Figure 4. Repeat size and HD risk

However, despite having performed additional histochemical staining, the authors failed to detect the presence of htt in the ubiquitin-positive neuronal intranuclear inclusions.

Although the detection of the htt stain is not absolutely necessary to establish the neuropathological diagnosis of HD (Jean Paul Vonsattel, MD, personal communication), arguably in this case it would be relevant to confirm the presence or absence of the "mutated" protein in the nucleus, as the subcellular localization of mutant htt is critical for the pathology of HD.

In agreement with this report, many recent experimental findings are shifting the focus from the polyglutamine expanded tract to other domains of the protein for toxicity (Zoghbi and Orr 2009). In other words, the notion that toxicity is simply a result of an expanded toxic polyglutamine tract that might escape the cellular degradation and quality control machinery becomes less likely.

The expansions of the polyglutamine tracts, although relevant, are not the only molecular mechanisms to cause toxicity. Many studies in spinobulbar muscular atrophy and HD indicate that protein domains outside the polyglutamine tract play a significant role in the selective neurotoxicity observed in these diseases (Graham, Deng et al., 2006).

The possibility of having HD with 29 CAG repeats would entail significant changes in genetic counselling of HD families, HD therapeutic trials and the current understanding of the molecular pathogenesis of the disease.

1.8. Neuropathologic Studies in HD

The brain regions initially affected in HD are the neocortex and the striatum, with the most extensive atrophy being demonstrated in these regions. The neurodegenerative process in the striatum occurs first in the caudate nucleus and then in the putamen (Vonsattel and DiFiglia, 1998; Gutekunst, Norflus et al., 2000). Pathological changes in the striatum develop in a seemingly topographically order, spreading along the caudal-rostral, dorsal-ventral and medial-lateral axes (Figure 5: Selective vulnerability of the striatum in HD).

The severity of post-mortem neuropathological features and cell loss has been graded using the scale described by Vonsattel, Meyers et al. 1985. HD pathology is divided into grades 0-4, designated in ascending order of severity. Neuropathological grade correlates with the severity of motor symptoms, mental state and CAG repeat length (Rosenblatt, Abbott, et al. 2003). Within affected brain regions, certain neurons degenerate whereas others remain relatively spared. In the striatum, medium spiny neurons, which receive dense innervation from the cortex, are preferentially affected, with the sparser interneurons, lacking cortical innervations being relatively spared (MacDonald and Halliday, 2002; Heinsen, Rub et al., 1999). Most of these medium spiny neurons become dysfunctional and are eventually lost in HD, thus disrupting the circuitry of the basal ganglia.

Substantial degeneration also occurs in the cerebral cortex of patients with HD. Volume reductions are seen in associative, frontal, temporal, parietal and primary somatosensory cortices. Large cortical pyramidal neurons, including those that project directly to the striatum, are selectively lost. The disruption of corticostriatal circuitry in HD is known to contribute to the motor symptoms of the disease and may also play a key role in cognitive symptoms.

Patients with HD have a heterogeneous presentation regarding their clinical condition and progression, which may well reflect individual variations in patterns of neurodegeneration. For example, even in early to mid stages of the disease, structural magnetic resonance imaging (MRI) studies reveal that regions other than the neocortex and striatum, notably hippocampus, globus pallidus, amygdale and cerebellum, can also show morphometric changes (Rosas, Koroshetz, et al., 2003). Pathology in the neocortex and hippocampus may be of particular significance with regard to the cognitive deficits seen in HD.

Reactive microglia has been detected in the cortex and striatum of post-mortem human HD brains. The density of activated microglia correlated with the degree of neuronal loss and Vonsattel grading. Markers of increased inflammatory gliosis were also identified post mortem in the putamen and frontal cortex, and together, these findings suggest that an inflammatory process may possibly be contributing to HD, although it is not known whether this has any impact on disease onset or progression.

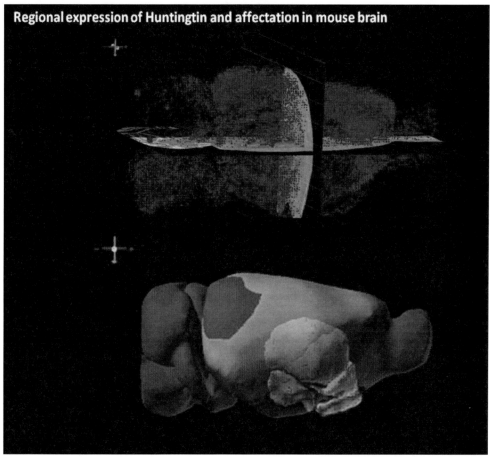

The genetic expression of Huntingtin can be observed through a tridimensional approach using "in situ hybridizations" performed with mRNAs in murine cerebral cortex slices (http://www.brain-map.org, Allen Institute for Brain Science. © 2004). The results of these hybridization assays are in concordance with the ones previously published in "Nature" (Site, Lein, E.S. et al., Genome-wide atlas of gene expression in the adult mouse brain, Nature 445: 168-176 (2007), doi:10.1038/nature05453). The bottom figure is a tridimensional reconstruction that highlights the regions showing selective vulnerability in HD, namely the striatum and cerebral cortex.

Figure 5. Selective vulnerability of the striatum in HD.

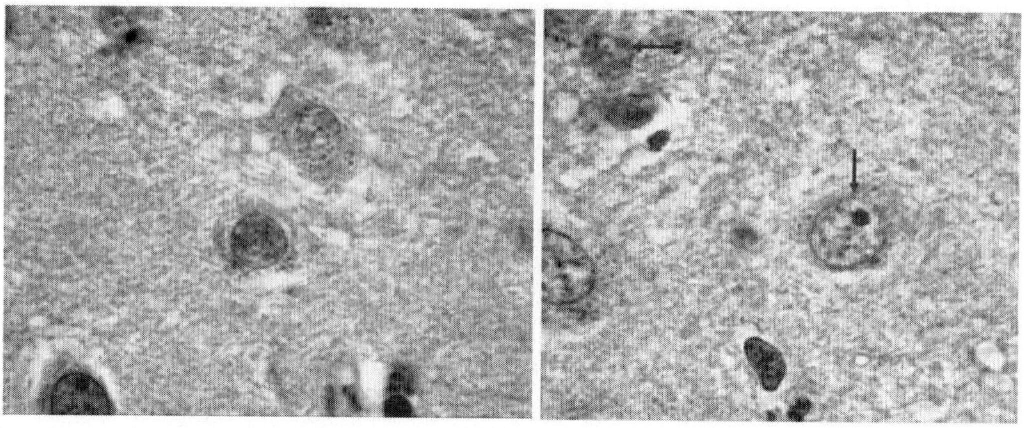

Control   **Affected**

One characteristic neuropathological feature, present in both human HD patients and transgenic HD mice, is the presence of protein aggregates, also known as neuronal inclusions, which are formed from the aggregation of expanded polyglutamine-cointaining Htt fragments as well as numerous other proteins.

Figure 6. Neuritic aggregates and perinuclear inclusions in HD

One characteristic neuropathological feature, present in both human HD patients and transgenic HD mice, is the presence of protein aggregates, also known as neuronal inclusions (Davies, Turmaine et al., 1997) (Figure 6: Neuritic aggregates and perinuclear inclusions in HD), which are formed from the aggregation of expanded polyglutamine-cointaining htt fragments, as well as numerous other proteins.

The aggregates appear in many different tissues, including the cortex and the striatum, and can be seen in either the cell nucleus or the cytoplasm.

These protein aggregates appear to disrupt cytoskeletal structures closely associated with protein and mRNA trafficking to synapses, possibly leading to secondary deficits in synaptic signaling.

However, abundant evidence suggests that aggregates may represent a neuroprotective cellular strategy (Menalled, Sison et al., 2003; Arrasate, Mitra et al., 2004), with polyglutamine containing fragments possibly exerting their toxic effects on cellular processes such as gene expression and protein transport prior to the formation of the large aggregates visible using light microscopy (Bodner, Outeiro et al., 2006; Bodner, Housman et al., 2006).

2. Molecular Pathogenesis of Huntington's Disease

2.1. HD Gene and Gene Product

The genetic defect responsible for HD is an expansion of a CAG repeat in the gene coding for the HD protein. This CAG repeat is an unstable triplet repeat DNA sequence, and its length is inversely correlated with the age at onset of disease, especially in juvenile HD cases, in which the repeat length is often >60 CAG units (Andrew, Goldberg et al., 1993) (Figure 7: A schematic diagram of human Huntingtin).

Expanded CAG repeats have been found in 10 other inherited neurodegenerative diseases, as well, including spinocerebellar ataxia (SCA) and spinobulbar muscular atrophy (SBMA) (Zohbi and Orr, 2000; Gatchel and Zoghbi 2005; Butler and Bates 2006) (see Tables 2 and 3). It is now clear that expansion of this repeat in various genes can cause distinct neurodegenerative pathology leading to different disorders.

The CAG repeat is translated into a polyglutamine (polyQ) domain in the disease proteins. Human htt is a large protein comprising 3144 amino acids. A normal polyQ domain, which in htt begins at amino acid position 18, tipically contains 11-34 glutamine residues in unaffected individuals, but this expands to more than 37 glutamines in HD patients. The length of the polyQ repeat varies among species. For example, mouse htt has 7 glutamines, whereas pufferfish htt contains only 4 (Harjes and Wanker 2003), which suggests that the polyQ domain may not be essential, but that it can regulate protein function. Consistently, deletion of the CAG repeat in the HD gene only results in subtle behavioral and motor phenotypes in mice.

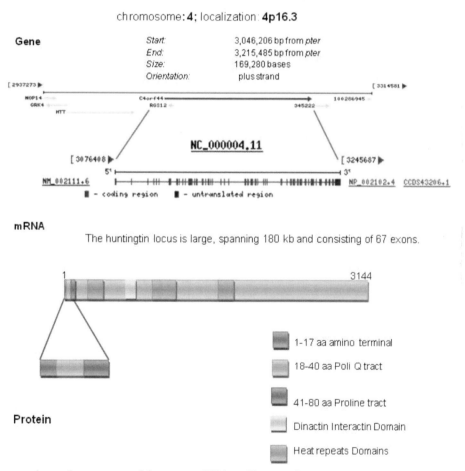

The diagram shows the structure of the gene, mRNA and htt protein.

Figure 7. A schematic diagram of human huntingtin

Htt is ubiquitously expressed in the brain and body and distributed in various subcellular regions (Gutenkunst, Levey et al., 1995). Its sequences do not show homology to other proteins of known function. One structural feature of htt, is the presence of HEAT repeats (Takano and Gusella 2002), which are sequences of ~40 amino acids that occur several times within a given protein and are found in a variety of proteins involved in intracellular transport and chromosomal segregation (Neuwald and Hirano 2000).

Several lines of evidence also suggest that htt is involved in intracellular trafficking and various cellular functions. As an example, htt is associated with a number of subcellular organelles (DiFiglia, Sapp et al., 1995; Gutenkunst, Levey, et al., 1995; Sharp, Love et al., 1995; Gutenkunst, Li et al., 1998).

In concordance with that, htt is known to interact with a variety of proteins that can be grouped according to whether they are involved in gene transcription, intracellular signaling, trafficking endocytosis, or metabolism (Harjes and Wanker 2003; Li and Li 2004).

Identification of these htt-interacting proteins suggests that htt may function as a scaffold involved in coordinating sets of proteins for signaling processes and intracellular transport.

The essential role of htt has been established using HD gene knockout mice. In this model, the absence of htt causes cell degeneration and embryonic lethality (Duyao, Auerbach et al., 1995; Nasir, Floresco et al., 1995; Zeitlin, Liu et al., 1995). Conditional knockout mice also show degeneration in adult cells (Dragatsis, Levine et al., 2000). These observations have led to the theory that a loss of htt function may contribute to the neuropathology of HD (Loss of function hypothesis) (Cattaneo, Rigamonti et al, 2001). However, there is more evidence to support the theory wherein mutant htt gains a toxic function (gain of function hypothesis). For example, heterozygous HD knock mice are known to live normally. Further, identification of the HD gene has allowed the generation of various animal models in which mutant htt is expressed in the presence of endogenous normal htt, and these transgenic mice still develop neurological symptoms and die early, even when endogenous normal htt is expressed at the normal levels (Davies, Turmaine et al., 1997; Schilling, Becher et al., 1999).

Moreover, mutant htt can rescue the embryonic lethal phenotype of htt-null mice (Hodgson, Smith et al., 1996), which also suggests the HD mutation can lead to neuronal toxicity, independent of the essential function of htt.

2.2. Correlation between Neuropathology and Pathogenesis in HD

Despite its widespread expression, htt causes selective neurodegeneration, which occurs preferentially and most prominently in the striatum and deep layers of the cortex in the early stages of HD (Vonsattel, Myers et al., 1985). In advanced stages, other brain regions, such as the hippocampus, hypothalamus, cerebellum, amygdala, and some thalamic nuclei, are also affected. Among these other brain regions, the lateral tuberal nucleus of the hypothalamus exhibits severe atrophy (Kremer, Roos et al., 1990).

The neurons that are most severely affected in HD are striatal projection neurons, which send their axons to different brain regions. These are the GABAergic medium sized spiny neurons (MSNs), which constitute 95% of all striatal neurons. MSNs receive abundant glutamatergic input from the cortex and primarily innervate the substantia nigra and globus pallidus. Thus, their preferential loss in HD is thought to be the result of glutamate excitotoxicity. Consistently, there is relative sparing of interneurons that colocalize somatostatin, neuropeptide Y, and NAPDH diaphorase, as well as of cholinergic interneurons and a subclass of GABAergic neurons that contain parvalbumin (Ferrante, Kowall et al., 1985; Graveland, Williams et al., 1985; Vonsattel, Myers et al., 1985). Another important pathological feature in the postmortem brains of HD patients is gliosis (Myers, Vonsattel et al., 1991; Singhrao, Thomas et al., 1998; Sapp, Kegel et al., 2001). Reactive glia or gliosis often occurs in response to neuronal injury. For example, neuronal degeneration is evidenced by a dramatic elevation in the density of large glia (Rajkowska, Selemon et al., 1998). Marked astrogliosis and microgliosis were observed in caudate and internal capsule samples of HD patients, but not in normal brain. In the striatum and cortex, reactive microglia also occurred in all grades of pathology, accumulated with increasing grade, and grew in density in relation to the degree of neuronal loss (Ferrante, Kowall et al., 1985; Graveland, Williams et al., 1985; Vonsattel, Myers et al., 1985). Thus, reactive microglia were considered to be an early response to change in neuropil (Sapp, Kegel et al., 2001). While reactive gliosis does represent an early neuropathological event in HD, glial pathology can also impact neuronal viability. Indeed, gliosis is a pathological feature in several HD mouse models that lack neuronal cell degeneration. These models include transgenic mice expressing N-terminal mutant htt (Ishiguro, Yamada et al., 2001; Yu, Li et al., 2003; Gu, Li et al., 2005) and "knock-in" mice that express full-length mutant htt (Reddy, Williams et al., 1998; Lin, Tallaksen-Greene et al., 2001).

Since the discovery of the HD gene, various antibodies and DNA aptamers to htt have been generated to characterize the distribution of mutant htt. Immunostaining of brains from transgenic mice that express mutant htt have revealed nuclear inclusions (Davies, Turmaine et al., 1997). Similar nuclear inclusions were identified in the brains of HD patients (DiFiglia, Sapp et al., 1997; Gutekunst, Li et al., 1999). Subsequently, the accumulation of expanded polyQ-containing proteins in the nucleus and nuclear inclusions were found to be common pathological features of other polyglutamine diseases (Zoghbi and Orr 2000; Gatchel and Zoghbi 2005; Butler and Bates 2006). The role of these nuclear inclusions in HD remains controversial, since their formation is correlated with disease progression, but is not associated with neuronal degeneration (Gutekunst, Li et al., 1999; Kuemmerle, Gutekunst et al., 1999; Slow, Graham et al., 2005). Moreover, several studies have shown that htt inclusions are protective against htt toxicity in cultured cells (Saudou, Finkbeiner et al., 1998; Arrasate, Mitra et al., 2004). Despite the controversy surrounding their exact role, htt inclusions reflect protein misfolding caused by an expanded polyQ domain and represent a pathological hallmark for the accumulation of toxic mutant htt. It is also remarkable that normal htt is predominantly localized in the cytoplasm, whereas mutant htt with its expanded polyQ domain accumulates in the nucleus. Hence, nuclear inclusions reflect the aberrant accumulation of mutant htt in the nucleus.

Importantly, HD also features abundant cytoplasmic aggregates localized in neuronal processes (neuropil aggregates) including axons and dendrites (Gutekunst, Li et al., 1999; Li, Li et al., 1999; Schilling, Becher et al., 1999; Li, Li et al., 2000; Li, Li et al., 2001; Menalled, Sison et al., 2003; Tallaksen-Greene, Crouse et al., 2005). In the early stages of the disease, the brains of HD patients contain more dystrophic neurites or neuropil aggregates than nuclear inclusions (DiFiglia, Sapp et al., 1997; Gutekunst, Li et al., 1999).

In addition, the progressive development of neuropil aggregates is strongly correlated with disease progression in transgenic mice (Li and Li 1998; Li, Li et al., 1999; Li, Li et al.,

2000; Li, Li et al., 2001). Moreover, the neuropil aggregates are associated with axonal degeneration in HD mouse brains (Li, Li et al., 2001; Yu, Li et al., 2003).

Taken together, the localization of htt aggregates in the nucleus and neuronal processes reveals that mutant htt elicits toxicity in both the nucleus and cytoplasm.

2.3. Models for the Pathogenesis of HD

A number of mouse models have provided in vivo evidence for the pathology of HD. Several transgenic mice were generated using either the human htt promoter or neuronal promoters. For example, transgenic mice R6/2 express exon 1 htt with 115-150 glutamine repeats (115-150Q) under the control of the human HD gene promoter (Davies, Turmaine et al., 1997). YAC (Yeast Artificial Chromosomes) transgenic mice use the human HD gene promoter to drive the expression of full-length mutant htt (Hodgson, Agopyan et al., 1999; Graham, Deng et al., 2006). N171-82Q transgenic mice express the first 171 amino acids with 82Q under the neuronal prion promoter (Schilling, Becher et al., 1999). These transgenic mice have been widely studied and found to have neurological and behavioral phenotypes similar to those observed in HD patients.

There are also HD repeat knock-in mouse models, which are generated by inserting an expanded repeat into the endogenous mouse HD gene (Shelbourne, Killeen et al., 1999; Wheeler, White et al., 2000; Lin, Tallaksen-Greene et al., 2001; Menalled, Sison et al., 2002). However, most HD mouse models do not show the overt neurodegeneration seen in human HD patients, even though some models display severe neurological symptoms and early death (Davies, Turmaine et al., 1997; Schilling, Becher et al., 1999).

It is possible that the short life span of the mouse does not allow sufficient time for the development of obvious neurodegeneration, although some earlier pathological events do occur.

HD mouse models also suggest that small fragments containing expanded polyQ are more toxic than larger fragments (HD and the toxic fragment hypothesis). This fits the finding that small N-terminal htt fragments are misfolded and form aggregates and inclusions in the brains of HD patients (DiFiglia, Sapp et al., 1997; Gutekunst, Li et al., 1999). It is obvious that proteolysis of htt generates multiple N-terminal htt fragments in HD repeat knock-in mice (Zhou, Cao et al., 2003). A great number of protease cleavage sites, including those for caspase-3, caspase-6, calpain, and unknown aspartic protease, have been found within the first 550 amino acids of htt (Kim, Yi et al., 2001; Gafni and Ellerby 2002; Lunkes, Lindenberg et al., 2002; Wellington, Ellerby et al., 2002; Graham, Deng et al., 2006).

In a recent study, the role of htt cleavage by caspase-3 and caspase-6 in disease was examined by expressing caspase-3 and caspase-6 resistant forms of mutant htt in mice. Controlling for expression levels, Graham, Deng et al., 2006, found that transgenic mice expressing polyglutamine-expanded htt with a mutated caspase-6 cleavage site did not manifest any behavioral deficits or neurodegeneration. In contrast, the caspase-3 resistant form of mutant htt remained fully pathogenic. Several assessments were performed to test the importance of the caspase-6 cleavage site in htt for disease. These included motor phenotypes (rotarod and open field) and neuropathology (brain weight, striatal volumen, and nuclear htt accumulation) of polyglutamine-expanded htt transgenic mice with a mutated cleavage site. Transgenic mice expressing polyglutamine expanded htt with a mutated caspase-6 cleavage site did not manifest behavioral deficits or neurodegeneration, even when the expression level of htt exceeded that in unmutated polyglutamine expanded htt transgenic mice. Also, htt mutated at caspase-6 cleavages sites had a significant delay in its nuclear translocation. Nuclear translocation is an early step in pathogenesis in an HD knock-in mouse model (Wheeler, Gutekunst et al., 2002) and is known to be required for neurotoxicity of other polyglutamine disease proteins.

These results demonstrate that sequences outside the polyglutamine tract are critical for pathogenicity and are consistent with cleavage of htt by caspase-6 as a critical event for pathology in HD. However, the importance of caspase-6 cleavage for HD needs to be confirmed through manipulation of caspase-6 activity in transgenic mice expressing full length mutant htt.

Most studies used transfected proteins to identify cleavage sites, and the nature of toxic N-terminal fragments generated organically in the HD brain is still being explored. It is likely that the proteolysis of full-length htt generates a number of N-terminal htt fragments. The decreased activities of the proteasomes and chaperones, which are responsible for clearing out misfolded and toxic peptides, promote the accumulation of htt fragments in aged neurons. In the meantime, an expanded polyglutamine tract causes them to misfold and aggregate in the nucleus and neuronal processes. The accumulation of mutant htt in the nucleus and neuronal processes therefore suggests that these subcellular regions are the primary sites for mutant htt to elicit its toxicity.

2.4. Nuclear Effect of Mutant Huntingtin

The nuclear inclusions of mutant htt led investigators to study the mechanisms for this phenomenon. Although some immunostaining and nuclear fractionation studies have shown that normal htt is also localized in the nucleus (Hoogeveen, Willemsen et al., 1993; Kegel, Meloni et al., 2002), it is clear that the majority remains in the cytoplasm. Moreover, nuclear htt aggregates can only be recognized by antibodies against the N-terminal region of htt (DiFiglia, Sapp et al., 1997; Gutekunst, Li et al., 1999). Furthermore, isolation of nuclear fractions from HD knock-in mice, which express full-length mutant htt under the endogenous mouse HD gene, provides evidence that multiple N-terminal htt fragments accumulate in the nucleus (Zhou, Cao et al., 2003). The association between nuclear accumulation of mutant htt and disease progression is clear from several HD mouse models. In HD knock-in mouse models, mutant htt accumulates preferentially in the nuclei of striatal neurons and forms more prominent aggregates as disease progresses (Wheeler, White et al., 2000; Lin, Tallaksen-Greene et al., 2001). A progressive phenotype is also associated with the nuclear accumulation of an amino-terminal cleavage fragment in a transgenic mouse model with inducible expression of full-length mutant huntingtin (Tanaka, Igarashi et al., 2006). Targeting mutant htt with nuclear localization sequences to direct mutant htt to the nucleus of mouse brains produces neurological phenotypes (Schilling, Savonenko et al., 2004; Benn, Landles et al., 2005). Moreover, prevention of htt cleavage by mutating caspase-6 site can alleviate neurological phenotypes and delay the nuclear accumulation of mutant htt in AC transgenic mice (Graham, Deng et al., 2006).

Studies of N-terminal htt fragments have failed to find whether these fragments contain nuclear localizations sequences. Thus, N-terminal htt fragments may passively enter the nucleus, but expanded polyQ repeats prevent their export from the nucleus (Cornett, Cao et al., 2005). The presence of mutant htt fragments in the nucleus and various cleavage sites in the N-terminal region of htt (Sun, Savanenin et al., 2001) also support the notion that proteolysis of htt leads to the generation of toxic htt fragments. Consistently, smaller N-terminal htt fragments appear to be more toxic than large-sized fragments in both cultured cells (Hackam, Singaraja et al., 1998) and transgenic animals (Davies, Turmaine et al., 1997; Schilling, Becher et al., 1999; Yu, Li et al., 2003).

The aberrant nuclear accumulation of mutant htt is likely to cause gene transcriptional dysregulations. Indeed, several nuclear transcription factors that bind htt have been found (Sugars and Rubinsztein 2003; Li and Li 2004). Of these, the coactivators cAMP response element-binding protein (CREB) and the specificity protein 1 (Sp1) are particularly important for neuronal function. Deletion of CREB in the brain causes selective neurodegeneration in the hippocampus and striatum (Mantamadiotis, Lemberger et al., 2002). Many neuronal genes that lack a TATA box require Sp1 for their transcription (Myers, Dingledine et al., 1999). Dysregulation of gene expression mediated by CBP (CREB-binding protein) and Sp1 have been found in HD mouse brains (Luthi-Carter, Hanson et al., 2002).

The interactions of mutant htt with transcription factors may occur at various binding sites. Many transcription factors contain a polyQ-rich domain. Since CBP is recruited into aggregates formed by different polyQ proteins, such as the androgen receptor (McCampbell, Taylor et al., 2000), the SCA3 (Chai, Wu et al., 2001), and the Dentatorubral-Pallidoluysian atrophy (DRPLA) proteins (Nucifora, Sasaki et al., 2001), it has been thought that the polyQ domain is the binding site to interact with other polyQ proteins. Supporting this idea, a number of transcription factors containing polyQ or proline –rich domains, including CBP (Nucifora, Sasaki et al., 2001; Steffan, Bodai et al., 2001), TBP (Huang, Faber et al., 1998; Perez, Paulson et al., 1998), and TAF130 (Shimohata, Nakajima et al., 2000), have been found in nuclear polyQ inclusions.

However, subsequent studies showed that the acetyltransferase domain in CBP interacts with htt (Chai, Wu et al., 2001; Steffan, Bodai et al., 2001), which led to the finding that inhibition of histone deacetylase (HDAC) or promotion of histone acetylation ameliorates neurodegeneration in cellular and fly models (Steffan, Bodai et al., 2001) and motor deficits in a mouse model of HD (Hockly, Richon et al., 2003).

These observations prompted a hypothesis whereby the pathogenic process was linked to the state of histone acetylation; specifically, mutant huntingtin induced a state of reduced histone acetylation and as a result altered gene expression. Support for this idea was obtained from the Drosophila HD model expressing an N-terminal fragment of huntingtin with an expanded polyglutamine tract in the eye. Administration of inhibitors of histone deacetylation (HDAC) arrested neurodegeneration and lethality (Steffan, Bodai et al., 2001). In 2002, Hughes et al. reported protective effects of HDAC inhibitors for other polyglutamine disorders, prompting the concept that at least some of the observed effects in polyglutamine-disorders are due to alterations in histone acetylation. This hypothesis has led to several preclinical studies using HDAC inhibitors (Ferrante and Kubilus 2003, Gardian, Browne et al., 2005).

Whether HDAC inhibitors, such as the FDA-approved SAHA, can be used in treating HD and other polyglutamine disorders remains to be proven. The evidence linking histone acetylation to polyglutamine pathogenesis is based, for the most part, on work performed using a fragment of the mutant polyglutamine protein. Thus, the biological relevance of this work depends on the extent to which the pathogenesis will prove to rest on the properties of the polyglutamine tract.

The colocalization of some transcription factors in nuclear polyQ inclusions also led to the idea that recruitment of transcription factors into polyQ inclusions reduces the level of these transcription factors. However, after examining several HD mouse models, researchers were unable to find decreased levels of CBP in symptomatic mouse brains (Yu, Li et al., 2002; Tallaksen-Greene, Crouse et al., 2005). In addition, altered expression of a number of genes was not necessarily associated with the formation of htt aggregates in HD mice (Luthi-Carter, Hanson et al., 2002) and may occur in cell models in the absence of nuclear inclusions (Kita, Carmichael et al., 2002; Sipione, Rigamonti et al., 2002). Thus, it is likely that soluble or misfolded htt may interact with transcription factors to alter transcriptional activity. This idea is further supported by the finding that soluble mutant htt reduces the binding of Sp1 to DNA (Dunah, Jeong et al., 2002; Li, Cheng et al., 2002).

In conclusion, there is ample evidence that mutant htt acts in the nucleus to affect gene transcription (Transcriptional dysregulation hypothesis)

2.5. Cytoplasmic Effect of Mutant Huntingtin

A. Axonal transport in HD

Earlier studies have reported that mutant htt not only increases caspase activity (Ona, Li et al., 1999; Sanchez, Xu et al., 1999; Chen, Ona et al., 2000; Li, Lam et al., 2000) but also affects various signaling pathways (Cepeda, Ariano et al., 2001; Song, Perides et al., 2002; Zeron, Hansson et al., 2002; Tang, Tu et al., 2003). These findings suggest that mutant htt also acts in the cytoplasm affecting cellular functions (Figure 8: Molecular and cellular pathways involved in HD pathogenesis).

Therefore, an intensive search to reveal the interactions between htt and cytoplasm proteins began. By means of the yeast two-hybrid screen and in vivo binding assays, a number of cytoplasmic proteins were found to interact with htt (Sharp, Loev et al., 1995; Harjes and Wanker 2003; Borrell-Pages, Zala et al., 2006). Out of them, htt associated protein 1 (HAP1) and htt-interacting protein 1 (HIP1) have been studied extensively. Both proteins may well be involved in intracellular trafficking. HAP1 binds more tightly to mutant htt than to normal htt (Li, Li et al., 1995) HAP1 also associates with both dynactin p150, which is involved in microtubule-dependent retrograde transport (Engelender, Sharp et al., 1997; Li, Gutekunst et al., 1998), and kinesin light chain 2 (McGuire, Rong et al., 2006), which is involved in anterograde transport. Several studies suggest that HAP1 participates in the trafficking or endocytosis of membrane receptors, including those for epidermal growth factor (Li, Chin et al., 2002), type 1 inositol (1,4,5) –triphosphate receptor (InsP3R1) (Tang, Tu et al., 2003), GABA (Kittler, Thomas et al., 2004), and nerve growth factor (NGF) (Rong, McGuire et al., 2006). Like htt, HAP1 is located at various subcellular sites, including microtubules and synaptic vesicles in axonal terminals (Gutekunst, Li et al., 1998). Mice lacking HAP1 often die at postnatal day 3 (Chan, Nasir et al., 2002; Li, Yu et al., 2003), which is likely due to neuronal degeneration in the hypothalamus (Li, Yu et al., 2003). The hypothalamic function of HAP1 appears to be critical for feeding behavior and metabolism (Sheng, Chang et al., 2006), and its dysfunction may contribute to hypothalamic pathology or degeneration in HD (Kremer, Roos et al., 1990; Li, Yu et al., 2003; Petersen, Gil et al., 2005; Sheng, Chang et al., 2006).

HIP1 is also important for assembly and function of the cytoskeleton and endocytosis (Kalchman, Koide et al., 1997) and binds clathrin and alpha-adaptin subunit AP-2 (Mishra, Agostinelli et al., 2001; Waelter, Scherzinger et al., 2001; Metzler, Li et al., 2003). The interactions of HIP1 with these proteins may constitute a protein complex involved in clathrin-mediated endocytosis. Unlike HAP1, HIP1 binds mutant htt weakly (Kalchman, Koide et al., 1997). This finding suggests that HIP1 requires interaction with htt for normal function, whereas dissociation from mutant htt may impair its function.

Although the interactions of htt with HAP1, HIP1, and other cytoplasmic proteins suggests that htt is involved in intracellular trafficking, more compelling evidence has come from the studies of trafficking function in cells that express mutant htt. Recent studies show that normal Drosophila htt functions in the axonal transport pathway and that polyQ expansion causes soluble htt to recruit more microtubule transporter proteins, thereby reducing the soluble pool of these proteins in axons (Gunawardena, Her et al., 2003). In cultured neurons, htt is involved in HAP1-associated axonal transport of brain-derived neurotrophic factor (BDNF), both anterogradely and retrogradely, which is disrupted by mutant htt (Gauthier, Charrin et al., 2004).

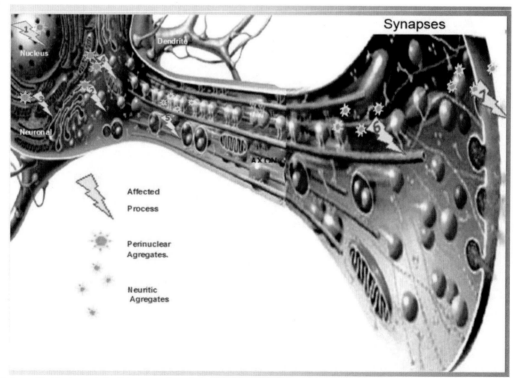

A hypothetical model of proposed molecular and cellular mechanisms involved in the pathogenesis of HD. The expanded polyglutamine in the mutant HD protein is suggested to disrupt key processes relating to cell signaling (4), gene transcription (1), protein trafficking (2, 5, 6), presynaptic and postsynaptic signaling (7), mitochondrial dysfunction (3), and protein-protein interactions.

Figure 8. Molecular and cellular pathways involved in HD pathogenesis

The presence of an expanded polyglutamine tract in huntingtin increases the interaction between huntingtin, HAP1 and p150 Glued, and thereby reduces the interaction of HAP1/p150 Glued with microtubules, which most likely accounts for the decrease in transport in the presence of mutant huntingtin. This mutant huntingtin-induced reduction in BDNF-containing vesicle transport reduces trophic support, causing neurotoxicity, which may contribute to the pathogenesis of HD. Recent studies (Colin, Zala et al., 2008) have demonstrated that Huntingtin is a positive regulatory factor for vesicular transport. Huntingtin is phosphorylated at serine 421 by the kinase Akt, and this event seems to be crucial to control the direction of vesicles in neurons. When phosphorylated, huntingtin recruits kinesin-1 to the dynactin complex on vesicles and microtubules, promoting anterograde transport. Conversely, when huntingtin is not phosphorylated, kinesin-1 detaches and vesicles are more likely to undergo retrograde transport. This also applies to other vesicles suggesting an essential role for huntingtin in the control of vesicular directionality in neurons.

Moreover, there are multiple indications that confirm that phosphorylation of htt at serine 421 (S421) restores its function in axonal transport. Using a strategy involving RNA interference and re-expression of various constructs, Zala, Colin et al. (2008) demonstrated that polyQ (polyglutamine)-htt is unable to promote transport of brain-derived neurotrophic factor (BDNF)-containing vesicles, but polyQ-htt constitutively phosphorylated at S421 is as effective as the wild-type (wt) as concerns transport of these vesicles. The S421 phosphorylated polyQ-htt displays the wt function of inducing BDNF release. Phosphorylation restores the interaction between htt and the p150 (Glued) subunit of dynactin and their association with microtubules in vitro and in cells. This is the first description of a mechanism that restores the htt function that got altered due to disease.

B. Mitochondrial/bioenergetic dysfunction and HD

Investigators have long thought that mitochondrial dysfunction plays an important role in neurodegenerative diseases. The strongest evidence that a mutant polyglutamine protein may have a direct effect on mitochondrial function is for HD. Such mitochondrial defects may be a direct effect of mutant huntingtin stems from studies demonstrating mitochondrial Ca+ defects in HD material (Tang, Slow, et al. 2005). Panov et al. found (2002) that mitochondria isolated from lymphoblasts of HD patients and mitochondria isolated from the brains of mice expressing full-length mutant huntingtin had a similar deficit in membrane potential and depolarization in response to Ca+ loading. In the mice, this deficit preceded onset of disease, and they found an N-terminal fragment of huntingtin at the mitochondrial membrane. Incubation of normal mitochondria with this N-terminal mutant huntingtin fragment induces the mitochondrial defects seen in HD patients and transgenic mice. In addition to altering calcium signaling at the mitochondrial membrane, Tang et al found that mutant huntingtin affects Ca+ signaling, leading to cytoplasmic Ca+ overload by sensitizing the InsP3R1 to activation by InsP3. Enhanced release of proapoptotic factors such as cytochrome C from mitochondria in medium spiny neurons was speculated to lead to apoptosis and HD.

A potentially important finding is the recent report of a relationship between the polyglutamine tract length and the cellular energy status (Seong, Ivanova, et al. 2005). In this study, the investigators first examined the ATP levels in a striatal cell line generated from mice carrying an expanded CAG inserted into the endogenous mouse SCA 1 gene, and found low mitochondrial ATP production and a decrease in the ability of the mitochondria to take up ADP. An analysis of the ATP/ADP ratio in 40 human lymphoblast cell lines generated from individuals with CAG repeats spanning in size from normal (9-34 repeats) to affected (35-70 repeats), alleles revealed an inverse association between the length of the longest repeat in an individual and the ATP/ADP ratio. This association extended throughout both the normal and mutant alleles, indicating that the length of the polyglutamine tract in huntingtin regulates in some fashion the energy status of a cell that may contribute to the increased susceptibility of striatal neurons in HD.

Recent studies (Lee, Ivanova, et al. 2007) were performed to test the prevailing hypothesis that huntingtin may directly affect the mitochondrion. By using comprehensive gene expression analysis, it was investigated whether the HD mutation may replicate the effects of 3-nitropropionic acid (3-NP), a compound known to inhibit mitochondria, with loss of striatal neurons. They found that, while mutant huntingtin and 3-NP both elicited energy starvation, the gene responses to the HD mutation, unlike the responses to 3-NP, did not highlight damage to mitochondria, but revealed effects on huntingtin-dependent processes instead. Thus, rather than direct inhibition, the polyglutamine tract size appears to modulate some normal activity of huntingtin that indirectly influences the management of the mitochondrion. Understanding the precise nature of this extra-mitochondrial process would critically guide efforts to achieve effective energy-based therapeutics in HD.

3. Conclusion

Huntington´s disease appears to involve disruption of numerous molecular and cellular processes, including protein-protein interactions, proteosomal actions, transcriptional regulation, protein trafficking, inter and intraneuronal signaling and synaptic plasticity. This disruption is expressed as an altered structure and function of networks of neurons, leading to motor, cognitive and psychiatric symptoms, prior to cell death. The role of other specific cellular processes, such as altered neurogenesis, may either be an epiphenomenon or a causative component of pathogenesis, but this remains one of the many issues yet to be fully elucidated. A key issue for further research is to distinguish cause and effect with respect to mechanisms of pathogenesis.

Other important issue facing researchers is how to sort out the major pathogenic pathways as targets for developing therapeutic strategies. For example, which is more critical for neuronal dysfunction and neurodegeneration, the nuclear or cytoplasmic effect of mutant htt? Answering this question would probably require a better understanding than what we currently have of mutant htt´s effects in different types of cells and at different stages of the disease.

Identification of molecular pathways altered by expanded repeats is beginning to reveal potential therapeutic interventions that can be tested in the existing animal models in preparation for eventual clinical investigations. The findings that motor, behavioral, and pathological phenotypes can be reversed in mouse models of HD upon silencing of the mutant transgene are quite exciting and provide hope that therapeutic interventions are likely to benefit not only presymptomatic individuals, but potentially individuals in early and midstages of the disease as well.

References

Al-Jader, L. N. & Harper, P. S. et al. (2001). "The frequency of inherited disorders database: prevalence of Huntington disease." *Community Genet, 4(3),* 148-57.

Andrade, M. A. & Bork, P. (1995). "HEAT repeats in the Huntington´s disease protein". *Nat Genet, 11,* 115-116.

Andrew, S. E. & Goldberg, Y. P. et al. (1993). "The relationship between trinucleotide (CAG) repeat length and clinical features of Huntington's disease." *Nat Genet, 4(4),* 398-403.

Arrasate, M. & Mitra, S. et al. (2004). "Inclusion body formation reduces levels of mutant huntingtin and the risk of neuronal death." *Nature, 431(7010),* 805-10.

Benn, C. L. & Landles C. et al. (2005). "Contribution of nuclear and extranuclear polyQ to neurological phenotypes in mouse models of Huntington's disease." *Hum Mol Genet, 14(20),* 3065-78.

Bertram, L. & Tanzi, R. E. (2005). "The genetic epidemiology of neurodegenerative disease." *J Clin Invest, 115(6),* 1449-57.

Bodner, R. A. & Housman, D. E. et al. (2006). "New directions for neurodegenerative disease therapy: using chemical compounds to boost the formation of mutant protein inclusions." *Cell Cycle, 5(14),* 1477-80.

Bodner, R. A. & Outeiro, T. F. et al. (2006). "Pharmacological promotion of inclusion formation: a therapeutic approach for Huntington's and Parkinson's diseases." *Proc Natl Acad Sci,* U S A *103(11),* 4246-51.

Borrell-Pages, M. & Zala D. et al. (2006). "Huntington's disease: from huntingtin function and dysfunction to therapeutic strategies." *Cell Mol Life Sci, 63(22),* 2642-60.

Brais, B., Bouchard, J. P., Xie, Y. G., Rochefort, D. L. & Chretien, N. et al. (1998). "Short GCG expansions in the PABP2 gene cause oculopharyngeal muscular dystrophy." *Nat. Genet.18,* 164-67. Erratum, 1998. *Nat. Genet., 19 (4),* 404.

Brinkman, R. R. & Mezei, M. M. et al. (1997). "The likelihood of being affected with Huntington disease by a particular age, for a specific CAG size." *Am J Hum Genet 60(5),* 1202-10.

Butler, R. and G. P. Bates (2006). "Histone deacetylase inhibitors as therapeutics for polyglutamine disorders." *Nat Rev Neurosci 7(10),* 784-96.

Cattaneo, E., D. Rigamonti, et al. (2001). "Loss of normal huntingtin function: new developments in Huntington's disease research." *Trends Neurosci 24(3),* 182-8.

Cepeda, C., Ariano M. A., et al. (2001). "NMDA receptor function in mouse models of Huntington disease." *J Neurosci Res 66(4),* 525-39.

Chai, Y., Wu L., et al. (2001). "The role of protein composition in specifying nuclear inclusion formation in polyglutamine disease." *J Biol Chem 276(48),* 44889-97.

Chan, E. Y., Nasir J., et al. (2002). "Targeted disruption of Huntingtin-associated protein-1 (Hap1) results in postnatal death due to depressed feeding behavior." *Hum Mol Genet 11(8),* 945-59.

Chen, M. & Ona V. O. et al. (2000). "Minocycline inhibits caspase-1 and caspase-3 expression and delays mortality in a transgenic mouse model of Huntington disease." *Nat Med, 6(7),* 797-801.

Colin, E., Zala, D. & Liot, G. et al., (2008). "Huntingtin phosphorylation acts as a molecular switch for anterograde/retrograde transport in neurons". *EMBO J. , 6,* 27(15),2124-34.

Conneally, P. M. & Haines, J. L. et al., (1989). "Huntington disease: no evidence for locus heterogeneity." *Genomics, 5(2)*, 304-8.

Cornett, J. & Cao F. et al., (2005). "Polyglutamine expansion of huntingtin impairs its nuclear export." *Nat Genet, 37(2)*, 198-204.

Davies, S. W. & Turmaine, M. et al., (1997). "Formation of neuronal intranuclear inclusions underlies the neurological dysfunction in mice transgenic for the HD mutation." *Cell, 90 (3)*, 537-48.

DiFiglia, M. & Sapp E. et al., (1995). "Huntingtin is a cytoplasmic protein associated with vesicles in human and rat brain neurons." *Neuron, 14(5)*, 1075-81.

Dragatsis, I. & Levine M. S. et al. (2000). "Inactivation of Hdh in the brain and testis results in progressive neurodegeneration and sterility in mice." *Nat Genet, 26(3)*, 300-6.

Duff, K. & Beglinger, L. J. et al., (2009). *"Cognitive deficits in Huntington's disease on the Repeatable Battery for the Assessment of Neuropsychological Status". J Clin Exp Neuropsychol., 29,1-9*

Dunah, A. W. & Jeong, H. et al. (2002). "Sp1 and TAFII130 transcriptional activity disrupted in early Huntington's disease." *Science, 296(5576)*, 2238-43.

Duyao, M. P. & Auerbach A. B. et al. (1995). "Inactivation of the mouse Huntington's disease gene homolog Hdh." *Science, 269(5222)*, 407-10.

Engelender, S. & Sharp, A. H. et al. (1997) "Huntingtin-associated protein 1 (HAP1) interacts with the p150Glued subunit of dynactin". *Hum Mol Genet, 6(13)*, 2205-2212.

Ferrante, R. J. & Kowall N. W. et al., (1985). "Selective sparing of a class of striatal neurons in Huntington's disease." *Science, 230(4725)*, 561-3.

Ferrante, R. J. & Kubilus, J. K. et al. (2003). "Histone deacetylase inhibition by sodium butyrate chemotherapy ameliorates the neurodegeneration phenotype in Huntington's disease mice". *J. Neurosci., 23*, 9418-27.

Gafni, J. & Ellerby, L. M. (2002). "Calpain activation in Huntington's disease." *J Neurosci, 22(12)*, 4842-9.

Gardian, G. & Browne, S. E. et al. (2005). "Neuroprotective effects of phenylbutyrate in the N171-82Q trangenic mouse model of Huntington's disease." *J. Biol. Chem., 280*, 556-63.

Gatchel, J. R. & Zoghbi, H. Y. (2005). "Diseases of unstable repeat expansion: mechanisms and common principles." *Nat Rev Genet, 6(10)*, 743-55.

Gauthier, L. R. & Charrin B. C. et al. (2004). "Huntingtin controls neurotrophic support and survival of neurons by enhancing BDNF vesicular transport along microtubules." *Cell, 118(1)*, 127-38.

Graham, R. K. & Deng Y. et al., (2006). "Cleavage at the caspase-6 site is required for neuronal dysfunction and degeneration due to mutant huntingtin." *Cell, 125(6)*, 1179-91.

Graveland, G. A. & Williams R. S. et al., (1985). "Evidence for degenerative and regenerative changes in neostriatal spiny neurons in Huntington's disease." *Science, 227(4688)*, 770-3.

Gu, X. & Li, C. et al., (2005). "Pathological cell-cell interactions elicited by a neuropathogenic form of mutant Huntingtin contribute to cortical pathogenesis in HD mice." *Neuron, 46(3)*, 433-44.

Gunawardena, S. & Her L. S. et al. (2003). "Disruption of axonal transport by loss of huntingtin or expression of pathogenic polyQ proteins in Drosophila." *Neuron, 40(1)*, 25-40.

Gusella, J. F. & Wexler N. S. et al. (1983). "A polymorphic DNA marker genetically linked to Huntington's disease." *Nature, 306(5940)*, 234-8.

Gutekunst, C. A. & Li, S. H. et al. (1999). "Nuclear and neuropil aggregates in Huntington's disease: relationship to neuropathology." *J Neurosci, 19(7)*, 2522-34.

Gutekunst, C. A. & Norflus F. et al. (2000). "Recent advances in Huntington's disease." *Curr Opin Neurol, 13(4)*, 445-50.

Gutekunst, C. A. & Levey A. I. et al. (1995). "Identification and localization of huntingtin in brain and human lymphoblastoid cell lines with anti-fusion protein antibodies." *Proc Natl Acad Sci, U S A, 92(19)*, 8710-4.

Gutekunst, C. A. & Li S. H. et al. (1998). "The cellular and subcellular localization of huntingtin-associated protein 1 (HAP1): comparison with huntingtin in rat and human." *J Neurosci, 18(19)*, 7674-86.

Hackam, A. S. & Singaraja R. et al. (1998). "The influence of huntingtin protein size on nuclear localization and cellular toxicity." *J Cell Biol, 141(5)*, 1097-105.

Harjes, P. & Wanker E. E. (2003). "The hunt for huntingtin function: interaction partners tell many different stories." *Trends Biochem Sci, 28(8)*, 425-33.

Harper, P. S. (1992). "The epidemiology of Huntington's disease." *Hum Genet, 89(4)*, 365-76.

Heinsen, H. & Rub U. et al., (1999). "Nerve cell loss in the thalamic mediodorsal nucleus in Huntington's disease." *Acta Neuropathol*, (Berl) *97(6)*, 613-22.

Hockly, E. & Richon V. M. et al. (2003). "Suberoylanilide hydroxamic acid, a histone deacetylase inhibitor, ameliorates motor deficits in a mouse model of Huntington's disease." *Proc Natl Acad Sci, U S A 100(4)*, 2041-6.

Hodgson, J. G. & Agopyan, N., et al. (1999). "A YAC mouse model for Huntington's disease with full-length mutant huntingtin, cytoplasmic toxicity, and selective striatal neurodegeneration." *Neuron, 23(1)*, 181-92.

Hodgson, J. G. & Smith D. J. et al. (1996). "Human huntingtin derived from YAC transgenes compensates for loss of murine huntingtin by rescue of the embryonic lethal phenotype." *Hum Mol Genet, 5(12)*, 1875-85.

Hoogeveen, A. T. & Willemsen R. et al. (1993). "Characterization and localization of the Huntington disease gene product." *Hum Mol Genet, 2(12)*, 2069-73.

Huang, C. C. & Faber P. W. et al. (1998). "Amyloid formation by mutant huntingtin, threshold, progressivity and recruitment of normal polyglutamine proteins." *Somat Cell Mol Genet, 24(4)*, 217-33.

Hughes, R. E. (2002). "Polyglutamine Disease: acetyltransferases away". *Curr. Biol., 12*, R 141-43.

Ishiguro, H. & Yamada, K. et al. (2001). "Age-dependent and tissue-specific CAG repeat instability occurs in mouse knock-in for a mutant Huntington's disease gene." *J Neurosci Res, 65(4)*, 289-97.

Kalchman, M. A. & Koide H. B. et al., (1997). "HIP1, a human homologue of S. cerevisiae Sla2p, interacts with membrane-associated huntingtin in the brain." *Nat Genet, 16(1)*, 44-53.

Kegel, K. B. & Meloni A. R. et al., (2002). "Huntingtin is present in the nucleus, interacts with the transcriptional corepressor C-terminal binding protein, and represses transcription." *J Biol Chem, 277(9)*, 7466-76.

Kenney C., Powell, S. & Jankovic, J. (2007) "Autopsy-proven Huntington's disease with 29 trinucleotide repeats". *Mov Disord*, 22, 127-130.

Kim, Y. J. & Yi Y. et al. (2001). "Caspase 3-cleaved N-terminal fragments of wild-type and mutant huntingtin are present in normal and Huntington's disease brains, associate with membranes, and undergo calpain-dependent proteolysis." *Proc Natl Acad Sci*, U S A 98(22), 12784-9.

Kita, H. & Carmichael J. et al. (2002). "Modulation of polyglutamine-induced cell death by genes identified by expression profiling." *Hum Mol Genet, 11(19)*, 2279-87.

Kittler, J. T. & Thomas, P. et al. (2004). "Huntingtin-associated protein 1 regulates inhibitory synaptic transmission by modulating gamma-aminobutyric acid type A receptor membrane trafficking." *Proc Natl Acad Sci*, U S A 101(34), 12736-41.

Kremer, B. & Squitieri, F. et al., (1993). "Molecular analysis of late onset Huntington's disease". *J. Med. Genet.*, 30, 991-995.

Kremer, H. P. & Roos R. A. et al., (1990). "Atrophy of the hypothalamic lateral tuberal nucleus in Huntington's disease." *J Neuropathol Exp Neurol, 49(4)*, 371-82.

Kuemmerle, S. & Gutekunst C. A. et al. (1999). "Huntington aggregates may not predict neuronal death in Huntington's disease." *Ann Neurol, 46(6)*, 842-9.

Lee J. M. & Ivanova E. V. et al., (2007). "Unbiased gene expression analysis implicates the huntingtin polyglutamine tract in extra-mitochondrial energy metabolism." *PLoS Genet., 3(8)*,e135.

Li S. H. & Gutekunst C. A., et al., (1998). " Interaction of huntingtin-associated protein with dynactin P150Glued." *J Neurosci*, 18, 1261-1269.

Li, H. & Li, S. H. et al. (1999). "Ultrastructural localization and progressive formation of neuropil aggregates in Huntington's disease transgenic mice." *Hum Mol Genet, 8(7)*, 1227-36.

Li, H. & Li, S. H. et al. (2000). "Amino-terminal fragments of mutant huntingtin show selective accumulation in striatal neurons and synaptic toxicity." *Nat Genet, 25(4)*, 385-9.

Li, H. & Li, S. H. et al. (2001). "Huntingtin aggregate-associated axonal degeneration is an early pathological event in Huntington's disease mice." *J Neurosci, 21(21)*, 8473-81.

Li, S. H. & Li, X. J. (1998). "Aggregation of N-terminal huntingtin is dependent on the length of its glutamine repeats." *Hum Mol Genet, 7(5)*, 777-82.

Li, S. H. & Li, X. J. (2004). "Huntingtin-protein interactions and the pathogenesis of Huntington's disease." *Trends Genet, 20(3)*, 146-54.

Li, S. H. & Cheng, A. L. et al. (2002). "Interaction of Huntington disease protein with transcriptional activator Sp1." *Mol Cell Biol, 22(5)*, 1277-87.

Li, S. H. & Gutekunst, C. A. et al. (1998). "Interaction of huntingtin-associated protein with dynactin P150Glued." *J Neurosci, 18(4)*, 1261-9.

Li, S. H. & Lam S. et al. (2000). "Intranuclear huntingtin increases the expression of caspase-1 and induces apoptosis." *Hum Mol Genet, 9(19)*, 2859-67.

Li, S. H. & Yu, Z. X. et al. (2003). "Lack of huntingtin-associated protein-1 causes neuronal death resembling hypothalamic degeneration in Huntington's disease." *J Neurosci, 23(17)*, 6956-64.

Li, X. J. & Li, S. H. et al. (1995). "A huntingtin-associated protein enriched in brain with implications for pathology." *Nature, 378(6555)*, 398-402.

Li, Y. & Chin, L. S. et al. (2002). "Huntingtin-associated protein 1 interacts with hepatocyte growth factor-regulated tyrosine kinase substrate and functions in endosomal trafficking." *J Biol Chem, 277(31)*, 28212-21.

Lin, C. H. & Tallaksen-Greene, S. et al. (2001). "Neurological abnormalities in a knock-in mouse model of Huntington's disease." *Hum Mol Genet, 10(2)*, 137-44.

Lipe H. & Bird, T. (2009). "Late onset Huntington Disease: clinical and genetic characteristics of 34 cases". *J Neurol Sci., 276(1-2),159-62*.

Lodi, R. & Schapira, A. H. et al., (2000). "Abnormal in vivo skeletal muscle energy metabolism in Huntington´s disease and dentatorubropallidoluysian atrophy". *Ann Neurol, 48*,72-76.

Lunkes, A. & Lindenberg, K. S. et al. (2002). "Proteases acting on mutant huntingtin generate cleaved products that differentially build up cytoplasmic and nuclear inclusions." *Mol Cell, 10(2)*, 259-69.

Luthi-Carter, R. & Hanson, S. A. et al. (2002). "Dysregulation of gene expression in the R6/2 model of polyglutamine disease: parallel changes in muscle and brain." *Hum Mol Genet, 11(17)*, 1911-26.

MacDonald, M. E. & Novelletto, A. et al. (1992). "The Huntington's disease candidate region exhibits many different haplotypes." *Nat Genet, 1(2)*, 99-103.

MacDonald, V. & Halliday, G. (2002). "Pyramidal cell loss in motor cortices in Huntington's disease." *Neurobiol Dis, 10(3)*, 378-86.

Mantamadiotis, T. & Lemberger, T. et al. (2002). "Disruption of CREB function in brain leads to neurodegeneration." *Nat Genet, 31(1)*, 47-54.

Marder, K. & Zhao, H. et al., (2009). "Dietary intake in adults at risk for Huntington disease: analysis of PHAROS research participants". *Neurology, 73(5)*, 385-92.

McCampbell, A. & Taylor, J. P. et al. (2000). "CREB-binding protein sequestration by expanded polyglutamine." *Hum Mol Genet, 9(14)*, 2197-202.

McGuire, J. R. & Rong J. et al. (2006). "Interaction of Huntingtin-associated protein-1 with kinesin light chain: implications in intracellular trafficking in neurons." *J Biol Chem, 281(6)*, 3552-9.

McNeil, S. M. & Novelletto, A. et al. (1997). "Reduced penetrance of the Huntington's disease mutation." *Hum Mol Genet, 6(5)*, 775-9.

Menalled, L. B. & Sison J. D. et al. (2003). "Time course of early motor and neuropathological anomalies in a knock-in mouse model of Huntington's disease with 140 CAG repeats." *J Comp Neurol, 465(1)*, 11-26.

Menalled, L. B. & Sison J. D. et al. (2002). "Early motor dysfunction and striosomal distribution of huntingtin microaggregates in Huntington's disease knock-in mice." *J Neurosci, 22(18)*, 8266-76.

Metzler, M. & Li, B. et al. (2003). "Disruption of the endocytic protein HIP1 results in neurological deficits and decreased AMPA receptor trafficking." *Embo J, 22(13)*, 3254-66.

Mishra, S. K. & Agostinelli, N. R. et al. (2001). "Clathrin- and AP-2-binding sites in HIP1 uncover a general assembly role for endocytic accessory proteins." *J Biol Chem, 276(49)*, 46230-6.

Myers, R. H. & Leavitt, J. et al. (1989). "Homozygote for Huntington disease." *Am J Hum Genet, 45(4)*, 615-8.

Myers, R. H. & MacDonald, M. E. et al. (1993). "De novo expansion of a (CAG)n repeat in sporadic Huntington's disease." *Nat Genet, 5(2)*, 168-73.

Myers, R. H. & Sax, D. S. et al. (1985). "Late onset of Huntington's disease." *J Neurol Neurosurg Psychiatry, 48(6)*, 530-4.

Myers, R. H. & Sax D. S. et al. (1991). "Factors associated with slow progression in Huntington's disease." *Arch Neurol, 48(8)*, 800-4.

Myers, R. H. & Vonsattel, J. P. et al. (1991). "Decreased neuronal and increased oligodendroglial densities in Huntington's disease caudate nucleus." *J Neuropathol Exp Neurol, 50(6)*, 729-42.

Myers, S. J. & Dingledine, R., et al. (1999). "Genetic regulation of glutamate receptor ion channels." *Annu Rev Pharmacol Toxicol, 39*, 221-41.

Nasir, J. & Floresco, S. B. et al. (1995). "Targeted disruption of the Huntington's disease gene results in embryonic lethality and behavioral and morphological changes in heterozygotes." *Cell, 81(5)*, 811-23.

Neuwald, A. F. & Hirano, T. (2000). "HEAT repeats associated with condensins, cohesins, and other complexes involved in chromosome-related functions." *Genome Res, 10(10)*, 1445-52.

Nucifora, F. C. & Jr., Sasaki, M. et al. (2001). "Interference by huntingtin and atrophin-1 with cbp-mediated transcription leading to cellular toxicity." *Science, 291(5512)*, 2423-8.

Ona, V. O. & Li, M. et al., (1999). "Inhibition of caspase-1 slows disease progression in a mouse model of Huntington's disease." *Nature, 399(6733)*, 263-7.

Panov A. V. & Gutekunst, C. A. et al., (2002) "Early mitochondrial calcium defects in Huntington's disease are a direct effect of polyglutamines". *Nat Neurosci., 5(8)*,731-6.

Peinemann, A. & Schuller, S. et al., (2005). "Executive dysfunction in early stages of Huntington's disease is associated with striatal and insular atrophy, A neuropsychological and voxel-based morphometric study". *J Neurol Sci., 239(1),11-9*.

Perez, M. K. & Paulson, H. L. et al. (1998). "Recruitment and the role of nuclear localization in polyglutamine-mediated aggregation." *J Cell Biol, 143(6)*, 1457-70.

Petersen, A. & Gil, J. et al. (2005). "Orexin loss in Huntington's disease." *Hum Mol Genet, 14(1)*, 39-47.

Rajkowska, G. & Selemon, L. D. et al. (1998). "Neuronal and glial somal size in the prefrontal cortex: a postmortem morphometric study of schizophrenia and Huntington disease." *Arch Gen Psychiatry, 55(3)*, 215-24.

Ranen N. G. & Stine C. O. et al. (1995). "Anticipation and instability of IT-15 (CAG)n repeats in parent-offspring pairs with Huntington's disease. *Am. J. Hum. Genet., 57*, 593-602.

Reddy, P. H. & Williams, M. et al. (1998). "Behavioural abnormalities and selective neuronal loss in HD transgenic mice expressing mutated full-length HD cDNA." *Nat Genet, 20(2)*, 198-202.

Rong, J. & McGuire, J. R. et al., (2006). "Regulation of intracellular trafficking of huntingtin-associated protein-1 is critical for TrkA protein levels and neurite outgrowth." *J Neurosci, 26(22)*, 6019-30.

Rosas H. D. & Koroshetz, W. J. et al. (2003). "Evidence for more widespread cerebral pathology in early HD: An MRI-based morphometric analysis". *Neurology, 60,* 1615-20.

Rosenblatt, A. & Abbott, M. H. et al., (2003). "Predictors of neuropathological severity in 100 patients with Huntington's disease". *Ann. Neurol., 54,* 488-93.

Rubinsztein, D. C. & Barton, D. E. et al. (1993). "Analysis of the huntingtin gene reveals a trinucleotide-length polymorphism in the region of the gene that contains two CCG-rich stretches and a correlation between decreased age of onset of Huntington's disease and CAG repeat number." *Hum Mol Genet, 2(10),* 1713-5.

Sanchez, I. & Xu, C. J. et al. (1999). "Caspase-8 is required for cell death induced by expanded polyglutamine repeats." *Neuron, 22(3),* 623-33.

Sapp, E. & Kegel, K. B. et al. (2001). "Early and progressive accumulation of reactive microglia in the Huntington disease brain." *J Neuropathol Exp Neurol, 60(2),* 161-72.

Saudou, F. & Finkbeiner, S. et al. (1998). "Huntingtin acts in the nucleus to induce apoptosis but death does not correlate with the formation of intranuclear inclusions." *Cell, 95(1),* 55-66.

Schilling, G. & Becher, M. W. et al. (1999). "Intranuclear inclusions and neuritic aggregates in transgenic mice expressing a mutant N-terminal fragment of huntingtin." *Hum Mol Genet, 8(3),* 397-407.

Schilling, G. & Savonenko, A. V. et al. (2004). "Nuclear-targeting of mutant huntingtin fragments produces Huntington's disease-like phenotypes in transgenic mice." *Hum Mol Genet, 13(15),* 1599-610.

Semaka, A. & Creighton, S. et al., (2006) "Predictive testing for Huntington disease: interpretation and significance of intermediate alleles". *Clin Genet., 70,* 283-94.

Seong, I. S. & Ivanova, E. et al. (2005)."HD CAG repeat implicates a dominant property of huntingtin in mitochondrial energy metabolism". *Hum Mol Genet., 14(19),2871-80.*

Sharp, A. H., Loev, S. J., et al. (1995). "Widespread expression of Huntington's disease gene (IT15) protein product." *Neuron, 14(5),* 1065-74.

Shelbourne, P. F. & Killeen, N. et al. (1999). "A Huntington's disease CAG expansion at the murine Hdh locus is unstable and associated with behavioural abnormalities in mice." *Hum Mol Genet, 8(5),* 763-74.

Sheng, G. & Chang, G. Q. et al. (2006). "Hypothalamic huntingtin-associated protein 1 as a mediator of feeding behavior." *Nat Med, 12(5),* 526-33.

Shimohata, T. & Nakajima, T. et al. (2000). "Expanded polyglutamine stretches interact with TAFII130, interfering with CREB-dependent transcription." *Nat Genet, 26(1),* 29-36.

Singhrao, S. K. & Thomas, P. et al. (1998). "Huntingtin protein colocalizes with lesions of neurodegenerative diseases: An investigation in Huntington's, Alzheimer's, and Pick's diseases." *Exp Neurol, 150(2),* 213-22.

Sipione, S. & Rigamonti, D. et al. (2002). "Early transcriptional profiles in huntingtin-inducible striatal cells by microarray analyses." *Hum Mol Genet, 11(17),* 1953-65.

Slow, E. J. & Graham, R. K., et al. (2005). "Absence of behavioral abnormalities and neurodegeneration in vivo despite widespread neuronal huntingtin inclusions." *Proc Natl Acad Sci, U S A 102(32),* 11402-7.

Song, C. & Perides, G. et al. (2002). "Expression of full-length polyglutamine-expanded Huntingtin disrupts growth factor receptor signaling in rat pheochromocytoma (PC12) cells." *J Biol Chem, 277(8),* 6703-7.

Steffan, J. S. & Bodai, L. et al. (2001). "Histone deacetylase inhibitors arrest polyglutamine-dependent neurodegeneration in Drosophila." *Nature, 413(6857)*, 739-43.

Strand, A. D., Aragaki, A. K., Shaw, D., Bird, T., Holton, J. & Turner, C. et al. (2005). "Gene expression in Huntington's disease skeletal muscle: a potential biomarker". *Hum Mol Genet*, 14,1863- 1876.

Stromme, P. & Mangelsdorf, M. E. et al,. (2002). "Mutations in the human ortholog of the Aristaless cause X-linked mental retardation and epilepsy. *Nat Genet., 30*, 441-45.

Sudarsky, L., Myers, R. H. & Walshe, T. M. (1983). "Huntington's disease in monozygotic twins reared apart". *J. Med. Genet., 20*, 408-411.

Sugars, K. L. & Rubinsztein, D. C. (2003). "Transcriptional abnormalities in Huntington disease." *Trends Genet, 19(5)*, 233-8.

Sun, Y. & Savanenin, A. et al. (2001). "Polyglutamine-expanded huntingtin promotes sensitization of N-methyl-D-aspartate receptors via post-synaptic density 95." *J Biol Chem, 276(27)*, 24713-8.

Takano, H. & Gusella, J. F. (2002). "The predominantly HEAT-like motif structure of huntingtin and its association and coincident nuclear entry with dorsal, an NF-kB/Rel/dorsal family transcription factor." BMC *Neurosci, 3*, 15.

Tallaksen-Greene, S. J. & Crouse, A. B. et al. (2005). "Neuronal intranuclear inclusions and neuropil aggregates in HdhCAG(150) knockin mice." *Neuroscience, 131(4)*, 843-52.

Tanaka, Y. & Igarashi, S. et al. (2006). "Progressive phenotype and nuclear accumulation of an amino-terminal cleavage fragment in a transgenic mouse model with inducible expression of full-length mutant huntingtin." *Neurobiol Dis, 21(2)*, 381-91.

Tang T. S. & Slow, E. et al., (2005). "Disturbed Ca2+ signaling and apoptosis of medium spiny neurons in Huntington's disease". *Proc Natl Acad, Sci U S A. 102(7)*, 2602-7.

Tang, T. S. & Tu, H. et al. (2003). "Huntingtin and huntingtin-associated protein 1 influence neuronal calcium signaling mediated by inositol-(1,4,5) triphosphate receptor type 1." *Neuron, 39(2)*, 227-39.

The Huntington's Disease Collaborative Research Group (1993). "A novel gene containing a trinucleotide repeat that is expanded and unstable on Huntington's disease chromosomes". *Cell, 72*, 971-983.

Turner, C. & Cooper, J. M. et al., (2007). "Clinical correlates of mitochondrial function in Huntington's disease muscle". *Mov Disord, 22(12)*, 1715-21.

Utsch, B. & Becker, K. et al., (2002)." A novel stable polyalanine (poly A) expansion in the HOXA 13 gene associated with hand-foot-genital syndrome: Proper function of poly (A)-harbouring transcription factors depends on a critical repeat length?". *Hum.Genet., 110*, 488-94.

Vonsattel, J. P. & DiFiglia, M. (1998). "Huntington disease." *J Neuropathol Exp Neurol 57(5)*, 369-84.

Vonsattel, J. P. & Myers, R. H. et al. (1985). "Neuropathological classification of Huntington's disease." *J Neuropathol Exp Neurol, 44(6)*, 559-77.

Waelter, S. & Scherzinger, E. et al. (2001). "The huntingtin interacting protein HIP1 is a clathrin and alpha-adaptin-binding protein involved in receptor-mediated endocytosis." *Hum Mol Genet, 10(17)*, 1807-17.

Wellington, C. L. & Ellerby, L. M. et al. (2002). "Caspase cleavage of mutant huntingtin precedes neurodegeneration in Huntington's disease." *J Neurosci, 22(18)*, 7862-72.

Wexler, N. S. & Young, A. B. et al. (1987). "Homozygotes for Huntington's disease." *Nature, 326(6109)*, 194-7.

Wheeler, V. C. & White, J. K. et al. (2000). "Long glutamine tracts cause nuclear localization of a novel form of huntingtin in medium spiny striatal neurons in HdhQ92 and HdhQ111 knock-in mice." *Hum Mol Genet, 9(4)*, 503-13.

Wheeler, V. C. & Gutekunst, C. A. et al., (2002). "Early phenotypes that presage late-onset neurodegenerative disease allow testing of modifiers in Hdh CAG knock-in mice". *Hum. Mol. Genet.*, 11,633-40.

Yu, Z. X. & Li, S. H. et al. (2003). "Mutant huntingtin causes context-dependent neurodegeneration in mice with Huntington's disease." *J Neurosci, 23(6)*, 2193-202.

Zala, D. & Colin, E. et al., (2008). "Phosphorylation of mutant huntingtin at S421 restores anterograde and retrograde transport in neurons". *Hum Mol Genet., 15*, 17(24), 3837-46

Zeitlin, S. & Liu J. P. et al. (1995). "Increased apoptosis and early embryonic lethality in mice nullizygous for the Huntington's disease gene homologue." *Nat Genet 11(2)*, 155-63.

Zeron, M. M. & Hansson, O. et al. (2002). "Increased sensitivity to N-methyl-D-aspartate receptor-mediated excitotoxicity in a mouse model of Huntington's disease." *Neuron, 33(6)*, 849-60.

Zhou, H. & Cao, F. et al. (2003). "Huntingtin forms toxic NH2-terminal fragment complexes that are promoted by the age-dependent decrease in proteasome activity." *J Cell Biol, 163(1)*, 109-18.

Zoghbi, H. Y. & Orr, H. T. (2000). "Glutamine repeats and neurodegeneration." *Annu Rev Neurosci, 23*, 217-47.

Zoghbi, H. Y. & Orr, H. T. (2009). "Pathogenic mechanisms of a polyglutamine-mediated neurodegenerative disease, spinocerebellar ataxia type 1". *J Biol Chem, 284(12)*,7425-9.

In: Encyclopedia of Neuroscience Research
Editors: Eileen J. Sampson and Donald R. Glevins
ISBN 978-1-61324-861-4
© 2012 Nova Science Publishers, Inc.

Chapter XXXIII

Huntingtin Interacting Proteins: Involvement in Diverse Molecular Functions, Biological Processes and Pathways

Nitai P. Bhattacharyya*, Moumita Datta, Manisha Banerjee, Srijit Das and Saikat Mukhopadhyay

Crystallography and Molecular Biology Division and Structural Genomics Section, Saha Institute of Nuclear Physics, 1/AF Bidhan Nagar, Kolkata 700 064, India.

Abstract

To gain insight into role of Huntingtin (HTT) interacting proteins in pathogenesis of Huntington's disease (HD), in the present review, using various databases and published data we analyzed 141 validated HTT interacting proteins out of 311 proteins identified as the interactors of HTT. Results revealed that (i) all 141 proteins express in brain, (ii) fifty three HTT interacting proteins are down regulated and 36 proteins are increased in caudate of HD patients (iii) 25 proteins preferentially interact with mutated HTT, 19 proteins have higher affinity towards wild type HTT and 33 proteins interact equally and (iv) 120 HTT interacting proteins interact with 1780 unique other proteins including 67 HTT interacting proteins and having 2998 interactions altogether. Altered expressions of HTT interacting proteins and their preferences for the wild type and mutated proteins are likely to alter the network of HTT interacting proteins in HD resulting in cellular dysfunctions observed in HD. Several interacting proteins are known to modulate HD pathogenesis in cell, animal models and HD patients. Significantly enriched HTT interacting proteins in various functional categories, biological processes and pathways compared to that coded by human genome indicates that these functions, processes and

---

* Correspondence: Nitai P. Bhattacharyya, Ph.D; Professor, Crystallography and Molecular Biology Division and Structural Genomics Section; 1/AF Bidhan Nagar, Kolkata 700 064, India; e.mail: nitaipada.bhattacharya@saha.ac.in; nitai.bhattacharyya@gmail.com; Telephone: 091 033 23375345, extension 1301; Fax: 091 033 23374637

pathways are altered in HD. All these data presented and reviewed in this chapter indicate that in spite of being a monogenic disease, HD is quite complex at molecular level. Understanding the molecular interactions and diverse pathways these interacting proteins participate are expected to help in devising therapeutic strategies to combat this devastating disease.

Introduction

Huntington's disease (HD, OMIM ID 143100) is a devastating autosomal dominant progressive neurodegenerative disorder named after George Huntington, who provided a classic systematic account of the conditions in HD (Huntington, 1872). Motor dysfunctions like involuntary purposeless motion, known as choreiform, cognitive impairment and psychiatric disturbances like anxiety, depression, aggression and compulsive behavior are the common symptoms in HD. The genetic basis of HD lies in the expansion of CAG repeats in exon1 of the gene, initially called IT15 (interesting transcript 15), now designated as *huntingtin (HTT)* at chromosome 4p16.3 (HDCRG, 1993). The *HTT* gene consists of 64 exons and codes for a protein of ~348kDa with unknown function(s). CAG repeat numbers vary from 7-36 among normal individuals. Repeat numbers greater than 36 however give rise to the disease. Appearance of the first symptoms, known as the age at onset, varies normally between 30 to 40 years, although early and late onset is also reported. Clinical features of HD progressively develop with an increase in choreic movements, dementia and other motor deficits like dystonia and rigidity. The disease is terminated in death within 10-20 years after the appearance of the initial symptoms. There is no cure for the disease at present. HD is highly prevalent among Caucasian and less frequent among Japanese, Chinese, Finnish and Indians (Harper, 1991). It has been proposed that prevalence of the disease is higher in diverse Indian populations than that of in Japan and China, while lower than that of Caucasian populations (Pramanik et al., 2000).

Extensive researches on the mechanism(s) by which mutant HTT exerts its toxicity to specific region of brain provide a wealth of information but the exact cause remains largely unknown. Autosomal dominant nature of the disease and other evidences indicate "gain of function" of the mutant HTT. However, reports for loss of function of wild type HTT allele together with other HTT interacting proteins (see the following sections) are also available. It is now believed that neurodegeneration in HD could arise from both gain of function and loss of function (reviewed by Ross, 2004). It is revealed that increased poly Q stretches alters the conformation of the protein leading to cytoplasmic and nuclear aggregates, also known as neuronal intranuclear inclusions (NII) through intermediate oligomers and finally results in various cellular dysfunctions observed in HD. It is still debatable whether NII is toxic or protective. Altered cellular and molecular abnormalities observed in various models of the disease and post mortem brains of patients include apoptosis, autophagy, excitotoxicity, dopamine toxicity, endocytosis, protein trafficking, oxidative stress, deregulation of transcription, mitochondrial function, protein degradation and chaperone assisted folding (for recent review see Imarisio et al., 2008). Various processes that are altered in HD are pictorially shown in the Figure 1.

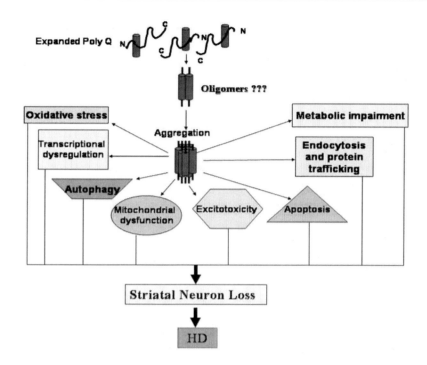

Figure 1 Various responses arising from the mutated HTT leading to neuronal death. Vertical bar represents the expanded poly Q stretch at the N-terminal HTT coded by the exon1 of the gene *HTT*.

Large numbers of proteins interact with HTT, however role of such interactions in the pathogenesis of the disease is not known fully. We cataloged 311 proteins from various sources that interact with HTT. Among these 311 proteins, 141 proteins are validated either by a second method than that initially used or shown to be involved in pathogenesis of HD. We analyzed 141 validated HTT interacting proteins to gather comprehensive information on (a) expressions of these interacting proteins in brain and other tissues, (b) altered expressions of the HTT interacting proteins in the caudate of HD patients, (c) preferential interaction of the proteins with wild type HTT or mutated HTT, (d) interacting partners of the HTT interacting proteins and (e) participation of HTT interacting proteins in diverse molecular functions, biological processes and pathways. Result of this analysis shows that the HTT interacting proteins are involved in various biological processes and pathways indicating that diverse cellular functions, processes and pathways are altered in HD.

Huntingtin Interacting Proteins

Evidences suggest that all cellular functions are carried out by a group of interacting proteins (Barabasi and Oltavai, 2004). This concept provides a framework for assigning function to a protein by identifying its interacting partners with known functions. In case of monogenic disease, mutation to the gene that causes disease may loose some of its interacting partners or interact with new partners resulting in the loss or gain of functions. These interacting partners of disease causing protein may provide clues for modulation of the disease onset, severity and progression. To identifying the function(s) of HTT, various interacting partners of HTT have been identified in yeast 2- hybrid (Y2H) assay and affinity pull down followed by mass spectrometry assay (reviewed in Li et al., 2007). In another approach, aggregates of mutant HTT are isolated, separated on gel and aggregate associated proteins are identified by mass spectrometry (Mitsui et al., 2002). Faber et al identified 13 proteins and named as huntingtin yeast two hybrid proteins (HYPs) in Y2H assay (Faber et al, 1998). Considering that Y2H predictions are subject to high rate of false positives (Von Mering et al., 2002), it is essential to establish the physical interaction of the protein with HTT. Some of these interactions are subsequently validated functionally to establish that the proteins can modulate the HD pathology at least in cultured cells or by a second method like immunoprecipitation (IP, for example see Raychaudhuri et al., 2008). In a similar Y2H assay, Goehler et al. identify 19 HTT interacting proteins, 4 of them was identified earlier. Most of theses new proteins were confirmed by additional IP assay (Goehler et al., 2004). Two hundred thirty four proteins are identified either in high stringent conditions of Y2H (104 proteins) or affinity pull down followed by mass spectrometry (130 proteins) using cell extracts from different types of cells. Only 4 proteins are common in these two methods. Among the interacting proteins thus obtained, 8 randomly chosen proteins are confirmed by IP out of 11 proteins tested. Thus, only ~63% of the proteins obtained by high throughput assay are true interactors. In an attempt to establish the functional role of the interacting proteins, randomly chosen proteins obtained using these two methods were tested in fly model of the disease. Either specific gene was over expressed together with mutant HTT or the mutant HTT was expressed in the flies with mutations in the homologous genes. About 80% of the proteins (48/60) modulate the toxicity in the fly model (Kaltenbach et al., 2007). We collated all these data from literatures and various databases and obtained 311 proteins that were identified to interact with wild type or mutated HTT. Given that ~40% of the proteins obtained in Y2H assay are false positive, we searched for the proteins that have been confirmed by another independent study, using a second method or shown to modulate the toxicity or aggregates in cell or animal models of the disease. Out 311 HTT interacting proteins only 141 proteins are validated by these criteria. Cytoscape, a bioinformatics tool for visualizing molecular interaction networks (http://www.cytoscape.org/) representation of these HTT interacting proteins is shown in the Figure 2. In further analysis, we concentrate only on 141 validated HTT interacting proteins.

Expressions of HTT Interacting Proteins in Various Tissues Including Brain

We argued that if HTT interacting proteins have any role in the pathogenesis of HD, they must also express together with the HTT in the same tissue. Since various in vitro assays, cells of non-human origin were used to identify the interacting proteins, we first identified the homologous genes in human and all subsequent analysis was carried out using human genes. We used mainly TiGER (Tissue-specific Gene Expression and Regulation, http://bioinfo.wilmer.jhu.edu/tiger/) database that describes tissue-specific expressed sequence tag (EST) data in 30 different tissues. These tissues are bladder, blood, bone marrow, brain, cervix, colon, eye, heart, kidney, larynx, liver, lymph node, mammary gland (breast), muscle, ovary, pancreas, peripheral nervous system, placenta, skin, small intestine, soft tissue, spleen, stomach, testis, tongue and uterus. We also searched GeneHub database (http://www.cgl.ucsf.edu/Research/genentech/genehub-gepis/genehub-gepis-search.html), although the tissues represent in this database were not the same as that of TiGER. In few cases, we also gather expression information provided in the Genecard (http://www.genecards.org/). In TiGER, ESTs for *HTT* gene was available for 25 tissues, except in heart, peripheral nervous system, soft tissue, spleen and thymus. In Gene card, expressions of *HTT* gene were shown in heart and thymus. EST for HTT gene was also present in spleen in GeneHub database. Taking together out of 30 tissues in TiGER, HTT gene expresses in 28 tissues, except in peripheral nervous system and soft tissue. Similarly, we cataloged the expressions of 141 validated HTT interacting proteins.

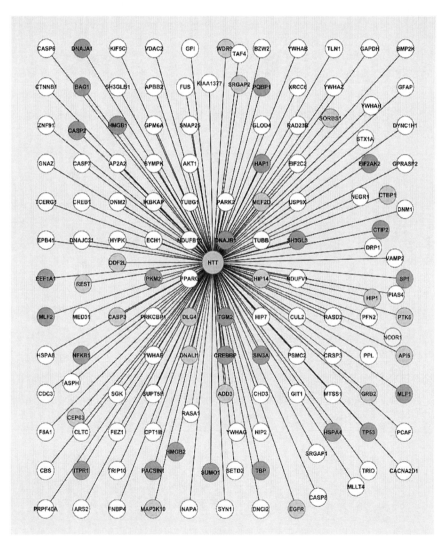

Figure 2. Cytoscape representation of the validated 141 HTT interacting proteins. Red colored 26 proteins (including EIF2AK2 that interacts with mutant HTT RNA) preferentially interacts with mutated HTT, green colored 19 proteins interact preferentially with the wild type HTT, and light yellow colored 33 proteins interact equally with wild and mutated HTT. Information for 61 proteins shown in white color is not known. Two proteins (NFYA and NFYC) interact with mutant HTT aggregates only are not shown in the picture.

It was evident that all 141 proteins are expressed in brain. Percent of interacting protein express in lung, eye, kidney, pancreas, muscle, liver, heart, spleen and thymus were 99%, 97%, 91%, 91%, 90%, 84%, 79%, 48% and 38% respectively. Striatal region is the major affected site in HD, however other organs are also affected resulting in risk of other diseases (reviewed in Sassone et al, 2009. van der Burg and Brundin, 2009). Absence of expression of the interacting partners may be cause of the lack of the effects of the HTT mutation in different tissues. Further studies are required to test the hypothesis.

Alterations of Expressions of the HTT Interacting Proteinsin Caudate of HD Patients

Deregulation of transcription in HD is thought to be one of the important mechanisms for contribution to pathological effects in HD (Sugar and Rubeinztein, 2003 Cha, 2007). Recruitment of transcription factors to mutant HTT aggregates and loss of their functions could be the simplest explanation for such deregulation

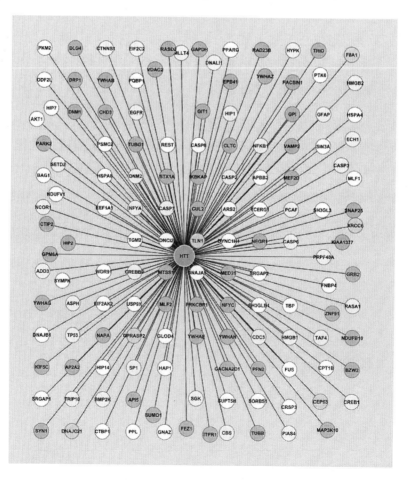

Figure 3. Cytoscape representations of the validated 141 HTT interacting proteins. Proteins (53) with red color are down regulated; green (36) colored proteins are increased, white colored proteins (46) are unaltered and expressions of proteins (6) with blue color are not known.

. Mutant HTT directly interacts with DNA sequences, changes conformation of DNA that allows transcription factors to bind to DNA easily and increase the activities of several transcription factors (Benn et al., 2008). In addition, several micro RNA that in general regulates the expressions of genes negatively in HD have been identified (reviewed in Johnson and Buckley, 2009). Thus, in HD several mechanisms for deregulation of genes might operate. However, it remains unknown whether the HTT interacting proteins are also deregulated in HD. Depending on the level of expression, protein-protein interaction may alter; causing cellular dysfunctions in HD.

To address whether HTT interacting proteins are altered in HD, we searched the published microarray data in caudate of HD patients. For identification of altered expressions, we followed the same criteria as described (Hodges et al., 2006). Out of 141 validated HTT interactors, expressions of 36 proteins were increased, 53 proteins were decreased significantly ($p < 0.002$), while the expressions of 46 genes are unaltered. Information of six genes is not available. Cytoscape representation of these proteins is shown in the Figure 3. It is to be noted that even these genes are observed in microarray assay, only few has been validated by another method, which is essential for the confirmation. SNAP25, which is decreased in caudate of HD patients, has shown to be depleted in sensory motor cortex of HD models indicating a role of this protein in HD pathogenesis (Freeman and Morton, 2004, Smith et al., 2007). Increased or decreased expressions of the HTT interacting protein is likely to alter the over all HTT interaction network and may result in the observed cellular dysfunctions in HD.

Influence of Poly Q Length at HTT for the Interactions with Other Proteins

HD is an autosomal dominant disease. One of the allele is mutated due to expansion of CAG repeats; the other allele is wild type in the diseased condition. HTT interacting proteins are identified in Y2H assay using either the wild type or mutated N-terminal HTT as baits. In these experiments as well in other directed experiments, it has been shown that some protein preferentially interact with wild type allele or mutant allele, some interacts equally with both the alleles. Biological consequences of interactions of HTT with its partners, if any, might be dependent not only on affinity of the interacting proteins but also on the amount of interacting proteins. We searched published literature to find out whether the interacting partners of HTT have any preference for the wild type or mutated HTT. Out of 141 validated HTT interactors, 33 proteins interact equally with mutant and wild type HTT, 25 proteins interact preferentially with mutant HTT, EIF2AK2 interacts with mutant RNA only and 19 proteins prefer the wild type HTT. Information on others is not known. Two proteins namely NFYA and NFYC interact with mutant HTT aggregates only. The proteins, which interact differentially, are also shown in the Figure 2. Thus, due to expansion of CAG repeats that causes HD, there might be new interactions with mutant allele and loss of some interactions with the wild type allele. HIP1 interacts preferentially with wild type HTT. In HD, due to mutation at HTT, freely available HIP1 is increased, which then interacts with HIPPI, a molecular partner of HIP1 and triggers apoptosis and/or gene expression (Gervais et al., 2002; Majumder et al., 2006, Bhattacharyya et al., 2008). Similar differential interaction of REST with wild type and mutant HTT may result in alterations of gene expressions (Zuccato et al., 2007). It is shown that wild-type HTT sequesters REST/NRSF in the cytoplasm thereby permitting activated transcription of the target gene such as BDNF. In the presence of mutant HTT, which interacts weekly with REST, REST accumulates in the nucleus resulting in decrease of BDNF transcription (Zuccato et al., 2003).

Interacting Partners of HTT Interacting Proteins

There are several databases that catalogue the interacting partners of proteins from various experimental data and literature searches. One of such databases is BioGRID (Stark et al., 2006), which includes data from Y2H, reconstituted complex (in vitro interactions with purified proteins), affinity capture-western blot or mass spectrometry and genetic interactions. In our search for interacting partners of HTT interacting proteins, we use only those proteins that are identified from first three methods. There are 21 proteins (SETD2, ASPH CPT1B, ZNF91, GPRASP2, ADD3, DLG4, GLOD4, F8A1, HYPK, CACNA2D1, NEGR1, NDUFB10, DNAJC21, CDH13, GPM6A, BMP2K, SRGAP1, HIP14, API5 and SRGAP2), which do not have any known partners presently. The other HTT partners have 1 (for example APBB2) to 214 interacting partners (p53). All together for 120 proteins, there are 2989 interactions of 1780 unique proteins including 67 HTT interacting proteins. As mentioned above, 25 proteins prefer the mutant HTT protein while EIF2AK2 prefers to interact with the mutant HTT RNA. From protein interaction network point of view, these are the new interactions in diseased condition. Twenty-six HTT interacting proteins (including one mutant RNA interacting protein), which preferentially interact with mutant HTT ineract with 748 proteins among which 36 proteins are HTT interacting proteins, altogether make 970 interactions. Such interactions are specific for the mutants HTT and thus are disease specific. It is important to know the expression of the partners of HTT interacting proteins in different tissues including brain to access role of the partners of HTT interacting proteins in HD and requires further studies.

Modulation of HD Pathogenesis by HTT Interacting Proteins

Biological implications of HTT interacting proteins in HD pathogenesis remain largely unknown. There are several approaches to find out the influences of other proteins on HD pathogenesis in different models of HD. In fly model of HD, mutant HTT aggregates and/or the toxicity are determined in the mutant flies where the fly homologue of specific protein is either mutated or the homologues are over expressed and compared with that in the wild type fly (Doumanis et al., 2009, Branco et al., 2008). Similar approach was in yeast model of HD (Willingham et al., 2003). Several drosophilae or yeast homologous genes are identified. Besides, in various cell models HTT interacting proteins are co-expressed with exon1 of the mutant HTT and the aggregates and toxicity/apoptosis are measured. Among 141 validated

HTT interacting proteins, we collected information for 60 proteins. Over expression of 25 proteins (CDC3, CLTC, CTNNB1, GAPDH, KIF5C, MEF2D, MLLT4, NDUFB10, PPARG, SUMO1, VDAC2, ADD3, DNAJC21, GIT1, TRIO, YWHAB, YWHAZ, CACNA2D1, GNAZ, GPM6A, ITPR1, NAPA, PSMC2, STX1A, YWHAE) enhance mutant HTT aggregates or the toxicity, while expressions of 35 proteins (CEP63, DNAJC11, DNCI2, DYNC1H1, FEZ1, GFAP, GPI, NEGR1, ODF2L, PPL, SORBS1, TLN1, BAG1, BMP2K, CREB1, CRSP3, CTIP2, DNAJB1, EEF1A1, HAP1, HIP1, HIP2, HMGB1, HMGB2, HSPA4, HSPA8, HYPK, MLF1, MLF2, PRKCBP1, TCERG1, VAMP2, PTK6, USP9X, ZNF91) are suppressed mutant HTT aggregates/toxicity. Depending on the function of HTT interacting protein, it may modulate the disease in different way. For example over expression of chaperones may involve in the early stage of aggregate formation by reversing the misfolded mutant HTT to normal state, while proteins in proteosomal degradation pathway would enhance the clearing stage of the damaged/misfolded proteins. On the basis several experimental result (Muchowski and Wacker 2005, McCampbell *et al.* 2000, Kazemi-Esfarjani and Benzer 2002) "sequestered poly Q hypothesis", that states over expression of any one of the HTT interacting proteins would reverse the effects of the mutant HTT, was put forwarded.

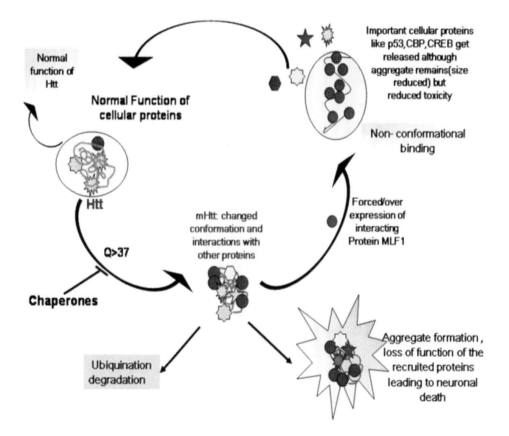

Figure 4. Pictorial presentation for "sequestered poly Q hypothesis" using MLF1 as a model. Over expression of MLF1 releases p53, CREB and CBP from the mutant aggregates. Released p53 increases the expression of GADD45A (Banerjee M and Bhattacharyya NP, unpublished observation).

This further implies that the recruited proteins to the HTT aggregate compromise their normal functions and contribute to the cellular abnormalities observed in HD. We recently obtained results with HTT interacting protein MLF1 that support the hypothesis (M. Banerjee and NP Bhattacharyya, unpublished observations). The pictorial presentation of this hypothesis is shown in the Figure 4. Other proteins, which alter the different functions, may increase the toxicity/mutant HTT aggregates. Thus, several HTT interacting proteins that are altered in HD as described above are capable modulating the pathogenesis.

Direct evidence that HTT interacting proteins alter the age at onset are also available. Imperfect CAGGCC repeat encoding Gln-Ala repeats in transcriptional coactivator CA150 gene explains the small but statistically significant variability in the age at onset in HD patients (Holbert et al., 2001) and subsequently confirmed in various populations with diverse genetic background (Chattopadhyay et al., 2003, Andresen et al., 2007). Analyzing 980 European HD patients, it is observed that that patients homozygous for the M441 genotype at HAP1, a HTT interacting protein delays the age at onset of the disease. Functional assays further show that homozygous M441-HAP1 interacts preferentially with mutant HTT that of T441-HAP1 reduces soluble HTT degraded products and protects against HTT-mediated toxicity (Metzger et al., 2008). Thus it is likely that variations HTT interacting proteins may alter the HD pathogenesis.

HTT Interacting Proteins: Their Molecular Functions

Validated HTT interacting proteins as well as the proteins, which were identified in Y2H or pull down assay only, classified on the basis of molecular functions, biological processes and pathways involved using Protein ANalysis Through Evolutionary Relationships (PANTHER). PANTHER provides a classification system that categorizes genes by their functions, biological processes and pathways using published scientific experimental evidence and evolutionary relationships to predict function even in the absence of direct experimental evidence (http://www.pantherdb.org/). HTT interacting proteins, classified into various categories, were then compared with that of proteins coded by human genome. In such analysis, we normalized the frequencies with only those proteins whose functions are predicted in PANTHER. There are no information in PANTHER for two proteins namely COX2 and COX3 having GENE IDs 4513 and 4514 respectively. Thus, only 139 proteins were analyzed. Validated HTT interacting proteins in functional categories chaperone (MF00077), cytoskeletal protein (MF00091) and membrane traffic protein (MF00267) are enriched significantly ($p < 0.01$, FDR <0.05) over that observed among proteins coded by human genome. Result of this analysis is shown in the Table 1. Among the HTT interacting proteins, which are not confirmed in another method, functional categories Chaperone (MF00077) and membrane traffic protein (MF00267) were also significantly enhanced. This analysis shows that even these proteins are not validated for their interactions with HTT, their functions are altered in HD.

Table 1. Significant enrichment of 141 validated HTT interacting proteins in different molecular functional categories

Moleculer Function (panther ID)	Gene ontology ID	Gene Name	Proteins in Human Genome (Normalized %)	HTT Interacting protein (Normalized %)	Level of Significance (p value)	FDR (Q Value)
Chaperone (MF00077)	Unfolded protein binding (GO:0051082)	YWHAB, DNAJB1, HSPA4 YWHAG, DNAJA1, HSPA8 YWHAE, YWHAH, YWHAZ	176 (1.21%)	9 (7.83%)	0.000	0
Cytoskeletal protein (MF00091)	Structural protein of cytoskeleton (GO:0005200)	ADD3,GFAP,CTNNB1 SYN1,TUBG1,TRIP10 TLN1,DNM2,DYNC1H1 MLLT4,TUBB,HIP1,DNALJ1 MTSS1,SORBS1,7-Sep, DNM1,KIF5C,DYNC1I2 FNBP4	878 (6.06%)	20 (17.39%)	0.000	0
Membrane traffic protein (MF00267)	No match	CLTC,SYN1,HAP1,STX1A PACSIN1,NAPA,SNAP25 VAMP2	359 (2.48%)	8 (6.96%)	0.003	0.0215

Among various interactors of HTT, chaperones play an important role in reduction of poly Q aggregates and toxicity (Raychaudhuri et al., 2008, Opal and Zoghbi, 2002). The effects of chaperones on the mutated HTT aggregates and toxicity have been extensively investigated in a diverse range of HD models including yeast, worms, flies and mice as well as *in vitro*. Chaperones may act on the aggregate formation by reversing the misfolded or conformation changed mutant HTT. HTT interacting protein GIT1 is a multi-domain protein and involves in diverse cellular processes. It is known that GIT1 and other member of the family of protein GIT2 traffic between three distinct cellular compartments namely cytoplasmic complexes, focal adhesions and the cell periphery (Hoefen and Berk 2006). In a recent review, it has been concluded that HTT regulates vesicular trafficking for the organelle transport along cellular cytoskeleton (Caviston and Holzbaur, 2009). Thus, 20 cytoskeleton proteins that interact with HTT are involved in the normal function of HTT; alterations of such interactions may contribute towards the pathogenesis of the disease.

HTT Interacting Proteins:
Their Involvement in Biological Processes

Out of 25431 genes in the human genome, only 14110 genes are predicted to participate in 30 biological processes. Among 141 validated HTT interacting proteins, 23 proteins are not assigned to any biological processes in PANTHER. Other HTT interacting proteins are classified into 25 different biological processes. Classifications of HTT interacting proteins on the basis of their involvement in various biological processes reveal that HTT interacting proteins are enriched significantly (p=0.00, FDR= 0.00) in apoptosis (BP00179), neuronal activities (BP00166), protein targeting and localization (BP00137), cell structure and motility (BP00285) and cell cycle (BP00203) compared to proteins coded by human genome (Table 2). For example, out of 118 HTT interacting proteins, which has been classified in different biological processes, 15 (12.71%) proteins involve in apoptosis, while there are 531 (3.76%) proteins coded by the human genome involve in apoptosis (p=0.00, FDR=0.00). Apoptosis has been implicated in HD (reviewed in Hickey Chesselet, 2003, Imarisio et al., 2008). Defects in axonal transport have been implicated in Drosophilae model of HD. One way to explain the HD pathogenesis is the disruption of axonal transport by mutant HTT protein (Gunawardena et al., 2003). HTT is known to interact with GASP2. As the GASP protein family plays a role in G protein-coupled receptor sorting, this suggests that HTT might influence receptor trafficking via the interaction with GASP2 (Horn et al., 2006). Transport of HTT from the cytoplasm to nucleus has been shown to involve in HD pathogenesis and recently reviewed (Truant et al., 2007). HIP1, a cytoskeleton protein, can act as a nucleo-cytoplasmic shuttling protein, a non-conventional function (Mills et al., 2005, Banerjee et al., 2009). HTT along with its interactor HAP1 accompanied by the dynein/dynactin microtubule-based motor complex is responsible for retrograde intracellular protein trafficking (BP00125) (Reviewed in Imarisio et al., 2008).

HTT Interacting Proteins Involve in Diverse Biological Pathways

Possible involvement of HTT interacting protein in diverse biological pathways is tested by analyzing validated 141 HTT interacting proteins using PANTHER. In PANTHER, only 2995 proteins coded by the human genome are classified in 153 different pathways. Thus, the majority of the proteins are not classified into any one of the pathways so far. It is also to be mentioned that same protein may participate in different pathways. Among the validated 141 proteins, only 64 proteins are classified into 76 pathways, 77 proteins are not classified into any pathway. Validated HTT interacting proteins are enriched significantly ($p<0.003$, FDR < 0.05) in 6 biological pathways in comparison with proteins coded by human genome (Table 3). Proportions of proteins in Huntington disease pathway (P00029), p53 pathway (P00059), FAS signaling pathway (P00020), apoptosis signaling pathway (P00006), muscarinic acetylcholine receptor 2 and 4 signaling pathway (P00043), EGF receptor signaling pathway (P00018) and Parkinson disease pathway (P00049) are enhanced.

Involvement of apoptosis in HD has been studied extensively (reviewed in Imarisio et al., 2008). Pro apoptotic proteins levels get increased and contribute in developing cell death in HD (Hickey and Chesselet, 2003). Moreover, caspase-8 directly interacts with HTT and regulates the apoptotic pathway (Sanchez et al., 1999). Mitochondrial dysfunction in neurons and glia are the common observation in HD. P53 interacts with mutant HTT, reduces the expression of its target genes $p21^{WAF1/CIP1}$ and *MDR-1* genes (Steffan et al., 2000). Level of p53 is increased in HD and can target the promoter of HTT gene (Feng et al., 2006) and influences the pathogenesis of HD in animal models (Ryan et al., 2006). Besides it interacts with 214 proteins and likely to be participated in diverse pathways of HD pathogenesis. Fas receptor is reduced in the caudate and putamen in HD brain. Moreover, Fas and Fas-L immunoreactivity is reduced in striatal neurons in HD (Ferrer et al., 2000). Involvement of EGF receptor signaling (Lievens et al., 2005), FGF signaling pathways (Jin et al., 2005), muscarinic acetylcholine receptor signaling (Lange et al., 1992, Vetter et al., 2003, Wastek and Yamamura, 1978) have been reported in several models of HD. All these experimental results indicate that HTT interacting proteins in these pathways are altered in HD.

It is interesting to note that non-validated HTT interacting proteins are also significantly enriched in several pathways. Among these pathways, synaptic vesicle trafficking (P05734), hypoxia response via HIF activation (P00030) and glycolysis (P00024), Axon guidance mediated by netrin (P00009), Opioid proopiomelanocortin pathway (P05917) and Opioid prodynorphin pathway (P05916) are known to involve in HD. Synaptic dysfunction in HD correlates with early clinical signs and symptoms of the disease. Abnormal association of mutant HTT with synaptic vesicles inhibits glutamate release resulting disorder in synaptic vesicle transport. Reduced association of HAP1, with synaptic vesicles is also observed in HD mouse brains due to aggregation formation (Li et al., 2003). Synaptic vesicle protein neurospecific phosphoprotein PACSIN1 interacts with HTT via its C-terminal SH3 domain and shows its mislocalization in neurons at early stage of HD. PACSIN1 interacts preferentially with mutant HTT and is sequestered into the aggregates (Modregger et al., 2002). In early HD striatum, selective defect in glycolysis is observed (Powers et al., 2007). Increase conversions of glucose into lactate as well as increased level of ATP in cytosolic extracts from the HD brain tissue are reported (Olah et al., 2008).

Table 2. Significantly enhanced HTT interacting proteins in different Biological Processes

Biological Process (panther ID)	Gene ontology ID	Gene Name	Proteins in Human Genome (Normalized %)	HTT interacting proteins (Normalized %)	Level of Significance (p Value)	FDR (Q Value)
Apoptosis (BP00179)	Apoptosis (GO:0006915)	AKT1, CASP6, CASP2, CASP7, CASP3, NFKB1, HIP1, SRGAP1, TP53, CASP8, SH3GLB1, SGK, SRGAP2, API5, SUMO1	531 (3.76%)	15 (12.71%)	0.000	0
Neuronal activities (BP00166)	Transmission of nerve impulse (GO:0019226)	FEZ1, SH3GL3, TRIO, CLTC, SYN1, FUS, HAP1, GPM6A, STX1A, SH3GLB1, APBB2, DLG4, ITPR1, CREB1	569 (4.03%)	14 (11.86%)	0.000	0
Protein targeting and localization (BP00137)	Protein targeting; GO:0008104 protein localization (GO:0006605)	YWHAB, IKBKAP, YWHAG, STX1A, CACNA2D1, YWHAE, WHAH, YWHAZ	253 (1.79%)	8 (6.78%)	0.000	0
Cell structure and motility (BP00285)	Cytoskeleton organization and biogenesis (GO:0007010)	ADD3, TRIO, GFAP, MAP3K10, TUBG1, TRIP10, TLN1, DNM2, DYNC1H1, MLLT4, TUBB, GPM6A, DNALI1, SRGAP1, MTSS1, GIT1, SORBS1, DNM1, KIF5C, ITPR1, SRGAP2, DYNC1I2	1148 (8.14%)	22 (18.64%)	0.000	0
Cell cycle (BP00203)	Cell cycle (GO:0007049)	YWHAB, AKT1, TUBG1, TRIP10, DYNC1H1, TUBB, YWHAG, SRGAP1, MTSS1, TP53, EGFR, SEPTIN7, YWHAE CUL2, YWHAH, SRGAP2, DYNC1I2, YWHAZ, SUMO1	1009 (7.15%)	19 (16.10)	0.000	0

Table 3. Involvement of HTT interacting proteins in different Biological Pathways

Biological Pathway (PANTHER ID)	Gene Name	Proteins in Human Genome (Normalized %)	HTT interacting protein (Normalized %)	Level of Significance (p value)	FDR (Q Value)
Huntington disease (P00029)	SH3GL3, AKT1, AP2A2, MAP3K10, CASP6, CASP3, HAP1, DYNC1H1, CREBBP, TUBB, HIP1, SIN3A, NCOR1, TP53, CASP8, PACSIN1, TAF4, DLG4, DYNC1I2	172 (5.74%)	19 (20.69%)	0.000	0
p53 pathway (P00059)	YWHAB, PCAF, AKT1, CREBBP, YWHAG, SIN3A, TP53, YWHAE, YWHAH, YWHAZ, SUMO1	136 (4.54%)	11 (17.19%)	0.000	0
FAS signaling pathway (P00020)	AKT1, CASP6, CASP7, CASP3, CASP8	36 (1.20%)	5 (7.81%)	0.000	0
Apoptosis signaling pathway (P00006)	AKT1, EIF2AK2, CASP7, CASP3, NFKB1, TP53, CASP8, HSPA8, CREB1	131 (4.37%)	9 (14.06%)	0.000	0
Muscarinic acetylcholine receptor 2 and 4 signaling pathway (P00043)	CPT1B, STX1A, CACNA2D1, SNAP25, VAMP2	59 (1.97%)	5 (7.81%)	0.001	0.0255
EGF receptor signaling pathway (P00018)	YWHAB, AKT1, YWHAG, EGFR, YWHAE, RASA1, YWHAH, YWHAZ, GRB2	150 (5.01%)	9 (14.06%)	0.002	0.0383
Parkinson disease (P00049)	YWHAB, PARK2, YWHAG, HSPA8, YWHAE, YWHAH, YWHAZ	106 (3.54%)	7 (10.94%)	0.002	0.0383

Induction of hypoxia inducible factor-1 (HIF-1) protects against mitochondrial dysfunction observed in a mouse chemical model of HD (Yang et al., 2005). Recently, inhibitor of HIF prolyl hydroxylase is shown to decrease neuronal death (Niatsetskaya et al., 2009) indicating functional role of HIF activation in HD. A significant reduction of dynorphin A1-8 concentration is observed in caudate nucleus, putamen, external globus pallidus and substantia nigra of postmortem brains of HD patients indicating prodynorphin opioid peptide system are affected in the basal ganglia in HD (Seizinger et al., 1986). These data supports the notion even though these HTT interacting proteins are not validated by another method, some of them might participate in HD pathogenesis.

Conclusion

Biological functions are carried out by macromolecular complexes composed of many proteins and regulated through concerted actions of many proteins interacting with each other. There are increasing evidences that onset and progression of genetic diseases arise from the interplay of these interconnected proteins. Here, we show that HTT interacts with at least 141 proteins, which again interacts with ~ 1000 other proteins. HTT interacting proteins are involved in HD pathogenesis. Effects of such interaction may depend on the level of expression of these proteins and affinity of these proteins towards wild type or mutated HTT. Protein-protein interaction network operating in normal condition may alter in diseased state. In some cases, the interaction is lost in HD, in other cases, new interactions take place with the mutated HTT. This may explain, at least partially, the loss of function as well as gain of functions of the network. Such alterations in turn may result in cellular dysfunctions observed in HD. It has been shown recently that among HD patients as well as in animal models of HD, various tissues other than striatum are affected by HTT mutation. Tissue specific expressions of the HTT interacting proteins and protein-protein interactions may contribute to such tissue specific affects of the HTT mutation. Further, in depth analysis of HTT interacting network, tissue specific expressions, and affinity towards wild type or mutated HTT are necessary to understand molecular alterations leading to pathological condition in brain and other tissues. HTT interacting proteins are enriched significantly in various pathways indicating that these pathways are altered in HD. This observation confirms the general notion that even though HD is a monogenic disease, at the molecular level it is quite complex. Understanding this molecular complexity is necessary to combat the devastating disease.

Acknowledgments

Mr. Srijit Das acknowledges the financial support for Junior Research Fellowship to C. S. I. R, Govt. of India.

References

Andresen, J. M., Gayán, J., Cherny, S. S., Brocklebank, D., Alkorta-Aranburu, G., Addis EA, US-Venezuela Collaborative Research Group, Cardon, L. R., Housman, D. E., Wexler, N. S. (2007). Replication of twelve association studies for Huntington's disease residual age of onset in large Venezuelan kindreds. *J. Med. Genet.* 44, 44-50.

Banerjee, M, Datta, M., Majumder, P., Mukhopadhyay, D., Bhattacharyya, N. P. (2009) Transcription regulation of caspase-1 by R393 of HIPPI and its molecular partner HIP-1. *Nucleic Acid Res.* (In Press)

Barabási, A., Oltvai, Z. N. (2004). Network biology: understanding the cell's functional organization. *Nature Reviews Genetics.* 5, 101-113

Benn, C.L., Sun, T. Sadri-Vakili. G., McFarland, K. N., DiRocco D. P., Yohrling, G. J., Clark, T. W., Bouzou, B., Cha, J. H. (2008). Huntingtin modulates transcription, occupies gene promoters in vivo, and binds directly to DNA in a polyglutamine-dependent manner. *J. Neurosci.*, 28, 10720-33.

Bhattacharyya, N. P., Banerjee, M., Majumder, P. (2008). Huntington's disease: roles of huntingtin-interacting protein 1 (HIP-1) and its molecular partner HIPPI in the regulation of apoptosis and transcription. *FEBS J.* 275, 4271-4279.

Branco, J., Al-Ramahi, I., Ukani, L., Pérez, A. M., Fernandez-Funez, P., Rincón-Limas, D., Botas, J. (2008). Comparative analysis of genetic modifiers in Drosophila points to common and distinct mechanisms of pathogenesis among polyglutamine diseases. *Hum. Mol. Genet.*, 17,376-390.

Caviston, J. P., Holzbaur, E. L. (2009). Huntingtin as an essential integrator of intracellular vesicular trafficking. *Trends Cell Biol.* 19, 147-155

Cha, J. H. (2007). Transcriptional signatures in Huntington's disease. *Prog. Neurobiol.* 83, 228-48.

Chattopadhyay, B., Ghosh, S., Gangopadhyay, P. K., Das, S. K, Roy, T., Sinha, K. K., Jha, D. K., Mukherjee, S. C., Chakraborty, A., Singhal, B. S., Bhattacharya, A. K., Bhattacharyya, N. P. (2003). Modulation of age at onset in Huntington's disease and spinocerebellar ataxia type 2 patients originated from eastern India. *Neurosci Lett.* 345, 93-6.

Doumanis, J., Wada, K., Kino, Y., Moore, A. W., Nukina, N. (2009). RNAi screening in Drosophila cells identifies new modifiers of mutant huntingtin aggregation *PLoS One* 4(9): e7275

Faber, P. W., Barnes, G. T, Srinidhi, J., Chen, J., Gusella, J. F., MacDonald, M. E. (1998). Huntingtin interacts with a family of WW domain proteins. *Hum. Mol. Genet.* 7, 1463-1474

Feng, Z., Jin, S., Zupnick, A., Hoh, J., de Stanchina, E., Lowe, S., Prives, C., Levine, A. J. (2006). p53 tumor suppressor protein regulates the levels of huntingtin gene expression. *Oncogene.* 25, 1-7.

Ferrer, I., Blanco, R., Cutillas, B., Ambrosio, S. (2000). Fas and Fas-L expression in Huntington's disease and Parkinson's disease. *Neuropathol. Appl. Neurobiol.* 26, 424-33.

Freeman, W., Morton, A. J., (2004). Regional and progressive changes in brain expression of complexin II in a mouse transgenic for the Huntington's disease mutation. *Brain Res. Bull.* 63, 45-55.

Gervais, F. G, Singaraja, R., Xanthoudakis, S., Gutekunst, C. A., Leavitt, B. R., Metzler, M., Hackam, A. S., Tam, J., Vaillancourt, J. P., Houtzager, V., Rasper, D. M., Roy, S., Hayden, M. R., Nicholson, D. W. (2002). Recruitment and activation of caspase-8 by the Huntingtin-interacting protein Hip-1 and a novel partner Hippi. *Nat. Cell Biol.* 4, 95-105.

Goehler, H., Lalowski, M., Stelzl, U., Waelter, S., Stroedicke, M., Worm, U., Droege, A., Lindenberg, K. S., Knoblich, M., Haenig, C., Herbst, M., Suopanki, J., Scherzinger, E., Abraham, C., Bauer, B., Hasenbank, R., Fritzsche, A., Ludewig, A. H., Büssow, K., Coleman, S. H., Gutekunst, C. A., Landwehrmeyer, B. G., Lehrach, H., Wanker, E. E. (2004). A protein interaction network links GIT1, an enhancer of huntingtin aggregation, to Huntington's disease. *Mol. Cell.* 15, 853-65.

Gunawardena, S., Her, L. S., Brusch, R. G., Laymon, R. A., Niesman, I. R., Gordesky-Gold, B., Sintasath, L., Bonini, N. M., Goldstein, L. S. (2003).Disruption of axonal transport by loss of huntingtin or expression of pathogenic polyQ proteins in Drosophila. *Neuron.* 40, 25-40.

Harper, P. S., (1991). The epidemiology of Huntington's disease. In Huntington's Disease, Second Edition (Ed. . Harper, P. S.) W.B. Saunders Co. London. pp 201-240

Hickey, M. A., Chesselet, M. F., (2003). Apoptosis in Huntington's disease. *Prog Neuropsychopharmacol. Biol. Psychiatry.* 27, 255-65.

Hodges, A., Strand, A. D., Aragaki, A. K., Kuhn, A., Sengstag, T., Hughes, G., Elliston, L. A., Hartog, C., Goldstein, D. R., Thu, D., Hollingsworth, Z. R., Collin, F., Synek, B., Holmans, P. A., Young, A. B., Wexler, N. S., Delorenzi, M., Kooperberg, C., Augood, S. J., Faull, R. L., Olson, J. M., Jones, L., Luthi-Carter R. (2006). Regional and cellular gene expression changes in human Huntington's disease brain. *Hum. Mol. Genet.* 15, 965-977

Hoefen, R. J., Berk, B. C. (2006). The multifunctional GIT family of proteins. *J. Cell Sci.* 119, 1469-1475.

Holbert, S., Denghien, I., Kiechle, T., Rosenblatt, A., Wellington, C., Hayden, M.R,. Margolis, R. L, Ross, C. A., Dausset, J., Ferrante, R. J. and Neri, C. (2001). The Gln-Ala repeat transcriptional activator CA150 interacts with huntingtin: neuropathologic and genetic evidence for a role in Huntington's disease. *Proc. Natl. Acad. Sci. USA* **98** 1811–1816

Horn, S.C., Lalowski, M., Goehler, H., Dröge, A., Wanker, E. E., Stelzl, U. (2006). Huntingtin interacts with the receptor sorting family protein GASP2. *J. Neural. Transm.* 113, 1081-90.

Huntington's Disease Collaborative Research Group. (1993). A novel gene containing a trinucleotide repeat that is expanded and unstable on Huntington's disease chromosomes. *Cell.* 72, 971-983

Huntington, G. (1872). The Medical and Surgical Reporter of Philadelphia, 26 (no.15), 317-321

Imarisio, S., Carmichael, J., Korolchuk, V., Chen, C. W, Saiki, S., Rose, C., Krishna, G., Davies, J. E., Ttofi, E., Underwood, B. R., Rubinsztein, D. C. (2008). Huntington's disease: from pathology and genetics to potential therapies. *Biochem. J.* 412, 191-209.

Jin, K., LaFevre-Bernt, M., Sun, Y., Chen, S., Gafni, J., Crippen, D., Logvinova, A., Ross, C. A., Greenberg, D. A., Ellerby, L. M. (2005). FGF-2 promotes neurogenesis and neuroprotection and prolongs survival in a transgenic mouse model of Huntington's disease. *Proc. Natl. Acad. Sci. U. S. A.* 102, 18189-94.

Johnson, R., Buckley, N. J. (2009). Gene dysregulation in Huntington's disease: REST, microRNAs and beyond. *Neuromolecular Med.,* 11, 183-199.

Kaltenbach LS, Romero E, Becklin RR, Chettier R, Bell R, Phansalkar A, Strand A, Torcassi C, Savage J, Hurlburt A, Cha GH, Ukani L, Chepanoske CL, Zhen Y, Sahasrabudhe S, Olson J, Kurschner C, Ellerby LM, Peltier JM, Botas J, Hughes RE (2007) Huntingtin interacting proteins are genetic modifiers of neurodegeneration. *PLoS Genet.* 3, e82.

Kazemi-Esfarjani, P. and Benzer, S. (2002). Suppression of polyglutamine toxicity by a Drosophila homolog of myeloid leukemia factor 1. *Hum Mol Genet,* **11,** 2657-2672.

Lange, K. W., Javoy-Agid, F., Agid, Y., Jenner, P., Marsden, C. D. (1992). Brain muscarinic cholinergic receptors in Huntington's disease. *J. Neurol.* 239, 103-104.

Li, X.J., Friedman, M., Li, S. (2007). Interacting proteins as genetic modifiers of Huntington disease. *Trends Genet.* 23, 531-533

Li, S. H., Yu, Z. X., Li, C. L., Nguyen, H. P., Zhou, Y. X., Deng, C., Li, X. J. (2003). Lack of huntingtin-associated protein-1 causes neuronal death resembling hypothalamic degeneration in Huntington's disease. *J. Neurosci.* 23, 6956-6964.

Liévens, J. C., Rival, T., Iché, M., Chneiweiss, H., Birman, S. (2005). Expanded polyglutamine peptides disrupt EGF receptor signaling and glutamate transporter expression in Drosophila. *Hum. Mol. Genet.* 14, 713-724.

Majumder, P., Chattopadhyay, B., Mazumder, A., Das, P., Bhattacharyya, N. P. (2006). Induction of apoptosis in cells expressing exogenous Hippi, a molecular partner of huntingtin-interacting protein Hip1. *Neurobiol. Dis.* 22, 242-256.

McCampbell, A., Taylor, J. P., Taye, A. A., Robitschek, J., Li, M., Walcott, J., Merry, D., Chai, Y., Paulson, H., Sobue, G., Fischbeck, K. H. (2000). CREB-binding protein sequestration by expanded polyglutamine. *Hum. Mol. Genet.* 9, 2197-2202

Metzger, S., Rong, J., Nguyen, H. P., Cape, A., Tomiuk, J., Soehn, A. S., Propping, P., Freudenberg-Hua, Y., Freudenberg, J., Tong, L., Li, S. H., Li, X. J., Riess, O. (2008). Huntingtin-associated protein-1 is a modifier of the age-at-onset of Huntington's disease. *Hum. Mol. Genet.* 17, 1137-1146.

Mills, I.G., Gaughan, L., Robson, C., Ross, T., McCracken, S., Kelly, J., Neal, D.E. (2005). Huntingtin interacting protein 1 modulates the transcriptional activity of nuclear hormone receptors. *J. Cell Biol.* 170, 191-200.

Mitsui, K., Nakayama, H., Akagi, T., Nekooki, M., Ohtawa, K., Takio, K., Hashikawa, T., Nukina, N. (2002). Purification of polyglutamine aggregates and identification of elongation factor-1alpha and heat shock protein 84 as aggregate-interacting proteins. *J. Neurosci.*, 22, 9267-9277

Modregger, J., DiProspero, N.A., Charles, V., Tagle, D.A., Plomann, M. (2002). PACSIN 1 interacts with huntingtin and is absent from synaptic varicosities in presymptomatic Huntington's disease brains. *Hum. Mol. Genet.* 11, 2547-2558.

Muchowski, P. J. and Wacker, J. L. (2005) Modulation of neurodegeneration by molecular chaperones. *Nat. Rev. Neurosci.* 6, 11-22.

Niatsetskaya, Z., Basso, M., Speer, R., McConoughey, S., Coppola, G., Ma, T. C., Ratan, R. (2009). HIF prolyl hydroxylase inhibitors prevent neuronal death induced by mitochondrial toxins: therapeutic implications for Huntington's disease and Alzheimer's disease. *Antioxid. Redox. Signal.* 2009 Aug 6 [Epub ahead of print].

Oláh, J., Klivényi, P., Gardián, G., Vécsei, L., Orosz, F., Kovacs, G. G., Westerhoff, H. V., Ovádi, J. (2008). Increased glucose metabolism and ATP level in brain tissue of Huntington's disease transgenic mice. *FEBS J.* 275, 4740-4755.

Opal, P., Zoghbi, H. Y., 2002. The role of chaperones in polyglutamine disease. *Trends Mol. Med.* 8, 232-236.

Powers, W. J., Videen, T. O., Markham, J., McGee-Minnich, L., Antenor-Dorsey, J. V., Hershey, T., Perlmutter, J. S. (2007). Selective defect of in vivo glycolysis in early Huntington's disease striatum. *Proc. Natl. Acad. Sci. U. S. A.* 104, 2945-2949

Pramanik, S., Basu, P., Gangopadhaya, P. K., Sinha, K. K., Jha, D. K, Sinha, S., Das, S. K., Maity, B. K, Mukherjee, S. C, Roychoudhuri, S., Majumder, P. P., Bhattacharyya, N. P. (2000). Analysis of CAG and CCG repeats in Huntingtin gene among HD patients and normal populations of India. *Eur. J. Hum. Genet.* 8, 678-682.

Raychaudhuri, S., Sinha, M., Mukhopadhyay. D., Bhattacharyya N. P. (2008). HYPK, a Huntingtin interacting protein, reduces aggregates and apoptosis induced by N-terminal Huntingtin with 40 glutamines in Neuro2a cells and exhibits chaperone-like activity. *Hum. Mol. Genet.* 17, 40-55.

Ross CA (2004) Huntington's Disease, *Cell.* 118, 4-7

Ryan, A. B., Zeitlin, S. O., Scrable, H. (2006). Genetic interaction between expanded murine Hdh alleles and p53 reveal deleterious effects of p53 on Huntington's disease pathogenesis. *Neurobiol. Dis.* 24, 419-427

Sánchez, I., Xu, C. J., Juo, P., Kakizaka, A., Blenis, J., Yuan, J. (1999). Caspase-8 is required for cell death induced by expanded polyglutamine repeats. *Neuron.* 22, 623-633.

Sassone, J., Colciago, C., Cislaghi, G., Silani, V., Ciammola, A. (2009). Huntington's disease: the current state of research with peripheral tissues, *Exp. Neurol.* 219, 385-397.

Seizinger, B. R., Liebisch, D. C., Kish, S. J, Arendt, R. M., Hornykiewicz, O., Herz, A. (1986), Opioid peptides in Huntington's disease: alterations in prodynorphin and proenkephalin system. *Brain Res.,* 378, 405-408

Smith, R., Klein, P., Koc-Schmitz, Y., Waldvogel, H. J., Faull, R. L., Brundin, P., Plomann, M., Li, J. Y. (2007). Loss of SNAP-25 and rabphilin 3a in sensory-motor cortex in Huntington's disease. *J. Neurochem.* 103, 115-123.

Stark, C., Breitkreutz, B. J., Reguly, T., Boucher, L., Breitkreutz, A., Tyers, M. (2006). BioGRID: A General Repository for Interaction Datasets. *Nucleic Acids Res.* 34:D535-539.

Steffan, J. S., Kazantsev, A., Spasic-Boskovic, O., Greenwald, M., Zhu, Y. Z., Gohler, H., Wanker, E. E., Bates, G. P., Housman, D. E., Thompson, L. M. (2000). The Huntington's disease protein interacts with p53 and CREB-binding protein and represses transcription. *Proc. Natl. Acad. Sci. U. S. A.* 97, 6763-6768.

Sugars, K. L., Rubinsztein, D. C. (2003). Transcriptional abnormalities in Huntington disease. *Trends Genet.* 19, 233-238.

Truant, R., Atwal, R. S., Burtnik, A. (2007). Nucleocytoplasmic trafficking and transcription effects of huntingtin in Huntington's disease. *Prog. Neurobiol.* 83, 211-227.

Valenza, M., et al., 2007. Progressive dysfunction of the cholesterol biosynthesis pathway in the R6/2 mouse model of Huntington's disease. *Neurobiol. Dis.* 28, 133-142.

van der Burg, J. M., Björkqvist, M., Brundin, P. (2009). Beyond the brain: widespread pathology in Huntington's disease. *Lancet Neurol.,* 8, 765-774

Vetter, J. M, Jehle, T., Heinemeyer, J., Franz, P., Behrens, P. F., Jackisch, R., Landwehrmeyer, G.B, Feuerstein, T.J. (2003). Mice transgenic for exon 1 of Huntington's disease: properties of cholinergic and dopaminergic pre-synaptic function in the striatum. *J. Neurochem.* 85, 1054-1063.

Von Mering, C., Krause, R., Snel, B., Cornell, M., Oliver, S.G., Fields, S., Bork. P. (2002). Comparative assessment of large-scale data sets of protein–protein interactions. *Nature.* 417, 399–403

Wastek, G. J., Yamamura, H. I., (1978). Biochemical characterization of the muscarinic cholinergic receptor in human brain: alterations in Huntington's disease. *Mol. Pharmacol.* 14, 768-780.

Willingham, S., Outeiro, T. F., DeVit, M. J., Lindquist, S. L, Muchowski, P. J. (2003). Yeast genes that enhance the toxicity of a mutant huntingtin fragment or alpha-synuclein. *Science.* 302, 1769-1772

Yang, H., et al., 2007. Ubiquitin ligase Hrd1 enhances the degradation and suppresses the toxicity of polyglutamine-expanded huntingtin. *Exp. Cell Res.* 313, 538-50.

Yang, Y. T., Ju, T. C., Yang, D. I. (2005). Induction of hypoxia inducible factor-1 attenuates metabolic insults induced by 3-nitropropionic acid in rat C6 glioma cells. *J. Neurochem.,* 93, 513-525

Zuccato, C., Belyaev, N., Conforti, P., Ooi, L., Tartari, M., Papadimou, E., MacDonald, M., Fossale, E., Zeitlin, S., Buckley, N., Cattaneo, E. (2007). Widespread disruption of repressor element-1 silencing transcription factor/neuron-restrictive silencer factor occupancy at its target genes in Huntington's disease. *J. Neurosci.*, 27, 6972-6983

Zuccato, C., Tartari, M., Crotti, A., Goffredo, D., Valenza, M., Conti, L., Cataudella, T., Leavitt, B. R., Hayden, M. R., Timmusk, T., Rigamonti, D., Cattaneo, E, (2003). Huntingtin interacts with REST/NRSF to modulate the transcription of NRSE-controlled neuronal genes. *Nat. Genet.* 35, 76–83

In: Encyclopedia of Neuroscience Research
Editors: Eileen J. Sampson and Donald R. Glevins

ISBN 978-1-61324-861-4
© 2012 Nova Science Publishers, Inc.

Chapter XXXIV

DNA Repair and Huntington's Disease

*Fabio Coppedè**
Department of Neuroscience, University of Pisa, Via Roma 67, 56126 Pisa, Italy

Abstract

Huntington's disease (HD) is a progressive neurodegenerative disorder resulting in cognitive impairment, choreiform movements and death which usually occurs 15-20 years after the onset of the symptoms. A CAG repeat expansion within exon 1 of the gene encoding for huntingtin (*IT15*) causes the disease. In the normal population the number of CAG repeats is maintained below 35, while in individuals affected by HD it ranges from 35 to more than 100, resulting in an expanded polyglutamine segment in the protein. HD age at onset is inversely correlated with the CAG repeat length; moreover the CAG repeat length seems to be related to the rate of progression of neurological symptoms and motor impairment. Somatic CAG repeat expansion in the huntingtin gene has been observed in several tissues of HD patients, but particularly in the striatum, the region most affected by the disease. An age-dependent somatic CAG repeat expansions was also observed in tissues of HD transgenic mice. Recently, it was found that somatic CAG repeat expansion is induced by oxidative stress in cultured HD fibroblasts and occurs during the repair of oxidized base lesions, dependent on a single base excision repair (BER) enzyme, the DNA glycosylase OGG1 which specifically removes oxidized guanine (8-oxo-G) from the DNA. It was therefore hypothesized that an age-dependent increase in 8-oxo-G formation in post-mitotic neurons could induce a DNA damage response mediated by OGG1 and give rise to a CAG repeat expansion that likely contributes to the onset and the progression of the disease. Further studies confirmed that somatic expansion of the CAG repeat tract in the brain is associated with an earlier HD age at onset. Preliminary results from our group suggest that a common OGG1 Ser326Cys polymorphism could contribute to CAG repeat number and disease age at onset. Several subsequent studies support a role for oxidative DNA damage and BER to somatic instability of CAG repeats. There is also evidence that another DNA repair pathway, the mismatch repair (MMR), is

---

* Corresponding author: Department of Neuroscience, University of Pisa, Via Roma 67, 56126 Pisa, ITALY., Current Address: Department of Human and Environmental Sciences, University of Pisa, Via S. Giuseppe 22, 56126 Pisa, ITALY. Phone: +39 050 2211028, E-mail: f.coppede@geog.unipi.it

required for the active mutagenesis of expanded CAG repeats. Since MMR is required for the repair of mismatched adenine opposite to 8-oxo-G, it remains to be seen if both BER and MMR co-operate physically in the process of CAG repeat expansion.

Introduction

Huntington's disease (HD) is a progressive neurodegenerative disorder resulting in cognitive impairment, choreiform movements and death which usually occurs 15-20 years after the onset of the symptoms. The disease is also characterized by psychiatric and behavioural disturbances. HD is an autosomal dominant disorder caused by a CAG repeat expansion within exon 1 of the gene encoding for huntingtin (*IT15*) on chromosome 4. In the normal population the number of CAG repeats is maintained below 35, while in individuals affected by HD it ranges from 35 to more than 100, resulting in an expanded polyglutamine segment in the protein. Huntingtin plays a role in protein trafficking, vesicle transport, postsynaptic signalling, transcriptional regulation, and apoptosis. Thus, a loss of function of the normal protein and a toxic gain of function of the mutant huntingtin contribute to the disruption of multiple intracellular pathways, ultimately leading to neurodegeneration [Gil and Rego, 2008]. Age at onset (AAO) of the disease is inversely correlated with the CAG repeat length; moreover the length of the expanded polyglutamine segment seems to be related to the rate of clinical progression of neurological symptoms and to the progression of motor impairment, but not to psychiatric symptoms [Furtado et al., 1996; Penney et al., 1997; Rosenblatt et al., 2006; Vassos et al., 2008].

The CAG repeat length accounts for 70% of the variability in HD AAO. However, 90% of individuals worldwide with expanded alleles possess between 40 and 50 CAG repeats. For these people, the size of their repeat only determines almost 50% of the variability in disease AAO [Gayán et al., 2008]. Significant variance remains in residual age at onset even after CAG repeat length is factored out, and studies in both HD patients and transgenic HD mice models suggest that other genetic and environmental factors can modify the onset and progression of motor symptoms, as well as physical, cognitive and social functions [Van Dellen et al., 2005; Nithianantharajah et al., 2008].

Errors in DNA replication are thought to underlie the lengthening of tracts of repeated DNA that occurs in HD and other neurodegenerative diseases [Mirkin, 2006]. However, recent findings indicate that mechanisms for repairing damaged DNA may also be responsible. Particularly, several studies support the notion that DNA repair contributes to somatic instability of CAG repeats, though other DNA associated processes, such as DNA replication, may also play a role [Pearson et al., 2005]. At least two repair processes appear to regulate *in vivo* somatic instability in the brain of HD mouse models. The mismatch repair process (MMR), whose role in somatic CAG repeat instability is well documented, and the DNA base excision repair (BER) pathway, which is specialized in the removal of damaged bases, such as oxidized bases [Manley et al., 1999; Kovtun et al., 2007]. This chapter describes in details the most recent findings linking the base excision repair of oxidatively modified bases to somatic CAG repeat expansion. A brief description of the contribution of the MMR pathway is given in the last section of the chapter.

Somatic CAG Repeat Instability and HD Age at Onset

Somatic CAG repeat expansion in the gene encoding for huntingtin has been observed in several tissues, including the blood, but particularly in the striatum of HD patients, the region most affected by the disease [Cannella et al., 2005, 2009; Shelbourne et al., 2007; Veitch et al., 2007]. A recent study was performed on cultured lymphoblasts obtained from HD subjects with various CAG repeat lengths. The study indicated that most of the cultures showed little or no repeat instability even after six or more months culture; however, in lymphoblasts with large expansion beyond 60 CAG repeats the mutation size and triplet mosaicism always increased during replication, implying that the repeat mutability for highly expanded mutations may quantitatively depend on the triplet expansion size. Therefore, these experiments suggest that the inherited CAG length of expanded alleles has a major influence on somatic repeat variation, with the longest triplet expansions showing wide somatic variations during cell replication [Cannella et al., 2009]. Similar results have been obtained in buccal cell DNA from HD subjects. Indeed, researchers observed that HD mutation length profiles in buccal cell DNA vary from individual to individual, and that the inherited CAG/CTG repeat length has a major influence on somatic CAG/CTG repeat length variation [Veitch et al., 2007]. Moreover, Kovtun et al. [2007] observed an age- and length-dependent somatic CAG repeat expansion in several tissues of HD transgenic mice; they also observed that CAG repeat expansion is induced by oxidative stress in cultured fibroblasts obtained from HD patients and that this phenomenon is caused by the process of removing oxidized base lesions, dependent on a single base excision repair enzyme, 7,8-dihydro-8-oxoguanine-DNA glycosylase (OGG1) [Kovtun et al., 2007]. Human OGG1 is an enzyme of the BER pathway which removes 8-oxoguanine (8-oxo-G) and nicks the DNA, leaving an abasic site, which is then completely repaired by the action of other enzymes [Klungland et al., 2007]. Somatic expansion does not require cell division and can occur in neurons after these cells are terminally differentiated and mitotic replication has ceased. Therefore, it was proposed a 'toxic oxidation cycle' model in which an age-dependent increase in oxidative lesions in post-mitotic neurons induces a DNA damage response and increases the need for OGG1 repair through a single strand breaks (SSB) mechanism, resulting in CAG repeat expansion that could contribute to HD onset and progression [Kovtun et al., 2007]. This study was published in *Nature* in 2007 and was the first paper linking the repair of oxidative DNA damage to HD [Kovtun et al., 2007].

A study performed with animal models of HD showed that region-specific CAG repeat mosaicism profiles are conserved between several mouse models of HD and therefore develop in a predetermined manner. Moreover, it was demonstrated that these synchronous, radical changes in CAG repeat size occur in terminally differentiated neurons. The neuronal population of the striatum was particularly distinguished by a high rate of CAG repeat allele instability and expression driving the repeat upwards and likely enhancing its toxicity [Gonitel et al., 2008]. Results from micro-dissected tissue and individual laser-dissected cells obtained from human HD cases and knock-in HD mice indicate that the CAG repeat is unstable in all cell types tested although neurons tend to have longer mutation length gains than glia. Mutation length gains occur early in the disease process and continue to accumulate as the disease progresses. In keeping with observed patterns of cell loss, neuronal mutation length gains tend to be more prominent in the striatum than in the cortex of low-grade human HD cases, less so in more advanced cases. Interestingly, neuronal sub-populations of HD mice appear to have different propensities for mutation length gains; in particular, smaller mutation length gains occur in nitric oxide synthase-positive striatal interneurons (a relatively spared cell type in HD) compared with the pan-striatal neuronal population [Shelbourne et al., 2007]. These data demonstrated that neuronal changes in HD repeat length can be at least as great, if not greater, than those observed in the germline, and the fact that significant CAG repeat length gains occur in non-replicating cells also argues that processes such as inappropriate DNA repair rather than DNA replication are involved in generating mutation instability in HD brain tissues [Shelbourne et al., 2007].

The link between somatic CAG repeat expansion and HD age at onset was supported in a recent study [Swami et al., 2009]. Researchers from the Center for Human Genetic Research of the Massachusetts General Hospital, Boston, quantified somatic instability in the cortex region of the brain from a cohort of HD individuals exhibiting phenotypic extremes of young and old disease onset as predicted by the length of their constitutive HD CAG repeat lengths. After accounting for constitutive repeat length, somatic instability was found to be a significant predictor of onset age, with larger repeat length gains associated with earlier disease onset. These data were consistent with the hypothesis that somatic CAG repeat length expansions in target tissues contribute to the HD pathogenic process, and supported pursuing factors that modify somatic instability as viable therapeutic targets [Swami et al., 2009].

The Cellular Defence against the Accumulation of Oxidized Bases

One of the major oxidative damage occurring in neuronal DNA is the formation of 8-oxo-G since, among DNA bases, guanine is the most susceptible to oxidative attack. Human cells possess several enzymes to counteract the mutagenic potential of 8-oxo-G accumulation into nucleic acids. These include the oxidized purine nucleoside triphosphatase MTH1 that hydrolyzes oxidized guanine in the nucleotide pools avoiding its incorporation into nucleic acids, OGG1 that removes 8-oxo-G from the DNA, and the humat MutY homolog (hMUTYH) that removes misincorporated adenine opposite to 8-oxo-G in DNA. The study performed by Kovtum et al [2007] suggested a direct role for OGG1 in mediating somatic CAG repeat expansion induced by oxidative stress in cultured fibroblasts obtained from HD patients. Further indication of a possible role for oxidative DNA damage and aberrant BER in somatic instability of CAG repeats is suggested by several studies (Table 1) described in the following sections. However, before describing all these studies in details I provide a brief description of the BER pathway.

The DNA base excision repair pathway is believed to be the major pathway for repairing DNA base modifications caused by oxidation, deamination and alkylation. Cells contain several DNA glycosylases, each of them exhibiting a specific substrate spectrum. DNA glycosylases, such as OGG1, catalyze the first step in the BER process by cleaving the N-glycosylic bond between the damaged base and the sugar moiety; after the cleavage the damaged base is released resulting in the formation of an abasic site which is then cleaved by an AP lyase activity or by the major mammalian apurinic/apyrimidinic endonuclease (APE1). Repair can then proceed through short or long-patch BER.

Table 1. Evidence Linking Oxidative DNA Damage and BER Activity to Somatic CAG Repeat Instability

ENZYME	FUNCTION	INVOLVMENT IN HD	REFERENCES
hMTH1 nucleoside triphosphatase	Removal of 8-oxo-G and other oxidized bases from nucleotide pools	Protects against oxidative stress induced HD-like symptoms	[De Luca et al., 2008]
OGG1 Oxoguanine DNA glycosylase	Removal of 8-oxo-G from the DNA	OGG1 has been suggested to cause somatic CAG repeat expansion. *In vitro* studies revealed reduced glycosylase activity in CAG repeats. The OGG1 Ser326Cys polymorphism has been linked to CAG repeats and HD age at onset.	[Kovtun et al., 2007] [Jarem et al., 2009] [Coppedè et al., 2009]
APE1 Endonuclease	Creation of abasic sites	Reduced activity in CAG repeats Abasic sites could favour somatic CAG instability	[Goula et al., 2009] [Völker et al., 2009]
POLβ DNA polymerase	Gap filling	Coordination between POLβ and FEN1 in long patch BER is required to prevent somatic CAG repeat instability	[Liu et al., 2009] [Goula et al., 2009]
DNA ligase I	DNA ligation	Impaired ligation might contribute to somatic CAG repeat instability	[López Castel et al., 2009]

In short-patch BER, which is the most common sub-pathway, a single nucleotide is incorporated into the gap by DNA polymerase β (POLβ) and ligated by the DNA ligase III/ X-ray repair cross-complementing group 1 (XRCC1) complex. However, when the 5'-terminal residue is resistant to POLβ liase activity, repair proceeds via long-patch sub-pathway In long-patch BER several nucleotides (two to seven-eight) are incorporated, followed by cleavage of the resulting 5' flap structure mediated by the 5'-flap endonuclease (FEN-1) and ligation. It has been suggested that after POLβ adds the first nucleotide into the gap, it is substituted by POLδ or POLε which continue long-patch BER. DNA ligase I completes the long-patch pathway. Several other proteins, including the proliferating cell nuclear antigen (PCNA), the RPA protein and FEN-1 participate in long-patch BER [Dianov et al., 2003]. BER takes places either in nuclei and mitochondria, and mitochondria have independent BER machinery encoded by nuclear genes. Indeed, several BER enzymes have been identified which have both nuclear and mitochondrial forms. For example, among DNA glycosylases, nuclear and mitochondrial adenine-DNA-glycosylases are generated by alternative spliced forms of the *MYH* gene. Similarly nuclear and mitochondrial uracil DNA glycosylases result from alternative splicing and transcription of the *UNG* gene. Two major OGG1 isoforms have been identified in human cells, the alpha and the beta form, arising as alternative splice products from the *OGG1* gene. There is indication suggesting that the beta form operates in mitochondria, while the alpha form operates in the nucleus as well as in mitochondria. Recent evidence suggests that XRCC1 acts as a scaffold protein in short-patch BER, regulating and coordinating the whole process. XRCC1 recruits DNA POLβ and DNA ligase III required for filling and sealing the damaged strand. Moreover, it also interacts with DNA glycosylases and APE1, mediating their exchange at the damaged site. XRCC1 also interacts with PARP-1, which is one of the cellular sensors of DNA damage. Several studies have been performed to investigate age-related changes in BER activity in neurons. We recently reviewed these studies and multiple evidence indicates that BER undergoes age-related and regional specific changes which are likely to contribute to the accumulation of oxidative DNA lesions in the aging brain [Coppedè and Migliore, 2010].

In Vitro Evidence Linking Oxidative DNA Damage and BER Activity to Somatic CAG Instability

Triplet repeat sequences, such as CAG/CTG, expand in the human genome to cause several neurological disorders, including HD. As part of the expansion process the formation of non-B DNA conformations, such as loops and hairpins, by the repeat sequence has been proposed [Lenzmeier and Freudenreich, 2003], and OGG1 has been recently implicated in the repeat expansion [Kovtun et al., 2007]. In a recent *in vitro* study researchers from the Department of Chemistry at the Brown University observed that the non-B conformation adopted by a tract containing 10 CAG repeats, a hairpin, is hyper-susceptible to DNA damage respect to duplex DNA, since a hot spot for DNA damage exists in this sequence. Specifically, researchers observed that a single guanine in the loop of the hairpin is susceptible to modification by peroxynitrite. Interestingly, it was observed that human OGG1 is able to excise 8-oxo-G from the loop of the hairpin substrate, but with a marked decrease in efficiency; indeed, the OGG1 enzyme removes 8-oxo-G from the loop of a hairpin with a rate that is approximately 700-fold slower than that observed for DNA duplex. Thus, while damage is preferentially generated in the loop of the hairpin, DNA repair is less efficient. The researchers suggested that these observed structure-dependent patterns of DNA damage and repair may contribute to the OGG1-dependent mechanism of trinucleotide repeat expansion [Jarem et al., 2009].

As previously observed, it has been suggested that CAG repeat expansion involves the transient formation of non-native slipped DNA structures within the triplet repeat domain [Lenzmeier and Freudenreich, 2003]. The observation that BER of oxidative damage at or near CAG repeats facilitates DNA triplet expansion [Kovtun et al., 2007; McMurray, 2008] suggests that the presence of abasic sites may influence the ability of repeat DNA sequences to form non-native slipped DNA structures. To assess this possibility, researchers from the Department of Chemistry and Chemical Biology of The State University of New Jersey evaluated the impact of abasic sites on the overall stability and conformational preferences of a CAG triplet repeat bulge loop structure that models slipped DNA states, observing that replacement within a triplet repeat bulge loop domain of a guanosine residue by an abasic site, the universal BER intermediate, increases the population of slipped/looped DNA structures relative to the corresponding lesion-free construct. Such abasic lesion-induced energetic enhancement of slipped/looped structures provides a linkage between BER and DNA expansion, potentially modulating levels of DNA expansion [Völker et al., 2009].

A recent study performed in mouse cell extracts showed a size-limited expansion of CAG repeats during repair of 8-oxo-G in wild-type cells. This expansion was deficient in extracts from cells lacking POLβ and HMGB1, which is a BER protein that is known to stimulate APE1 5'-incision activity and FEN1 cleavage activity. Authors demonstrated that CAG expansion is mediated through POLβ multinucleotide gap-filling DNA synthesis during long-patch BER. Unexpectedly, FEN1 promotes expansion by facilitating ligation of hairpins formed by strand slippage [Liu et al., 2009]. Researchers proposed that this alternate role of FEN1 and the POLβ multinucleotide gap-filling synthesis are the result of uncoupling of the usual coordination between POLβ and FEN1. HMGB1 probably promotes expansion by stimulating APE1 and FEN1 in forming single strand breaks and nicks, respectively [Liu et al., 2009]. The study by Liu and co-workers demonstrates that that disruption of POLβ and FEN1 coordination during long-patch BER results in CAG repeat expansion, and overall these *in vitro* studies suggest that several proteins and intermediates of the BER pathway might participate in the somatic instability of CAG repeats.

Stidies in HD Transgenic Mice Linking Oxidative DNA Damage and BER Activity to Somatic CAG Instability

A recent study performed on transgenic HD mice provided a nice complement to the *in vitro* studies described in the previous section. Authors observed that in HD mice oxidative DNA damage abnormally accumulates at CAG repeats in a length-dependent, but age- and tissue-independent manner, indicating that oxidative DNA damage alone is not sufficient to trigger somatic instability [Goula et al., 2009]. It was also observed that OGG1 and APE1 are unable to efficiently incise hairpin substrates, supporting the idea that oxidative DNA lesions accumulate at CAG repeats because they can form secondary structures that are refractory to processing by BER enzymes [Goula et al., 2009]. Protein levels and activities of major BER enzymes were compared between striatum and cerebellum of HD mice. FEN1 and the BER cofactor HMGB1 were significantly lower in the striatum than in the cerebellum, and POLβ was specifically enriched at CAG expansions in the striatum, but not in the cerebellum of HD mice [Goula et al., 2009]. Authors suggested that their results support a role for oxidative DNA damage and BER in somatic CAG instability in the striatum of HD mice. Particularly, their observations suggest that in the cerebellum an optimal cooperation between gap filling by POLβ and 5'-flap excision by FEN1 during the repair of oxidative DNA lesions via long-patch BER prevents the formation of slipped strand structures at CAG repeats. On the contrary, in the striatum, a poor cooperation between these two enzymes leads to the formation of complex structures that could promote somatic expansion [Goula et al., 2009].

Studies in Mammalian Cell Cultures Linking Oxidative DNA Damage and BER Activity to Somatic CAG Instability

DNA ligation is an essential step common to replication and repair, both potential sources of CAG repeat instability. Using DNA ligase I deficient human cells, López Castel and co-workers [2009] assessed the effect of ligase I activity, overexpression, and its interaction with proliferating cell nuclear antigen (PCNA) upon the ability to replicate and repair trinucleotide repeats. Compared with ligase I wild-type (+/+) cells, replication progression through repeats was poor, and repair tracts were broadened beyond the slipped-repeat for all mutant extracts. Increased repeat instability was linked only to ligase I over-expression and expression of a mutant ligase I incapable of interacting with PCNA. The endogenous mutant version of ligase I with reduced ligation activity did not alter instability. Authors distinguished the DNA processes through which ligase I contributes to trinucleotide instability. The highest levels of repeat instability were observed under the ligase I over-expression and were linked to reduced slipped-DNAs repair efficiencies [López Castel et al., 2009]. Therefore, this paper suggests that the replication-mediated instability can partly be attributed to errors during replication but also to the poor repair of slipped-DNAs formed during this process. The addition of purified proteins suggests that disruption of ligase I and PCNA interactions influences trinucleotide repeat instability [López Castel et al., 2009]. Overall, the paper by López Castel and co-workers, which is based on the use of mammalian cell lines impaired for ligase I, extends the number of steps of the BER process which are relevant for somatic CAG repeat instability by suggesting that the coordination with the downstream ligation step is also crucial.

Studies in Humans Linking the OGG1 Ser326Cys Polymorphism and Huntington's Disease

Further indication for a possible contribution of the BER pathway to CAG repeat expansion comes from a recent study by us [Coppedè et al., 2009]. We screened 91 HD subjects for the presence of the common OGG1 Ser326Cys polymorphism and searched for association between the polymorphism and both CAG repeat length in peripheral lymphocytes and disease age at onset. Interestingly, we observed that heterozygous Ser/Cys and mutant Cys/Cys subjects tend to have a mean increased CAG repeat number and an earlier age at onset of the disease compared to Ser/Ser wild type HD patients [Coppedè et al., 2009]. The OGG1 326Cys variant has been associated with a reduced enzyme activity under conditions of oxidative stress [Bravard et al., 2009]. Particularly, in order to investigate whether the variant 326Cys allele encodes a protein with altered glycosylase activity, Bravard et al. [2009] compared the 8-oxo-G repair activity both *in vivo* and in cell extracts of lymphoblastoid cell lines established from individuals carrying either Ser/Ser or Cys/Cys genotypes, observing that cells homozygous for the Cys variant display increased genetic instability and reduced *in vivo* 8-oxo-G repair activity. The analysis of the redox status of the OGG1 protein in the cells showed that the lower activity of OGG1-Cys326 is associated with the oxidation of Cys326 to form a disulfide bond, suggesting that individuals bearing the OGG1-Cys326 variant could more readily accumulate mutations under conditions of oxidative stress [Bravard et al., 2009]. Several authors observed increased oxidative stress in HD subjects and in pre-symptomatic individuals carrying the HD expanded gene, suggesting that oxidative stress occurs early during the progression of the disease, prior to the manifestation of disease symptoms [Saft et al., 2005; Chen et al., 2007; Klepac et al., 2007]. Therefore, our observation of an association between the OGG1 Ser326Cys polymorphism and both HD age at onset and CAG repeat length led us to formulate the hypothesis that under condition of oxidative stress, such as those occurring in HD subjects and in pre-symptomatic carriers of the expanded HD gene, the presence of the Cys326 aminoacid in OGG1 might render the enzyme less efficient in the repair of 8-oxo-G occurring in CAG repeats, thus enhancing size-expansion of the CAG tract [Coppedè et al., 2009]. Although our working hypothesis is that the OGG1 Ser326Cys polymorphism might contribute to CAG repeat instability in somatic tissues and to HD age at onset, our results should however be taken with caution since we only had DNA from blood cells of 91 HD subjects to search for a correlation between the mean number of CAG repeats and the OGG1 genotype. Even if we identified a significant correlation between the OGG1 genotype and the mean number of CAG repeats in blood cells, the results should be confirmed in other tissues, such as the striatum in the brain, where the CAG expanded tract is much more instable than in the blood.

Additional Evidence Linking Oxidative DNA Damage to Huntington's Disease

The accumulation of 8-oxo-G in the DNA can occur either through direct oxidation of DNA guanine or via incorporation of the oxidized nucleotide during replication. Hydrolases that degrade oxidized purine nucleoside triphosphates normally minimize this incorporation. hMTH1 is the major human hydrolase and recent findings suggest that hMTH1 expression in embryonic HD mice protects progenitor striatal cells from HD-like symptoms, thus providing additional evidence that 8-oxo-G accumulation in the DNA has a role in HD pathogenesis [De Luca et al., 2008]. The researchers constructed a transgenic mouse in which the human MTH1 (hMTH1) enzyme is expressed. hMTH1 expression protected embryonic fibroblasts and mouse tissues against the effects of oxidants. Particularly, wild-type mice exposed to 3-nitropropionic acid developed neuropathological and behavioural symptoms that resemble those of HD, while hMTH1 transgene expression conferred protection against these HD-like symptoms. In a complementary approach, an in vitro genetic model for HD was also used. hMTH1 expression protected progenitor striatal cells containing an expanded CAG repeat of the huntingtin gene from toxicity associated with expression of the mutant huntingtin. These findings provide additional evidence that the formation of 8-oxo-G is a critical step in the neuropathological features of HD [De Luca et al., 2008].

The Mismatch Repair (MMR) Pathway and Somatic CAG Instability

The mismatch repair pathway corrects mismatches and small insertions or deletions during DNA replication, thus eliminating potentially pre-mutagenic bases. Repair involves recognition of the mismatch by MutSα (MSH2 and MSH6 proteins), or by MutSβ (MSH2 and MSH3 proteins) in the case of small insertions/deletions (1-10 nucleotides). MutLα (a heterodimer of MLH1 and PMS2 proteins) is then recruited and serves to coordinate the process that involves, among others, the PCNA protein for strand discrimination and exonuclease 1, DNA POLδ and a DNA ligase, for DNA repair [Kunkel and Erie, 2005]. Hsieh and Yamane [2008] have recently provided a comprehensive review of the changes of MMR with aging concluding that the role of the MMR pathway in aging is not yet clearly discernible. Although several studies suggest that the capacity for MMR is diminished as a function of aging possibly due to the loss of key MMR proteins, it is still not clear whether correlations between aging and loss of MMR reflect a direct role for MMR in preventing aging or are indirect consequences of the aging process. These studies have been mainly performed on peripheral blood cells of donors at different ages [Hsieh and Yamane, 2008].

There is evidence that MMR proteins are required for the active mutagenesis of expanded CAG repeats, but the biological mechanism is still unclear. Studies in transgenic mice containing (CAG)n tracts have shown that certain MMR proteins are required for the somatic increase in repeat length. Particularly, in mice deficient for MSH2, MSH3 or PMS2, repeats are somatically stabilized, whereas deficiencies in MSH6 either lead to no change in somatic instability or else an increase in expansions. Moreover, deficiency of MSH2 or MSH3 resulted in CAG repeats instability in all tissues tested (for recent reviews see [McMurray, 2008; Slean et al., 2008]). Slean et al. [2008] suggested that there may be at least two distinct pathways for CAG repeat instability: one that is OGG1-dependent and independent of MSH2–MSH3, and another that is MSH2–MSH3-dependent, but independent of OGG1. However, since the failure to repair 8-oxo-G prior to DNA replication can lead to the formation of base mismatches, it remains to be seen if both OGG1 and MMR co-operate physically in the process of CAG repeat expansion [Slean et al., 2008].

Conclusion

Increasing evidence suggests a role for oxidative DNA damage and impaired DNA repair in the onset of several age-related neurodegenerative diseases, including Alzheimer's disease, Parkinson's disease, Amyotrophic Lateral Sclerosis, Huntington's disease, and many others [Coppedè and Migliore, 2010]. In the present chapter I have reviewed recent evidence suggesting that oxidative DNA damage and impairments in the DNA base excision repair pathway might be involved in somatic CAG repeat expansion in HD subjects, thus affecting disease symptoms and age at onset. Table 1 describes all the steps of the BER pathway which have been involved in somatic CAG instability. All these studies have been performed in the last two-three years, therefore we are at the beginning of the understanding of this process. Moreover, since most of the studies have been performed *in vitro* or in animal models of the disease, further investigation is required to confirm that something similar occurs in the striatum of HD subjects. Overall, the BER pathway is an emerging and exciting candidate to explain the origin of somatic CAG instability, but another DNA repair system, the mismatch repair, seems also to be involved [Slean et al., 2008]. Further studies are required to address the mechanism of action of the MMR pathway leading to somatic CAG instability and its possible interaction with the BER pathway, if any.

References

Bravard, A., Vacher, M., Moritz, E., Vaslin, L., Hall, J., Epe, B. & Radicella J. P. (2009). Oxidation status of human OGG1-S326C Polymorphic variant determines cellular DNA repair capacity. *Cancer Res.*, *69*, 3642-9.

Cannella, M., Maglione, V., Martino, T., Ragona, G., Frati, L., Li, G. M. & Squitieri, F. (2009). DNA instability in replicating Huntington's disease lymphoblasts. *BMC Med Genet*, *10*, 11.

Cannella, M., Maglione, V., Martino, T., Simonelli, M., Ragona, G. & Squitieri F. (2005). New Huntington disease mutation arising from a paternal CAG34 allele showing somatic length variation in serially passaged lymphoblasts. *Am J. Med Genet B Neuropsychiatr Genet*, 133B, 127-30.

Chen, C. M., Wu, Y. R., Cheng, M. L., Liu, J. L., Lee, Y. M., Lee, P. W., Soong, B. W. & Chiu, D. T. (2007). Increased oxidative damage and mitochondrial abnormalities in the peripheral blood of Huntington's disease patients. *Biochem Biophys Res Commun.*, *359*, 335-40.

Coppedè, F. & Migliore, L. (2010). DNA Repair in Premature Aging Disorders and Neurodegeneration. *Curr. Aging Sci.*, *3*, 3-19.

Coppedè, F., Migheli, F., Ceravolo, R., Bregant, E., Rocchi, A., Petrozzi, L., Unti, E., Lonigro, R., Siciliano, G. & Migliore, L. (2009). The hOGG1 Ser326Cys Polymorphism and Huntington's disease. *Toxicology*, in press. E-pub 2009, Oct. 24.

De Luca, G., Russo, M. T., Degan, P., Tiveron, C., Zijno, A., Meccia, E., Ventura, I., Mattei, E., Nakabeppu, Y., Crescenzi, M., Pepponi, R., Pèzzola, A., PoPoli, P. & Bignami, M. (2008). A role for oxidized DNA precursors in Huntington's disease-like striatal neurodegeneration. *PLoS Genet.*, *4*, e1000266.

Dianov, G. L., Sleeth, K. M., Dianova, I. I. & Allinson, S. L. (2003). Repair of abasic sites in DNA. *Mutat Res.*, *531*, 157-63.

Furtado, S., Suchowersky, O., Rewcastle, B., Graham, L., Klimek, M. L. & Garber, A. (1996). Relationship between trinucleotide repeats and neuropathological changes in Huntington's disease. *Ann Neurol.*, *39*, 132-6.

Gayán, J., Brocklebank, D., Andresen, J. M. & Alkorta-Aranburu, G., US-Venezuela Collaborative Research Group, Zameel Cader, M., Roberts, S. A., Cherny, S. S., Wexler, N. S., Cardon, L. R. & Housman, D. E. (2008). Genomewide linkage scan reveals novel loci modifying age of onset of Huntington's disease in the Venezuelan HD kindreds. *Genet Epidemiol.*, *32*, 445-53.

Gil, J. M. & Rego, A. C. (2008). Mechanisms of neurodegeneration in Huntington's disease. *Eur J. Neurosci.*, *27*, 2803-20.

Gonitel, R., Moffitt, H., Sathasivam, K., Woodman, B., Detloff, P. J., Faull, R. L. & Bates, G. P. (2008). DNA instability in postmitotic neurons. *Proc Natl Acad Sci U S A.*, 105, 3467-72.

Goula, A. V., Berquist, B. R., Wilson, D. M. 3rd, Wheeler, V. C., Trottier, Y. & Merienne, K. (2009). Stoichiometry of base excision repair proteins correlates with increased somatic CAG instability in striatum over cerebellum In Huntington's disease transgenic mice. *PLoS Genet.*, *5*, e1000749.

Hsieh, P. & Yamane, K. (2008). DNA mismatch repair: molecular mechanism, cancer, and ageing. *Mech Ageing Dev.*, *129*, 391-407.

Jarem, D. A., Wilson, N. R. & Delaney, S. (2009) Structure-dependent DNA damage and repair in a trinucleotide repeat sequence. *Biochemistry*, *48*, 6655-63.

Klepac, N., Relja, M., Klepac, R., Hećimović, S., Babić, T. & Trkulja, V. (2007). Oxidative stress parameters in plasma of Huntington's disease patients, asymptomatic Huntington's disease gene carriers and healthy subjects : a cross-sectional study. *J Neurol.*, *254*, 1676-83.

Klungland, A. & Bjelland, S. (2007). Oxidative damage to purines in DNA: role of mammalian Ogg1. *DNA Repair (Amst). 6*, 481-8.

Kovtun, I. V., Liu, Y., Bjoras, M., Klungland, A., Wilson, S. H. & McMurray, C. T. (2007). OGG1 initiates age-dependent CAG trinucleotide expansion in somatic cells. *Nature*, *447*, 447-52.

Kunkel, T. A. & Erie, D. A. (2005). DNA mismatch repair. *Annu Rev Biochem.*, *74*, 681-710.

Lenzmeier, B. A. & Freudenreich, C.H. (2003). Trinucleotide repeat instability: a hairpin curve at the crossroads of replication, recombination, and repair. *Cytogenet Genome Res.*, *100*, 7-24.

Liu, Y., Prasad, R., Beard, W. A., Hou, E. W., Horton, J. K., McMurray, C. T. & Wilson, S. H. (2009). Coordination between Polymerase beta and FEN1 can modulate CAG repeat expansion. *J Biol Chem, 284*, 28352-66.

López Castel, A., Tomkinson, A. E. & Pearson, C. E. (2009). CTG/CAG repeat instability is modulated by the levels of human DNA ligase I and its interaction with proliferating cell nuclear antigen: a distinction between replication and slipped-DNA repair. *J Biol Chem., 284*, 26631-45.

Manley, K., Shirley, T. L., Flaherty, L. & Messer, A. (1999). Msh2 deficiency prevents in vivo somatic instability of the CAG repeat in Huntington disease transgenic mice. *Nat Genet., 23*, 471-3.

McMurray, C. T. (2008). Hijacking of the mismatch repair system to cause CAG expansion and cell death in neurodegenerative disease. *DNA Repair (Amst). 7*, 1121-34.

Mirkin, S. M. (2006). DNA structures, repeat expansions and human hereditary disorders. Curr Opin *Struct Biol., 16*, 351-8.

Nithianantharajah, J., Barkus, C., Murphy, M. & Hannan, A. J. (2008). Gene-environment interactions modulating cognitive function and molecular correlates of synaptic plasticity in Huntington's disease transgenic mice. *Neurobiol Dis., 29*, 490-504.

Pearson, C. E., Nichol Edamura, K. & Cleary, J.D. (2005). Repeat instability: mechanisms of dynamic mutations. *Nat Rev Genet., 6*, 729-42.

Penney, J. B., Vonsattel, J. P., MacDonald, M. E., Gusella, J. F. & Myers R. H. (1997). CAG repeat number governs the development rate of pathology in Huntington's disease. *Ann Neurol., 41*, 689-92.

Rosenblatt, A., Liang, K. Y., Zhou, H., Abbott, M. H., Gourley, L. M., Margolis, R. L., Brandt, J. & Ross, C. A. (2006). The association of CAG repeat length with clinical progression in Huntington disease. *Neurology, 66*, 1016-20.

Saft, C., Zange, J., Andrich, J., Müller, K., Lindenberg, K., Landwehrmeyer, B., Vorgerd, M., Kraus, P. H., Przuntek, H. & Schöls, L. Mitochondrial impairment in patients and asymptomatic mutation carriers of Huntington's disease. *Mov Disord., 20*, 674-9.

Shelbourne, P. F., Keller-McGandy, C., Bi, W. L., Yoon, S. R., Dubeau, L., Veitch, N. J., Vonsattel, J. P., Wexler, N. S., US-Venezuela Collaborative Research Group, Arnheim, N. & Augood, S. J. (2007). Triplet repeat mutation length gains correlate with cell-type specific vulnerability in Huntington disease brain. *Hum Mol Genet., 16*, 1133-42.

Slean, M. M., Panigrahi, G. B., Ranum, L. P. & Pearson, C. E. (2008). Mutagenic roles of DNA "repair" proteins in antibody diversity and disease-associated trinucleotide repeat instability. *DNA Repair (Amst) 7*, 1135-54.

Swami, M., Hendricks, A. E., Gillis, T., Massood, T., Mysore, J., Myers, R. H. & Wheeler, V. C. (2009). Somatic expansion of the Huntington's disease CAG repeat in the brain is associated with an earlier age of disease onset. *Hum. Mol. Genet., 18*, 3039-3047.

Van Dellen, A., Grote, H. E. & Hannan, A. J. (2005). Gene-environment interactions, neuronal dysfunction and pathological plasticity in Huntington's disease. *Clin Exp Pharmacol Physiol., 32*, 1007-19.

Vassos, E., Panas, M., Kladi, A. & Vassilopoulos, D. (2008). Effect of CAG repeat length on psychiatric disorders in Huntington's disease. *J Psychiatr Res., 42*, 544-9.

Veitch, N. J., Ennis, M., McAbney, J. P., US-Venezuela Collaborative Research Project, Shelbourne, P. F. & Monckton, D. G. (2007). Inherited CAG.CTG allele length is a major modifier of somatic mutation length variability in Huntington disease. *DNA Repair (Amst).*, *6*, 789-96.

Völker, J., Plum, G. E., Klump, H. H. & Breslauer, K. J. (2009). DNA repair and DNA triplet repeat expansion: the impact of abasic lesions on triplet repeat DNA energetics. *J Am Chem Soc.*, *131*, 9354-60.

In: Encyclopedia of Neuroscience Research
Editors: Eileen J. Sampson and Donald R. Glevins

ISBN 978-1-61324-861-4
© 2012 Nova Science Publishers, Inc

Chapter XXXV

Putting Together Evidence and Experience: Best Care in Huntington's Disease

Zinzi Paola[1], Jacopini Gioia[1], Frontali Marina[2] and Anna Rita Bentivoglio[3]

[1]Consiglio Nazionale delle Ricerche, Istituto di Scienze e Tecnologie della Cognizione , Rome (Italy)
[2]Consiglio Nazionale delle Ricerche, Istituto di Neurobiologia e Medicina Molecolare, Rome (Italy)
[3]Università Cattolica del Sacro Cuore, Istituto di Neurologia, Rome (Italy)

Abstract

Huntington's Disease (HD) is a late onset autosomal dominant neurodegenerative disease leading to movement disorders, dementia and psychiatric manifestations. The management of these patients, for whom there is still no cure, has been for long time neglected by clinicians.

The identification of HD mutation in 1993 has raised a new interest in research as well as in clinical field and experimental data on patients care have begun to heap up. Among pharmacological therapies, data have become available on management of motor disorders with different types of neuroleptics or other classes of drugs while for the treatment of psychiatric and parkinsonian symptoms the available evidence is not specific for HD.

Among non pharmacological therapies, rehabilitation has recently provided good results supported by measurable outcomes.

Although these advances in management are encouraging, HD poses so many problems on so many different levels, that the way to attain an evidence based best care is still long.

In this chapter we provide some discussion as to the best care in HD as well as to some ethical problems posed by the disease. We take into consideration either the advances attained in the different types of therapies or the knowledge acquired outside an experimental setting, during many years clinical practice with HD patients.

Introduction

Care for Huntington's Disease (HD) patients has been for long time neglected. The rarity of the disorder (with a prevalence of 5-10/100.000) and the difficulty to have large and homogeneous cohorts of patients for clinical trials were certainly some of the causes, but other factors have to be considered. The progressive cognitive deterioration of these patients, their psychiatric symptoms (including aggressiveness, depression, schizophrenia) and different motor disabilities (be it bradychinesia and rigidity or choreic movements) have for a long time determined their confinement in asylums or nursing homes and justified their exclusion from the interest of clinicians and researchers.

Further aspects of the disease making it difficult for patients to reach research-oriented health centres are: the age at onset (around 40), in the mid of productive life, causing the loss of the job and economic difficulties; the possible presence of more than one patient in the same nuclear family at the same time due to the autosomal dominant transmission and to the symptoms onset anticipation over the generations; the length of disease duration (around 20 years) and the consequent economic burden quite often determining the progressive descent of family's social level and dramatic changes in caregiver's and family members standards of life.

In 1993, the identification of the genetic cause of the disease, the expansion of a CAG repeat in gene IT15, has given impulse to the investigation on the functional role of normal and mutated huntingtin, the protein coded by the gene. These investigations have shown the analogies with the neurodegenerative processes of other disorders, polyglutamine disorders and, in general, disorders due to misfolded proteins, and determined a new scientific interest in the disease and in the care of patients.

In the review of Bonelli et al. [1] of 218 papers published between 1965 and 2005, about the pharmacological management of HD, only three drugs were considered "possibly useful" for the treatment of chorea: haloperidol, fluphenazine, and olanzapine. In the last years a growing body of studies, according to the rules of evidence-based medicine, has emerged and the number of medications commonly used in the management of HD have dramatically increased [2].

However, when translating data into clinical practice some problems may arise either because the impact of treatments on patients' quality of life or daily activities is not monitored, or because sample size are small and patients are not always randomized according to the length of CAG expansion or because the modifying effect of self medications, often practiced by patients, has not been considered.

On the ground of non pharmacologic therapies, physiotherapy has quit only in the last years an anecdotal way of reporting its effects to enter the realm of measurable outcomes, while nutrition lacks trials on the best way to overcome the weight reduction often accompanying the disease.

This chapter will provide, whenever possible, evidence for HD best care and, when experimental data are lacking, the authors' experience will be used, acquired outside an experimental setting, in about twenty years clinical practice with HD patients. Surgical treatments will not be discussed, as the evidence on their effectiveness and safety is insufficient at present, and they are still to be considered experimental procedures. Considerations about psychological and ethical problems posed by the disease will also be expressed.

The Diagnosis

Caring for HD patients begins during the diagnostic process. Once non-genetic choreas have been excluded [3] and the hypothesis of HD has become possible or probable, clinicians usually ask for a confirmatory genetic test. The time and ways of communicating the genetic test result and the clinical diagnosis should be accurately planned according to the patient's psychiatric status and the familial and social context, in order to lead the affected individual and the family members, whenever possible, towards a gradual awareness and knowledge of the disease. In fact, the early phases of the disease are characterized by patients' and family's fragile equilibrium that can be easily broken with possible catastrophic effects (suicidal ideations or attempts) [4].

The diagnosis of HD has implications for the whole family and particular attention should be devoted to the conflicts that could arise between the desire of the patient to maintain the confidentiality of the diagnosis and the right of relatives to know their risk. The decision to communicate it or not to all family members should be agreed with the patient in the context of a counseling session, where the pros and cons of maintaining the confidentiality could be analyzed. Genetic counselling should be available also for family members, to provide either information on genetic risk and natural history of the disease or support in the process of coping with the many burdensome effects of the disease, such as social isolation and discrimination, loss of work and social role, disclosing children the truth. Contacts with lay organizations are invaluable for all these aspects.

In case at risk relatives would apply for the presymptomatic test, it should be performed following the ad hoc international guidelines published in 1994 [5]. Whenever an individual at risk results to be a presymptomatic gene-carrier, psychological support and clinical follow up should be offered with the aim of reassuring the testee about the possible presence of anxiety, depression and sleep disorders, which can often characterize the post test period, independently from the disease.

Based on data and clinical experience with HD, as well as with other movement disorders (Parkinson's Disease, dystonia, epilepsy, akathisia and so on), we will adopt a symptom-oriented approach to the pharmacological management of the classical features of HD: movement disturbances, cognitive decline and behavioural problems. The medications commonly used as best treatments in different stages of the disease are summarized in Table 1 while Table 2 provides some advices on how to take therapeutic decisions.

Pharmacologic Therapy of Signs and Symptoms

Table 1. Pharmacological and care intervention according to the disease's stage

Stage	Disease Burden	Pharmacologic Intervention	Non pharmachologycal intervention
Pre-symptomatic	Anxiety, Difficulty in personal and family planning	None, if not necessary, or: Benzodiazepines Serotonin Reuptake Inhibitors	Psychological counseling Genetic counseling Psychological support, Psychotherapy
Early (I-II)	Chorea: minimal to moderate	No treatment or: Tetrabenazine Atypical Neuroleptics	Physical therapy, Encourage active life.
	Minimal akinetic signs	Amantadine	
	Minimal to moderate psychiatric symptoms: mood depression, anxiety	Benzodiazepines Serotonin Reuptake Inhibitors Valproate, Gabapentin	Psychological support, Psychotherapy
	Insomnia	Melatonine, Mirtazapine	
	Forgetfulness, difficulty in planning strategies		Day-to-day strategies (memory notes, planning charts) Cognitive rehabilitation
	Weight loss	Dietary supplementation	High calorie diet
Middle (II)	Chorea	Tetrabenazine, Neuroleptics	Rehabilitation therapy
	Parkinsonism	Levodopa, Dopamine-agonists, Amantadine	
	Psychiatric Symptoms	Atypical Neuroleptics, mood stabilizers; (injectable potent neuroleptics only for emergencies)	Psychological support
	Focal dystonia/contracture	Botulinum toxin injection	
	Weight loss	Dietary supplementation	High calorie diet
	Cognitive decline		Cognitive rehabilitation (Discuss with patient and family on future choices in terms of life support in advanced stages)
Advanced (IV-V)	Chorea Hypokinesia Rigidity Anxiety or Depression Obsessions, Psychosis	As above, adjust doses.	Mobilization Assisted gait
	Dysphagia	Dietary supplementation (proteins, vitamins)	Dietary advice. For severe dysphagia consider naso-gastric tube or PEG to prevent inhalation.
	Dysartria		Speech and language therapy, Respiratory exercises
	Balance, Stability, Gait disorders, Frequent falls		Fall prevention
	Wheel-chaired or bedridden		Pressure sore prevention

* The Shoulson–Fahn Scale measures independence in activities of daily living. Functional capacity ranges from Stage 1 to Stage 5, with Stage 1 representing the most independent level of function. (Shoulson I, Fahn S. Huntington disease: clinical care and evaluation. Neurology. 1979 Jan;29(1):1–3)

Table 2. Criteria to guide the choice of drug therapy

1.	Start the decisional process on therapeutic approach, after a detailed anamnestic interview to patient and caregivers, widely oriented to all disease aspects (mentation, behaviour, mood, affectivity, sleep, weight loss, appetite, impulse control, verbal and physical aggressivity, suicidal ideation).
2.	Discuss with patient and caregivers the therapeutic options, elucidate which advantages is reasonable to expect, alert them on the expected side effects. This is a respectful approach and improves the compliance.
3.	Treat only the symptoms affecting patient's quality of life, tailoring drugs and doses on the patient.
4.	Even when hyperkinesias are the sole manifestation of HD, it is advisable to avoid potent D2-blockers, as these drugs precipitate /worsen/induce parkinsonism and other side effects such as anaffectivity, anedonia, lack of motivation, catatonia, and should be taken into consideration only to treat severe psychotic symptoms or in the emergency.
5.	In the course of the disease the clinical picture changes: test the need for antipsychotic and antidyskinetic drugs along the time, and stop what is unnecessary.
6.	The decision to treat the patient should be taken according to the status of HD. Some times the best thing is not to use any drug, particularly in the latest stages of the disease
7.	Not all that can be done, must be done: the choice on how far should the assistance be pushed during the latest stage of the disease should belong to the patient. Therefore, all these medical-ethical aspects regarding which kind of end-of –life support should be used, have to be discussed with the patient and caregivers in advance, never in the emergency room, so as to give patient and family the opportunity to express their will and to choose.

Dyskinesias

Typical neuroleptics such as haloperidol, fluphenazine and sulpiride have been extensively used in the past. However, their adverse effects such as parkinsonism, tardive dyskinesias, apathy, excessive sedation, together with the availability of less harmful drugs have drastically limited their use. More recently, depletors of dopamine presynaptic vesicles proved to be highly effective and safe in controlling choreic movements. In particular Tetrabenazine, a reversible and selective blocker of central monoamine transporter, has been extensively studied [6, 7, 8, 9, 10].

When chorea is prominent, tetrabenazine at low doses (12.5 mg/day) can be started and slowly titrated until hyperkinesias are reasonably reduced. The most serious potential adverse effect of tetrabenazine is depression, that should not be neglected as it might be very severe.

When tetrabenazine is insufficient or contraindicated (e.g. previous or present diagnosis of depression), atypical neuroleptics such as olanzapine and risperidone should be considered.

Amantadine, an antiglutammatergic drug, also resulted to improve motor performance [11, 12] without deleterious effects on parkinsonism and cognition.

Parkinsonism

Early, in the course of the rigid-akinetic form (Westphal variant) or late, in the last stages of the most common hyperkinetic form of HD, akinesia, bradykinesia, and tremor may be, sometimes, more invalidating then the chorea itself.

As a first step, particular care should be devoted in assessing whether the appearance of parkinsonism is due to the use of antidopaminergics, in which case the dosages should be reduced, or typical neuroleptics should be abandoned in favour of the atypical ones.

One should be aware that rigidity and akinesia are part of the phenotypic expression of HD, and this awareness should guide the choice of drugs since the disease onset, by avoiding potent D2-blockers for the management of dyskinesias and minor psychiatric symptoms.

Amantadine can be positively used to alleviate parkinsonism while anticholinergics should not be used, as they might exacerbate chorea.

Other Movement Disorders

Dystonic movements or postures are common among the spectrum of movement disorders affecting HD patients. Dystonia is poorly responsive to drugs, tetrabenazine and anticholinergics included; the latter ones should not be used, anyway, because they may worsen chorea. Focal dystonia, contractures and bruxism, may be improved by local injection of botulinum toxin. Myoclonus, a rare but potentially disabling feature, improves with valproate [13], which may also have a beneficial effect on chorea and mood, as well as with Clonazepam.

HD patients may present tremor of different types. Often a high frequency, low amplitude, postural tremor responds to propanonol or benzodiazepines and, when resistant, to antiepileptics such as topiramate and levetiracetam. Resting tremor is ameliorated by antiparkinsonian drugs. Akathisia is a relatively common side effect of antipsichotics and, albeit uncommonly, it is also a symptom of HD. It can be improved by lowering the dose of the antipsichotics or shifting to an atypic neuroleptic.

When patients do not improve, tetrabenazine, amantadine, or levetiracetam, can also be tried.

Epilepsy

Associated with mental decline and rigidity, epilepsy is more common among juvenile HD patients. Myoclonic epilepsy often responds to valproate, lamotrigine, clonazepam. Also levetiracetam should be considered, as it might also reduce dyskinesias [14]. The status epilepticus should be treated according to the currently shared guidelines.

Spasticity is an uncommon feature of HD, mostly complicating the latest stages of the disease. Diazepam and other benzodiazepines may be used to reduce it.

Psychiatric Symptoms

Although mood disorders, anxiety and depression are particularly frequent in HD patients, rigorous trials on best pharmacologic treatment of this population are lacking.

Tryciclic antidepressants (amitriptiline, nortriptiline) may lead to anticholinergic effects and no data are available supporting the use, in HD patients, of newer antidepressant drugs, such as paroxetine, fluoxetine, sertraline, or mirtazapine and venlafaxine. In a very recent study on 26 HD patients with diagnosis of major depression Venlafaxine XR (extended release) was demonstrated to produce significant ameliorations of symptoms of depression with evident clinical improvement. However, relevant side effects were reported in five patients and further exploration is needed in the future [15]. Among psychiatrists there is some debate on the potential increase of suicidal ideation in patients with mood disorders, especially the young patients [16], after the introduction of antidepressants. In our experience, such an increase has not been observed, but higher degree of irritability and aggressiveness have been reported sometimes by patients and caregivers. In any case, when an antidepressant is prescribed, patients observation should be warranted, and the drug should be introduced and slowly titrated without concomitant adjunct of benzodiazepines, as they may mask side effects such as agitation, irritability, excitation. According to our experience, mood stabilizers such as valproate, carbamazepine or atypical antipsychotic (quetiapine), may be useful to treat depression with concomitant anxiety and sleep disorders. When psychosis is prominent, clozapine and quetiapine (unfrequently causing extrapyramidal side effects) should be considered, while haloperidol should be left as a second choice. There is a general agreement that Lithium is the best choice to treat suicide risks in bipolar disorders and so is clozapine in schizophrenic patients [17] while there is no specific evidence for HD [18].

Acute psychosis with agitation and aggressive behaviours can be treated with risperidone or olanzapine, [19] or, when really severe, with i.m. haloperidol and concomitant use of valproate or benzodiazepine [19].

Cognition

Rivastigmine has been recently reported to have a positive effect on MMSE score in HD patients, [20] but more data are needed to confirm this result. Memantine was found beneficial in 27 HD subjects with reported no decline for 2 years in both cognitive and motor parameters [21] A pilot open label carried out in 12 patients was not confirming effects on cognitive domain [22]. A larger placebo controlled trial of this drug for use in Huntington's disease primarily testing for cognitive signs is presently underway.

Non Pharmacological Treatment

A growing body of experimental evidence is available on the beneficial effects of an enriched environment in rodent models of HD [23, 24, 25, 26]. These studies strongly encourage to promote through exercise an active, stimulating and autonomous life as long as possible, in patients too.

Rehabilitation Therapy

Bilney et al [27] and Busse et al [28] in reviewing literature data on rehabilitation therapies in HD found that physiotherapy is generally reported as beneficial, but the studies have a poor methodology, small sample sizes, and insufficient information about the protocols and selection criteria.

Recently, however, the effects of an intensive multidisciplinary rehabilitation program were assessed through motor and functional scales in a sample of 40 HD patients [29]. The results showed that the treatment has a positive, short-medium term effect on motor and functional performance in patients with early- and middle-stage HD and no severe psychiatric symptoms. In addition no decline in motor, functional, emotional and cognitive scores was evident in a subsample of 11 patients who underwent rehabilitation treatment for two years [29]. New valid, reliable and responsive tools to measure walking activity, mobility, risk of fall and balance in HD patients, have been recently proposed [30, 31, 32] which will be used to provide further and accurate evidence on the effects of rehabilitation.

In the early stages of the disease, when minimal impairments (chorea, impaired balance, reduced flexibility) and functional limitations (difficulties in performing activities necessary to fully participate in daily life at work or at home) are present, balance and core stability training may be useful to prevent falls and to delay the onset of mobility restrictions. Speech and respiratory therapy can jointly help to maintain lung capacity and induce relaxation.[29]. Exercise aimed at maintaining cardiovascular fitness are important to prevent additional limitations or impairments. Patients should be encouraged to maintain their usual activity for as long as possible. Occupational therapy may play a relevant role in this, helping patients to adapt to new conditions, promoting safety and delaying the loss of social role at work and at home.

At these early stages, subtle cognitive deficits such as attention and memory deficits, reduced speed of processing, decreased cognitive flexibility and decision making ability are also usually present. Retraining or stimulation exercise focused on attention and memory can be useful and compensation and adaptation strategies should be suggested to overcome deficits in more complex cognitive abilities [33, 27].

With the progression of HD functional limitations increase. Difficulty in postural changes, impaired balance, weakness of stabilizing muscles, dystonia, chorea, contribute to lower mobility, increase the risk of falls and impair everyday activities (personal hygiene, dressing oneself, eating, house cleaning, cooking, etc). At this stage of the disease interventions should be aimed at increasing the strength of stabilizing muscle and at suggesting strategies to perform at best the personal and household activities.

Environmental modifications at home are frequently needed and aids can be used to reduce the risks of falls (walking aids) and to protect vulnerable parts (e.g hip protectors, knee and elbow pads, etc). Speech and respiratory therapy is essential for maintaining communication, safe eating and expectorating.

In the later stages functional limitations are severe. The rehabilitative approach should be mainly devoted to avoid secondary complications such as muscle contractures, pressure sores and chest infections and at maximizing comfort, maintaining contact with external environment and easing caregivers' burden as a full time caring is needed. The physiotherapist should be working closely with the speech and language therapist in order to develop program and strategies to minimize the risk of choking and aspiration of the fluid into the lungs and, ultimately, chest infections.

Nutrition

Although weight loss and increased energy expenditure have been repeatedly documented in HD patients [34, 35, 36, 37], very little or no evidence is available on best diet to overcome this problem. A hypercaloric diet, particularly in early and middle stages of the disease, is an obvious empirical treatment. The amount of calories to be added should be assessed on the base of the weight loss and the actual food intake. An addition of 473kc/day to the usual diet was found to be a sufficient supplement [38]. Oral hygiene should be particularly cared for and dentist visits planned 2-3 times/year in order to prevent the worsening of chewing problems.

A randomized, placebo-controlled, double-blind study has shown that highly unsaturated fatty acids seem to be beneficial to patients with Huntington's disease [39]. Creatine, a guanidine compound produced endogenously and acquired exogenously through diet, seems to be a relevant component in maintaining much cellular energy.

Strong evidence exists for early oxidative stress in HD, coupled with mitochondrial dysfunction, exacerbating each other and leading to an energy deficit. [40]. The addition of antioxidants to the diet could be beneficial, although no data are available at present.

As dysphagia progresses and nutrition becomes more problematic dry food should be avoided in favour of wet food, cut in small pieces or grinded, and liquids with jellying agents should be offered. It may be necessary to offer individual assistance to help the patient to complete the meal. In many institutional settings, feeding tubes are used routinely. In our opinion this choice should be taken into consideration only when patient's swallowing deteriorates more rapidly that other functions of the body. In fact, the meal represents an opportunity for the patient to receive time and attention, it is a social interaction with an affective value and until the patient can manage the meal with pureed food we think the effort to avoid feeding tubes can be worthwhile.

Although artificial nutrition may improve life span, some patients refuse to be fed artificially and are upset when such solution is proposed. The use of life supports should be discussed with patients in advance, at the early stages of the disease, when they are able to understand medical explanations, when speech and cognition are still strong and it is possible to make meaningful choices, so as to be prepared if and when the progression of disease poses the question. [41,42].

Conclusion

HD is a long-duration, neurologic, dementing, and psychiatric disorder with an autosomal dominant inheritance. All these aspects have far-reaching effects, meaning that patients and families will need care and assistance for years, generation after generation. The ideal model of caring for HD patients is to set up a team with a physician receiving input from many health care professionals (neurologists, psychiatrists, rehabilitation therapists, psychologists, genetic counsellors, nurses, social workers and other specialists) and from family members in order to develop a care-plan. The best approach to patient care is being flexible: although the symptoms and functional difficulties follow some common patterns, their combination widely vary between individuals. In addition, the course of the disease and the patients' quality of life throughout the disease is highly dependent on their personality characteristics (e.g. good-humour, ability to cope, previous interests, friendliness) as well as on the social context (family background, type of work, meaningful social relationships, community resources, and external sources of help) which have influence on behaviour through the disease. Because of the genetic implications of HD, the team should also be concerned for the care and support of the whole family, as the social and emotional needs of family members are very closely linked to those of the patient.

Referrals to lay organizations is a further resource to be considered either for support groups and sharing experience or for bringing the needs of people with HD to the attention of institutional bodies and society at large.

Acknowledgments

We thank Dr. Michele Raja for sharing his experience on the management of the psychiatric burden of the disease, and Dr. Paolo Zappata for his expertise in HD patient multidisciplinary rehabilitation. We are indebted to the three lay associations AICH-Roma-Napoli-Milano for their joint effort in publishing an Italian Consensus Handbook on Huntington Disease which has been a guide in preparing this paper**. Finally we thank all HD patients and their relatives from whom we drew our caring experience.

References

[1] Bonelli, RM, Wenning, GK. Pharmacological Management Of Huntington's Disease: An Evidence-Based Review. *Curr Pharm Des*, 2006, 12, 2701-2720.

[2] Phillips, W; Shannon, KM; Barker, RA. The Current Clinical Management Of Huntington's *Disease. Mov Disord.*, 2008, 23(11), 1491-504.

[3] Cardoso, F. Chorea: Non Genetic *Causes Curr Opin Neurol*, 2004, 17, 433-436.

[4] Paulsen, JS; Hoth, KF; Nehl, C; Stierman, L. Critical Period Of Suicide Risk In Huntington's Disease. *Am J Psychiatry*, 2005, 162, 725-731.

[5] International Huntington Association (IHA) And The World Federation Of Neurology (WFN) Research Group On Huntington's Chorea. Guidelines For The Molecular *Genetics Predictive Test In Huntington's Disease. Neurology*, 1994, 44, 1533-1536.

[6] Huntington Study Group. Tetrabenazine As Antichorea Therapy In Huntington Disease: A Randomized Controlled Trial. *Neurology,* 2006, 66, 366-372.

[7] Frank, S; Ondo, W; Fahn, S; Et Al. A Study Of Chorea After Tetrabenazine Withdrawal In Patients With Huntington's Disease. *Clin Neuropharmacol*, 2008, 31, 127-133.

[8] Mclellan, DL; Chalmers, RJ; Johnson, RH. A Double-Blind Trial Of Tetrabenazine, Thiopropazate, And Placebo In Patients With Chorea. *Lancet*, 1974, 1, 104-107.

[9] Jankovic, J. Treatment of Hyperkinetic Movement Disorders With Tetrabenazine: A Doubleblind Cross-Over Study. *Ann Neurol*, 1982, 11, 41-47.

[10] Hayden, MR; Leavitt, BR; Yasothan, U; Kirkpatrick, P. Tetrabenazine. *Nat Rev Drug Discov.*, 2009, 8, 17-18.

[11] Lucetti, C; Del Dotto, P; Gambaccini, G; Et, Al. IV Amantadine Improves Chorea In Huntington's Disease: An Acute Randomized, *Controlled Study. Neurology*, 2003, 60, 1995-1997.

[12] Verhagen Metman, L; Morris, MJ; Farmer, C; Et Al. Huntington's Disease: A Randomized, Controlled Trial Using The NMDA-Antagonist Amantadine. *Neurology*, 2002, 59, 694-699.

[13] Saft, C; Lauter, T; Kraus, PH; Przuntek, H; Andrich, J. Dose-Dependent Improvement Of Myoclonic Hyperkinesia Due To Valproic Acid In Eight Huntington's Disease Patients: A Case Series. *BMC Neurol*, 2006, 6, 11.

[14] Woods, SW; Saksa, JR; Baker, CB; Cohen, SJ; Tek, C. Effects Of Levetiracetam On Tardive Dyskinesia: A Randomized, Double-Blind, Placebo-Controlled Study. *J Clin Psychiatry.*, 2008, 69, 546-554.

[15] Holl, AK; Wilkinson, L; Painold, A; Holl, EM; Bonelli, RM. Combating Depression In Huntington's Disease: Effective Antidepressive Treatment With Venlafaxine XR. *Int Clin Psychopharmacol.*, 2010, 25(1), 46-50.

[16] Hetrick, S; Merry, S; Mckenzie, J; Sindahl, P; Proctor, M. Selective Serotonin Reuptake Inhibitors (Ssris) For Depressive Disorders In Children And Adolescents. *Cochrane Database Syst Rev.*, 2007, 3, CD004851.

[17] Pompili, M; Lester, D; Innamorati, M; Tatarelli, R; Girardi, P. Assessment And Treatment Of Suicide Risk In Schizophrenia. *Expert Rev Neurother.*, 2008, 8, 51-74.

[18] Pompili, M; Rihmer, Z; Innamorati, M; Lester, D; Girardi, P; Tatarelli, R. Assessment And Treatment Of Suicide Risk In Bipolar *Disorders Expert Rev Neurother.*, 2009, 9, 109-136.

[19] Wilhelm, S; Schacht, A; Wagner, T. Use Of Antipsychotics And Benzodiazepines In Patients With Psychiatric Emergencies: Results Of An Observational Trial. *BMC Psychiatry.*, 2008, 22, 61

[20] De Tommaso, M; Difruscolo, O; Sciruicchio, V; Specchio, N; Livrea, P. Two Years' Follow-Up Of Rivastigmine Treatment In Huntington Disease. *Clin Neuropharmacol.*, 2007, 30, 43-46.

[21] Beister, A; Kraus, P; Kuhn, W; Dose, M; Weindl, A; Gerlach, M. The N-Methyl-D-Aspartate Antagonist Memantine Retards Progression Of Huntington's Disease. *J Neural Transm Suppl.*, 2004, (68), 117-22.

[22] Ondo, WG; Mejia, NI; Hunter, CB. A Pilot Study Of The Clinical Efficacy And Safety Of Memantine For Huntington's Disease. *Parkinsonism Relat Disord.*, 2006, 13, 453-4

[23] Van Dellen, A; Blakemore, C; Deacon, R; York, D; Hannan, AJ. Delaying The Onset Of Huntington's In Mice. *Nature*, 2000, 404, 721-722.

[24] Carter, RJ; Hunt, MJ; Morton, AJ. Environmental Stimulation Increases Survival In Mice Transgenic For Exon 1 Of The Huntington's Disease Gene. *Mov Disord*, 2000, 15, 925-937.

[25] Hockly, E; Cordery, PM; Woodman, B; Et Al. Environmental Enrichment Slows Disease Progression In R6/2 Huntington's Disease Mice. *Ann Neurol*, 2002, 51, 235-242.

[26] Spires, TL; Grote, HE; Varshney, NK; Et, Al. Environmental Enrichment Rescues Protein Deficits In A Mouse Model Of Huntington's Disease, Indicating A Possible Disease Mechanism. *J Neurosci.*, 2004, 24, 2270-2276.

[27] Bilney, B; Morris, ME; Perry, A. Effectiveness of Physiotherapy, Occupational Therapy, And Speech Pathology For People With Huntington's Disease: A Systematic Review. *Neurorehabil Neural Repair*, 2003, 17, 12-24.

[28] Busse, ME; Rosser, AE. Can Directed Activity Improve Mobility In Huntington's Disease? *Brain Res Bull.*, 2007, 72, 172-174.

[29] Zinzi, P; Salmaso, D; De Grandis, R; Graziani, G; Maceroni, S; Bentivoglio, A; Zappata, P; Frontali, M; Jacopini, G. Effects Of Intensive Rehabilitation Programme On Patients With Huntington's Disease: A Pilot Study. *Clinical Rehabilitation*, 2007, 21, 603-613.

[30] Rao, AK; Muratori, L; Louis, ED; Moskowitz, CB; Marder, KS. Clinical Measurement Of Mobility And Balance Impairments In Huntington's Disease: Validity And Responsiveness. *Gait & Posture*, 2009, 29, 433-436.

[31] Busse, ME; Van Deursen, RW; Wiles, CM. Real-Life Step And Activity Measurement: Reliability And Validity. *J Med Eng Technol.*, 2009, 33, 33-41.

[32] Busse, ME; Wiles, CM; Rosser, AE. Mobility And Falls In People With Huntington's Disease. *J Neurol Neurosurg Psychiatry.*, 2009, 80, 88-90.

[33] Rosenblatt, A; Ranen, NG; Nance, MA; Paulsen, JS. *A Physician's Guide To The Management Of Huntington's Disease*. Second Edition. Huntington's Disease Society Of America 1999.

[34] Pratley, RE; Salbe, AD; Ravussin, E; Caviness, JN. Higher Sedentary Energy Expenditure In Patients With Huntington's Disease. *Ann Neurol*, 2000, 47, 64-70.

[35] Trejo, A; Tarrats, RM; Alonso, ME; Boll, MC; Ochoa, A; Velásquez, L. Assessment of The Nutrition Status Of Patients With Huntington's Disease. *Nutrition.*, 2004, 20, 192-196.

[36] Gaba, A; Zhang, K; Moskowitz, CB; Boozer, CN; Marder, K. Harris-Benedict Equation Estimations of Energy Needs As Compared To Measured 24-H Energy Expenditure By Indirect Calorimetry In People With Early To Mid-Stage Huntington's Disease. *Nutr Neurosci.*, 2008, 11, 213-218.

[37] Aziz, NA; Van Der Burg, JM; Landwehrmeyer, GB; Brundin, P; Stijnen, T; EHDI Study Group; Roos, RA. Weight Loss In Huntington Disease Increases With Higher CAG Repeat Number. *Neurology.*, 2008, 71, 1506-1513.

[38] Trejo, A; Boll, MC; Alonso, ME; Ochoa, A; Velásquez, L. Use Of Oral Nutritional Supplements In Patients With Huntington's *Disease. Nutrition.*, 2005, 21, 889-894.

[39] Vaddadi, KS; Soosai, E; Chiu, E; Dingjan, P. A Randomised, Placebo-Controlled, Double Blind Study Of Treatment Of Huntington's Disease With Unsaturated Fatty Acids. *Neuroreport.*, 2002, 13, 29-33. Erratum In: Neuroreport 2002; 13: Inside Back Cover.

[40] Stack, EC; Matson, WR; Ferrante, RJ. Evidence Of Oxidant Damage In Huntington's Disease: Translational Strategies Using Antioxidants. *Ann N Y Acad Sci.*, 2008, 1147, 79-92.

[41] Simpson, S. Late Stage Care In Huntington's Disease. *Brain Res Bull.*, 2007, 72, 179-181

[42] Klager, J; Duckett, A; Sandler, S; Moskowitz, C. Huntington's Disease: A Caring Approach To The End Of Life. *Care Management Journals*, 2008, 9, 75-81.

In: Encyclopedia of Neuroscience Research
Editors: Eileen J. Sampson and Donald R. Glevins
ISBN 978-1-61324-861-4
© 2012 Nova Science Publishers, Inc.

Chapter XXXVI

Oral Health Care for the Individual with Huntington's Disease

Robert Rada

Department of Oral Medicine and Diagnostic Sciences,
University of Illinois College of Dentistry, Chicago, Illinois, USA

With the many advances in medical care, the dentist will be quite likely to be asked to treat patients with special health care needs. The dentist may be asked to recommend preventive regimens to maintain good oral health or provide comprehensive treatment of advanced dental disease. The patient with Huntington's disease is but one example of the debilitating effects systemic disease can have upon the oral cavity. There are no direct adverse affects on the oral cavity due to Huntington's disease; however complications associated with the disease increase the risk for dental caries and periodontal disease. As Huntington's disease progress, the person's ability to cooperate will diminish as functional and cognitive abilities decline. This will require the development of realistic treatment plans and easily maintainable dental restorations. Caregivers will need to be involved throughout the process as oral complications are likely throughout this long and difficult disease.

Huntington characterized the disease by a triad of symptoms to include gradual personality changes, dementia and choreiform movements. Other symptoms include dysarthria, disturbances of gait and oculomotor dysfunction. The dementia is characterized by forgetfulness, slowness of thought and altered personality with apathy or depression. The patient can also become moody, irritable and incapable of dealing with the routine details of life. Subtle personality changes often become apparent before any motor symptoms arise. [1-4] These conditions can all play a role in maintaining a healthy mouth. The dentist is likely to be among the first to notice a deterioration in oral health status.

Oral Health and Systemic Health

Both direct effects from the disease and side effects from the medications used in treating this illness can have a negative impact on one's quality of life. A person with a chronic disease can suffer from depression. Depression can cause them to have a lack of concern over oral health and diet. The individual may not enjoy eating. [5] An inability to chew effectively, due to a broken down, painful dentition, could have life threatening consequences. Dental infection and choking are serious potential problems. [6] Secondary illnesses, such as pneumonia, are often the cause of death. [7] Of greatest concern would be aspiration resulting in a life-threatening pulmonary abscess. These risks can all be increased by poor oral health.[8,9].

Oral complications will negatively impact the health and comfort of these individuals. The dentist should be a necessary partner in the health care team. Of paramount importance is preventive therapy to maintain good health for the oral tissues. Dental consultation early in the disease process is essential. As the disease progresses, dysphagia is a significant problem. A sound and functioning dentition is essential in this process for anything but a pureed diet.[10] The lack of controlled body movements makes oral hygiene difficult Poor oral hygiene will result in carious teeth and periodontal inflammation. Many prescribed medications cause xerostomia further complicating the situation. The poor oral hygiene and advanced oral disease can result in loss of teeth.[9].

Continuous movement of the orofacial musculature may induce orofacial pain, temporomandibular joint discomfort, cracked and excessively worn teeth. Bruxism has been reported in patients with Huntington's disease, but it is not a common symptom. Botulinum toxin has been used for those patients requiring relief of severe bruxism associated with the disease.[11].

Dysphagia and Aspiration Pneumonia

Difficulty swallowing is a common symptom and may be associated with fatal complications if food is aspirated and pulmonary abscess results.[9] The result of this dysphagia can also be nutritional compromise. Patients have been classified into hyperkinetic and rigid-bradykinetic swallowing abnormalities of deglutition. In the first group lingual chorea, swallowing incoordination and inability to stop respiration were among the noted characteristics. In the second group mandibular rigidity, coughing on foods and choking on liquids were noted. In both groups pharyngeal space retention and aspiration were also identified.[12] Dentists can work with speech therapists in managing the dysphagia aspect.[13]

For severely debilitated patients, enhanced oral care may reduce the risk of illness and even death. There is an increased colonization of bacteria in persons enterally fed due to a lack of mastication and swallowing which aid in mechanically cleansing the mouth. When a specialized suction toothbrush is used, pathogenic bacteria are significantly reduced, decreasing the risk of infection and respiratory illness. [8]

Of especially serious consequence to the patient with Huntington's disease is the susceptibility to aspiration pneumonia with consequent colonization of the lungs with gram negative anaerobes from inflamed gingival tissues. Microorganisms can travel into the lungs through aspiration of bacteria rich saliva due to poor oral health. The risk for aspiration pneumonia increases when high levels of bacteria are present in the mouth. There is a demonstrated relationship between aspiration pneumonia and periodontal disease, dental caries and poor oral hygiene.[14, 15]

Medication Induced Xerostomia

Medications that inhibit parasympathetic activity and stimulate sympathetic activity are most likely to diminish salivary output. Agents that modulate nerve transmission within the central nervous system, such as antidepressants, inhibit autonomic stimulation of salivary glands. Anxiety, depression or stress may also give rise to subjective symptoms of dry mouth. Patients who are taking more than one xerogenic medication appear to have a higher incidence of dry mouth. Antianxiety agents, antipsychotics, antidepressants, anti-Parkinsonism and antihypertensive medications are all known to cause xerostomia.[16,17] Many of these medications are taken by individuals with Huntington's disease.[18]

Asking several questions of the individual will help to determine whether the patient has xerostomia. An affirmative response to one of these questions correlates with a decrease in saliva: "Does your mouth usually feel dry? Does your mouth feel dry when eating a meal? Do you have difficulty swallowing dry foods? Do you sip liquids to aid in swallowing dry foods? Is the amount of saliva in your mouth too little most of the time?"[19]

Saliva aids in both chewing and swallowing. Saliva provides oral irrigation and contains proteins and electrolytes which inhibit the growth of microorganisms and buffer oral acids. [16,17,19] Caregivers must be cautious as to not allow the patient to eat too much soft, cariogenic foods and drink sugar containing beverages to help stimulate salivary flow. One of the most detrimental complications of xerostomia is dental caries. This usually occurs near the gumline and causes rapid destruction of the teeth.[17]

Lack of saliva can also contribute to opportunistic fungal overgrowth of Candida albicans. Oral mucositis, erythema, white patches and inflamed fissures at the corner of the mouth, called angular chelitis, are all clinically evident indicators. The patient will frequently complain of generalized oral discomfort or a burning feeling. [16,17,19]

Xerostomia will also cause difficulties with mastication, taste, maintaining a comfortable fit for oral appliances as well as difficulty speaking or sleeping. Traumatic oral lesions may arise when the dry mucosa rubs against a sharp tooth or restoration. Halitosis and gingivitis are also possible complications. [16,17,19].

The treatment of xerostomia is primarily palliative. Individuals suffering from xerostomia are frequently seen carrying a bottle of water with them to take an occasional sip. More debilitated patients, who may have significant difficulty swallowing, can be offered ice chips to slowly melt in the mouth. Occasionally patients can be placed on an alternative medication which may have a lesser side-effect. The physician may also modify the dosage or instruct the patient to take the medicine at a different time of the day.

A variety of over the counter products are available to soothe the discomfort of dry mouth. The products are manufactured as gels, sprays, mouth rinses and toothpastes. Individuals frequently go through a process of trial and error to determine which works best. Many well-advertised mouthwashes contain alcohol, which can desiccate and dehydrate the oral mucosa. Individuals with xerostomia should purchase mouthwashes which are alcohol free. Petroleum jelly should be avoided because it dries out the lips.[16,17,19]

Occupational Therapy and Oral Care

Occupational therapists "enable patients to participate in their everyday life activities in their desired roles, context and life situations." [20] Everyday life activities, also known as activities of daily living, include personal oral hygiene. Occupational therapists can nicely complement the efforts of the dentist or dental hygienist. Although not licensed to clean teeth, occupational therapists can teach and assist patients to perform oral self-care.

Occupational therapists are perfect individuals to fill in the gap of patient education, specifically related to maintaining oral health, for individuals with Huntington's disease. Dental hygienists frequently do not have the time to provide the activities to patients that occupational therapists can offer. Occupational therapists include demonstration, modeling, return demonstration, positive reinforcement and corrective feedback in self-care. Occupational therapy sessions on dental hygiene followed by brief weekly reinforcement sessions for three months significantly reduced dental plaque in dependent and cognitively impaired long term care residents. For an occupational therapist with minimal exposure to oral health education, a one to two hour training session with a dental hygienist should be sufficient to effectively work with patients.[21] Oral home telecare, through videoconferencing on the internet, has even been used to provide enhanced communication of the importance and techniques for personal oral hygiene to homebound patients.[21] Occupational therapists can provide suggestions to dentists for fabrication of special handles or supports to make it easier for individuals with Huntington's disease to hold a toothbrush or floss.

The Dental Appointment

Although actual treatment times may need to be kept short, adequate time is necessary for the patient to formulate and express their thoughts.[6] Aggressive behavior and agitation can be part of Huntington's disease. However these can also be due to frustration from difficulties with communication. The patient may become frustrated at their ability to stay still and cooperate with dental treatment. The dentist should attempt to use questions that require a simple yes/no response. Procedures should be explained in brief sentences while maintaining eye contact with the patient. Reduction of excessive noise and distractions within the treatment room will help the patient's concentration.[22]

Patients with Huntington's disease will often have difficulty keeping their mouth open. Bite blocks, especially the Molt type, can be very useful. Many patients cannot be treated while leaned back due to a diminished cough reflex and inability to swallow. A rubber dam can prevent debris from falling into the oropharynx and potentially aspirated. The dental assistant must be especially meticulous in suctioning the oral cavity while retracting the soft tissues, and possibly helping to stabilize the patient.[22]

Protective stabilization may be necessary to provide safe and effective care while treating patients with Huntington's disease.[6] The sudden, uncontrolled movements may result in injury to the patient or dental health care worker when using sharp instruments or the dental drill in the mouth. Frequently this can be managed with simple hand holding and head support offered by family members or caregivers. If full body, uncontrolled movements are present, then a papoose board or full body wrap can be useful.

Informed consent must be obtained prior to using any type of restraint. This would involve a description of the method of restraint to be used as well as the amount of time and reasons that protective stabilization is necessary. It is also a good idea to discuss these issues with both the patient a family member.[23] Without consensus misunderstandings can easily arise. Alternatives of sedation and general anesthesia may need to be discussed.[24] Part of the dialogue will include the consequences of no treatment.

When a patient undergoes extensive oral rehabilitation a one week follow-up is appropriate. Frequently these patients are best placed on a three month recall program, at which a prophylaxis polish and topical fluoride treatment is performed. Even the patient who exhibits significant choreiform movements can have regular preventive maintenance appointments. With frequent breaks and light restraint, such as hand holding and head stabilization by a dental assistant and family member, a thorough job is possible. The dental office staff should continually attempt to reinforce oral hygiene techniques. However effecting any change, if the patient has been lax in maintaining good oral hygiene in the past can be difficult. This will require much positive reinforcement. For example having the patient understand the benefits of an electric toothbrush, and encouraging assistance from caregivers can help. In addition, educating the role a healthy mouth plays on overall health cannot be understated.

Advanced Anesthesia Needs for Dental Treatment

Uncontrolled movements, anxiety disorders and advanced oral disease may require that treatment be completed under general anesthesia. General anesthesia does bring some risks even to a healthy patient receiving dental care. In the Huntington's disease patient there are some additional special concerns. Of paramount importance in these patients is airway protection. [25] Even minimal post-operative bleeding in the mouth from tooth extractions can be especially serious. There is a potential for aspiration related to the Huntington's patient's lack of oropharyngeal muscle control. A thorough preanesthesia work-up is essential in successful management of these patients. [24]

Timely treatment of the diseased oral cavity is an absolute medical necessity in maintaining overall heath. Short procedures may be completed with in-office intravenous sedation. However the potential for aspiration is again a concern and multiple appointments may be necessary to complete lengthy treatment plans. The key to minimizing the potential for post-operative complications is very close, virtually one-on-one monitoring. Airway monitoring and maintenance must be of highest priority, Suctioning of oral fluids during the recovery period eliminates the potential for aspiration. The importance of close monitoring of the patient after discharge should also be communicated to caregivers whenever any type of advanced anesthesia techniques are used.[24,26]

Restorative Dentistry Considerations

Routine restorative and endodontic therapy can be difficult for the patient and dentist because of the uncontrolled movements.[27] Removable appliances can be difficult to retain because of the choreiform movements. Excessive movements of the jaw and tongue might cause dislodgement of a partial denture. Even inserting and removing these appliances can be difficult for patients in advanced stages of the disease. If the patient has never worn a removable partial denture, adaptation to the appliance in the presence of Huntington's disease may be very difficult. The dentist should watch the patient use his hands to handle objects or sign his name. If it is evident that the patient has poor manual dexterity, the partial may be frustrating to insert. The dentist must also be cognizant of what level of caregiver support the individual has prior to recommending a removable appliance.

Implant therapy is a possibility for patients able to afford this form of treatment.[28] For the fully edentulous patient, dental implants would be an ideal adjunct, if the patient and his caregivers are able to maintain acceptable oral hygiene. Stability of a lower full denture can be a challenge for any patient. Implant retained dentures have made it possible for many people to adapt quite easily to a removable prosthesis. Fitting implants for patients with severe neurological diseases requires highly flexible interdisciplinary decision-making, and a willingness to reconsider conventional contraindications.[28]

The dentist can also choose to place restorative materials which can help prevent recurrent decay. Glass ionomer materials appear to inhibit secondary caries *in vivo* and data from *in vitro* studies support this effect. Glass ionomers not only release fluoride onto the adjacent tooth structure, they can be recharged by topically applied fluoride.[29,30] In vitro studies have shown zones of caries inhibition 0.25mm from the tooth restoration interface.[31] These materials are frequently used to restore teeth damaged by root caries due to xerostomia. Whenever possible, glass ionomers should be included in the treatment options for the high-caries risk patient.

Disease Prevention

Prevention of dental disease is of utmost importance in maintaining the overall health of a patient with Huntington's disease. The modalities will differ between individuals and may vary as the disease progresses. For some, 3 month recall appointments will be necessary, others may require a specialized or mechanical toothbrush, many will benefit from topically applied chemotherapeutic agents to reduce bacterial plaque. The cornerstone of preventive therapy will be oral hygiene evaluation, instruction and reinforcement as a significant part of each dental appointment.[6]

Caregivers may be incapable, unwilling, or too overwhelmed with multiple responsibilities to provide effective daily care. It is important to ensure that caregivers have the knowledge, skill and willingness to help with oral healthcare. For example, in studies of nurses providing care to patients who have had a stroke, the nurses were relatively unaware of the impact a patient's oral health has on overall health.[32] An oral health teaching program should be part of training for all types of caregivers. Quality of life can be enhanced if the discomfort and potential hazard of poor oral health are reduced.[33]

Publications can be quite useful in providing information on maintaining good oral hygiene in a special needs adult. Dental Care Every Day: A Caregiver's Guide [34] distributed by the National Institute of Dental and Craniofacial Research (NIDCR) describes brushing and flossing techniques for caregivers as well as practical tips for making toothbrush and floss handles easier to hold and use for an individual who can clean his/her own teeth. Specialized brushes are available which can cover multiple tooth surfaces with one stroke. Listed below are three popular manual type brushes.

Benefit Toothbrush   Benedent Corporation (www.benedent.com)

Collis-Curve Toothbrush   (www.colliscurve.com)

Surround Toothbrush   Specialized Care Company (www.specializedcare.com)

Battery operated mechanical toothbrushes are available in either a rotating brush head or a sonic – oscillating style. Both have been shown to be effective and the choice is generally up to individual preference and personal cleaning effectiveness. The dentist or dental hygienist can help with recommending a particular type and offering personalized instructions for use. Mechanical toothbrushes have been shown to be an effective choice for patients with disabilities.[35] In addition the NIDCR has produced a brochure on the causes of xerostomia and ways to manage the associated discomfort.[36] The Importance of Dental Care[37] and Eating and Swallowing Difficulties[38] are two of many publications distributed by the Huntington's Disease Association. These provide valuable information for patients, caregivers and health care workers regarding oral health challenges and ways to overcome the challenges associated with Huntington's disease.

The fundamentals of maintaining good oral health involve plaque removal through brushing and flossing, the use of fluoridated toothpaste, drinking and cooking with fluoridated water, reducing the intake of fermentable carbohydrates and regularly scheduling professional oral care. Frequently, brushing and flossing will not be enough to prevent the development of dental caries or advancement of periodontal disease. In these situations adjunctive therapies become necessary. In these cases fluoride application and chlorhexidine rinses become the next step. Protocols for use of these agents for the prevention of dental disease in community settings for people with special needs have been researched and developed. [39]

Topical Fluorides

Topically applied fluorides have become universally accepted for the anti-caries activity. Topical fluorides are easy to use, effective, inexpensive and safe. Fluorides function to inhibit tooth demineralization, enhance remineralization and inhibit bacterial plaque.[40] The fluoride products available in the United States are sodium fluoride, acidulated phosphate fluoride, stannous fluoride and sodium monofluorophosphate. Clinician will chose the particular product based on caries risk, anticaries objectives and the patient's ability to comply with treatment recommendations.[41]

Fluoride mouth rinses, purchased over the counter, are 230 ppm sodium fluoride. They are intended for an individual who can safely rinse and expectorate without ingestion. If the individual cannot rinse or spit out the solution, it can be applied with a cotton swab or sponge applicator twice daily. [41]

High concentration fluoride toothpastes or gels are beneficial for individuals suffering from medication induced xerostomia. High concentration fluoride products need to be prescribed by a dentist. Their use may need to be overseen by a caregiver depending on the ability of the patient. These products contain 5000 ppm sodium fluoride gel and can be toxic if excessive amounts are swallowed. The toothpastes are brushed on to the teeth at bedtime and the excess spit out, leaving the residual paste on the teeth while sleeping. Water should not be used for rinsing. Custom trays can also be fabricated by the dentist into which the fluoride gel can placed. The patient can then give themselves a daily fluoride treatment. [41]

Fluoride varnishes are much faster to apply and for some individuals may be better tolerated. These agents contain 5 percent sodium fluoride which is painted on each tooth as a thin adherent film with a small brush. Application of fluoride varnish is regulated by state law, so it must be applied by a healthcare professional. Those allowed to apply the varnish varies from state to state. Fluoride varnish requires removal of food or debris from the teeth prior to application. One protocol recommends application of fluoride varnish three times per week, once per year, or applying a single coat of fluoride varnish once every six months.[41] Fluoride varnishes offer a benefit of low total fluoride exposure and high tooth surface concentration.

Chlorhexidine Rinses

Chlorhexidine has been safely and effectively used as a dental plaque reducing rinse in two intraoral concentrations; 0.12% in the United States and 0.2% in Europe. Chlorhexidine is an effective agent in breaking down oral biofilms. The chemical acts to disrupt the cell membrane so that exposed bacteria are unable to maintain their osmotic balance in the oral environment. [43] Chlorhexidine is an approved and effective treatment of gingivitis. It is also used in varying prescriptions for caries management strategies. Chlorhexidine is formulated as an alcohol based and alcohol free product. The patient suffering with xerostomia and mucosal sensitivity may be better prescribed the alcohol free product. One suggested regimen is to rinse with a half-ounce of chlorhexidine solution for 1 minute twice daily for two weeks. The regimen should be repeated 4 times per year. A dental professional should be involved in determining whether chlorhexidine rinses would be useful and in monitoring the efficacy of treatment. [42]

Oral Care and Palliative Care

As the disease progresses into the end stage of life the dentist may be consulted to play a role in providing assistance in palliative care. Dentists infrequently interact with dying patients, yet dentists can play an integral role in the palliative care team. Dental professional should be called upon to offer assistance to the patient in the end stages of Huntington's disease. Dental professionals can provide training for palliative care professionals and manage specific oral problems.

The chief complaints of palliative care patients are xerostomia and symptoms from poorly fitting dentures. Candidiasis is often present, but it is not usually the most frequently verbalized complaint.[43] The dentist is an expert in diagnosis and treatment of these areas. The dentist may even be able to excavate tooth decay bedside and place a sedative filling to eliminate a toothache. Water-soluble lubricants should be used to lubricate the oral tissues. Biotene Oral Balance gel (GlaxoSmithKline Consumer Healthcare) is an excellent water-soluble agent and an alternative to the typical lubricants as it contains lactoperoxidase, lysozyme, glucose oxidase, lactoferrin and no glycerin. Nursing staff should be instructed to apply the product thinly all around the mouth using a foam brush.[44]

End of life mouth care may be unfamiliar to medical practitioners. One literature review has highlighted a number of inconsistencies in both the knowledge of mouth care and its implementation by nursing staff.[45] The goals of oral health palliative care are simple; eliminate pain and infection, keep the mouth moist and clear from dental plaque or food debris.[46] Increased awareness by all health care professionals of palliative oral care would go a long way in providing relief, comfort, and consolation to terminally ill patients and their families.

Summary

Huntington's disease exemplifies the importance of cooperation between physician and dentist to positively impact the overall health of an individual through improved oral health. Frequently oral health problems in severely medically compromised individuals are dismissed as trivial. This should not be the case. Oral health professionals should be involved at every stage of this disease. It is essential that dentists understand medical terminology and the pathophysiology of disease to be able to manage these medically compromised patients. Medical technology will continue to advance and there will be no shortage of these types of patients arriving at dental offices in the future. It is essential that individuals with Huntington's disease be able to enjoy optimum oral health throughout the course of this illness. Good oral health will undoubtedly impact their quality of life.

References

[1] Bonelli, RM; Wenning, GK. Kapfhammer. Huntington's disease: present treatments and future therapeutic modalities. *Int Clin Psychopharmacol*, 2004, 19(2), 51-62.

[2] Lauterbach, EC; Cummings, JL; Duffy, J; Coffey, CE; Kaufer, D; Lovell, M; Malloy, P; Reeve, A; Royall, DR; Rummans, TA; Salloway, SP. Neuropsychiatric correlates and treatment of lenticulostriatal diseases: a review of the literature and overview of research opportunities in Huntinton's, Wilson's and Fahr's diseases. *Journal of Neuropsychiatry and Clinical Neurosciences*, 1998, 10, 249-266.

[3] Leori, I; Michalon, M. Treatment of the psychiatric manifestations of Huntington's disease: a review of the literature. *Can J Psychiatry*, 1998, 43, 933-40.

[4] Naarding, P; Kremer, HP; Zitman, FG. Huntington's disease: a review of the literature on prevalence and treatment of neuropsychiatric phenomena. *Eur Psychiatry*, 2001, 16, 439-45.

[5] Tran, P; Mannen, J. Improving oral heathcare: improving the quality of life for patients after a stroke. *Spec Care Dentist*, 2009, 29, 218-221.

[6] Lewis, D; Fiske, J; Dougall, A. Access to special care dentistry, part 7. Special care dentistry services: seamless care for people in their middle years- part 1. *Br Dent J*, 2008, 205, 305-317.

[7] Cardoso, F. Huntington disease and other choreas. *Neurol Clin*, 2009, 27, 719-736.

[8] Ferozali, F; Johnson, G; Cavagnaro, A. Health benefits and reductions in bacteria from enhanced oral care. *Spec Care Dent*, 2007, 27, 168-176.

[9] Moline, DO; Iglehart, DR. Huntington's chorea: review and case report. *Gen Dent*, 1985, 33, 131-133.

[10] Hamakawa, S; Koda, C; Umeno, H; Yoshida, Y; Nakishima, T; Asaoka, K; Shoji, H. Oropharyngeal dysphagia in a case of Huntington's disease. *Auris Nasus Larynx*, 2004, 31, 171-176.

[11] Nash, MC; Ferrell, RB; Lombardo, MA; Williams, RB. Treatment of bruxism in Huntington's disease with botulinum toxin. *J Neuropsychiatry Clin Neurosci*, 2004, 16(3), 381-82.

[12] Kagel, MC; Leopold, NA. Dysphagia in Huntington's disease: a 16-year retrospective. *Dysphagia*, 1992, 7, 106-114.

[13] Bilney, B; Morris, ME; Perry, A. Effectiveness of physiotherapy, occupational therapy, and speech therapy for people with Huntington's disease: a systematic review. *Neurorehabil Neural Repair*, 2003, 17, 12-24.

[14] Scannapieco, FA; Bush, RB; Paju, S. Associations between periodontal disease and risk for nosocomial bacterial pneumonia and chronic obstructive pulmonary disease. A systemic review. *Annual Periodontal*, 2003, 8, 54-69.

[15] Terpenning, MS; Taylor, GW; Lopatin, DE; Kerr, CK; Dominguez, BL; Loesche, WJ. Aspiration pneumonia: dental and oral risk factors in an older veteran population. *J Am Geriatr Soc*, 2001, 49, 557-563.

[16] Moore, PA; Guggenheimer, J. Medication induced hyposlivation: etiology, diagnosis and treatment. *Compend Contin Educ Dent.*, 2008, 29(1), 50-5.

[17] Turner, MD; Ship, JA. Dry mouth and its effects on the oral health of elderly people. *JADA*, 2007, 138(9 supplement):15S-20S.

[18] Bonelli, RM; Wenning, GK. Pharmacological management of Huntington's disease: an evidence based review. *Current Pharaceutical Design*, 2006, 12(21), 2701-2720.

[19] Guggenheimer, J; Moore, PA. Xerostomia: etiology, recognition and treatment. *JADA*, 2003, 61-69.

[20] American Occupational Therapy Association. Scope of Practice. *Am J Occup Ther*, 2004, 58, 673-7.

[21] Yuen, HK; Pope, C. Oral home telecare for adults with tetraplegia: a feasibility study. *Spec Care Dent*, 2009, 29, 204-209.

[22] Friedlander, AH; Mahler, M; Norman, KM; Ettinger, RL. Parkinson disesase: systemic and orofacial manifestations, medical and dental management. *JADA*, 2009, 140, 658-669.

[23] Karibe, H; Umezu, Y; Hasegawa, Y; Ogihara, E; Masuda, R; Ogata, K; Shirase, T; Kawakami, T; Warita, S. Factors affecting the use of protective stabilization in dental patients with cognitive disabilities. *Spec Care Dent*, 2008, 28, 214-220.

[24] Rada, RE. Comprehensive dental treatment of a patient with Huntington's disease: literature review and case report. *Spec Care Dentist.*, 2008, 28, 131-5.

[25] Nagele, P. Hammerle. Sevoflurane and mivacurium in a patient with Huntington's chorea. *Br J Anaesth*, 2000, 85, 320-21.

[26] Cangemi, CF; Miller, RJ. Huntington's disease: review and anesthetic case management. *Anesth Prog*, 1999, 45, 150-153.

[27] Braddord, H; Britto, LR; Leal, G; Katz, J. Endodontic treatment of a patient with Huntington's disease. *J Endod*, 2004, 30, 366-369.

[28] Jackowski, J; Andrich, J; Kappeler, H; Zollner, A; Johren, P; Muller, T. Implant supported denture in a patient with Huntington's disease: interdisciplinary aspects. *Spec Care Dent*, 2001, 21, 15-20.

[29] Burgess, JO. Fluoride releasing materials and their adhesive characteristics. *Compend Contin Educ Dent*, 2008, 29(2), 82-91.

[30] Randall, RC; Wilson, NHF. Glass-ionomer restoratives: a systematic review of a secondary caries treatment effect. *J Dent Res*, 1999, 78, 628-637.

[31] Tantbirojn, D; Rusin, RP; Bui, HT; Mitra, SB. Inhibition of dentin demineralization adjacent to a glass-ionomer / composite sandwich restoration. *Quintessence Int*, 2009, 40, 287-294.

[32] Preston, AJ; Punekar, S; Gosney, MA. Oral care of elderly patients: nurses' knowledge and views. *Postgrad Med J*, 2000, 76, 89-91.

[33] Locker, D; Matear, D; Stephens, M; Jokovic, A. Oral health-related quality of life of a population of medically compromised elderly people. *Community Dent Health.*, 2002, Jun, 19, 90-97.

[34] National Institute of Dental and Craniofacial Research. Dental care every day: a caregiver's guide. 2009. http://www.nidcr.nih.gov/OralHealth/Topics/ Developmental Disabilities/DentalCareEveryDay.htm. (accessed December 10, 2009).

[35] Bozkurt, FY; Fentoglu, O; Yetkin, Z. The comparison of various oral hygiene strategies in neuromuscularly disabled individuals. *J Contemp Dent Pract*, 2004, 15, 5(4), 23-31.

[36] National Institute of Dental and Craniofacial Research. Dry mouth. 2009. http://www.nidcr.nih.gov/OralHealth/Topics/DryMouth/DryMouth.htm (accessed December, 10, 2009).

[37] The Huntington's Disease Association . Fact Sheet 8. The importance of dental care. 2008. http://www.hda.org.uk/download/fact-sheets/HD-Dental-Care.pdf. (accessed Deceber 10, 2009).

[38] The Huntington's Disease Association . Fact Sheet 6. Eating and swallowing difficulties. 2008. http://www.hda.org.uk/download/fact-sheets/HD-Eating-Swallowing.pdf. (accessed December 10, 2009).

[39] Glassman, P; Anderson, M; Jacobsen, P; Schonfeld, S; Weintraub, J; White, A; Gall, T; Hammersmark, S; Isman, R; Miller, CE; Noel, D; Silverstein, S; Young, D. Practical protocols for the prevention of dental disease in community settings for people with special needs: the protocols. *Spec Care Dentist*, 2003, 23, 160-164.

[40] Featherstone, JDB. The science and practice of caries prevention. *JADA*, 2000, 131, 887-899.

[41] Jacobsen, P; Young, D. The use of topical fluoride to prevent or reverse dental caries. *Spec Care Dentist*, 2003, 23, 177-179.

[42] Anderson, M. Chlorhexidine and xylitol gum in caries prevention. *Spec Care Dentist*, 2003, 23, 173-176.

[43] Wiseman, MA. Palliative care dentistry. *Gerodontology*, 2000, 17(1), 49-51.

[44] Wiseman, MA. The treatment of oral problems in the palliative patient. *J Can Dent Assoc*, 2006, 72, 453-8.

[45] Gillam, JL; Gillam, DG. The assessment and implementation of mouth care in palliative care: a review. *J R Soc Promot Health.*, 2006, 126(1), 33-7.

[46] Saini, R; Marawar, PP; Shete, S; Saini, S; Mani, A. Dental expression and role in palliative treatment. Indian J Palliat Care [serial online] 2009, 15, 26-9. http://www.jpalliativecare.com/text.asp?2009/15/1/26/53508 (accessed December 12, 2009).

In: Encyclopedia of Neuroscience Research
Editors: Eileen J. Sampson and Donald R. Glevins
ISBN 978-1-61324-861-4
© 2012 Nova Science Publishers, Inc.

Chapter XXXVII

Suicidal Ideation and Behaviour in Huntington's Disease

*Tarja-Brita Robins Wahlin**
Department of Neurobiology, Care Sciences and Society,
Karolinska Institutet, Stockholm, Sweden and
The University of Queensland School of Medicine, Brisbane, Australia

Abstract

Huntington's Disease (HD) is an autosomal-dominant neuropsychiatric disorder characterized by irreversible physical and mental deterioration, personality change, mental disorder, and increased susceptibility to suicidal ideation and suicide. Typically the disease has a very long course to run before death. There is little that the affected or HD persons who find out their carrier status can do, although if they come early for genetic testing, they can make reproductive choices. If the diagnosis or the predictive information has little therapeutic value to the patient, especially if there is no family, one possible available action is planning of a suicide. Higher suicide rates are reported for HD patients compared to the general population. It is also well known that persons with HD have an increased propensity to psychiatric dysfunction and elevated rates of catastrophic social events. Furthermore, surveys of attitudes about likelihood of attempted suicide have indicated that 5-29% of at risk individuals would contemplate suicide if they received a result indicating a carrier status. The prevalence of suicide and suicide attempts is difficult to estimate because of methodological problems. Suicide rates of HD populations vary among countries, and they are affected by factors such as socio-economic status, age, sex, and prevalence of HD. It appears that cases of suicide vary from 1.6% to 13.8% in HD populations. In this chapter, suicidal behaviour, psychological affect of predictive testing, suicidal ideation and behaviour before and after predictive testing, and depression in HD, are discussed. At present, the problems

---

* Address for communication:Tarja-Brita Robins Wahlin, Department of Neurobiology, Care Sciences and Society, Karolinska Institutet, Novum plan 5, 141 86 Stockholm, Sweden, Tel. +46 8 5858 9397, Fax: +46 8 5858 5470, E-mail: Tarja-Brita.Robins.wahlin@ki.se

associated with suicidal ideation and suicide in HD populations, worldwide, are much the same as two decades ago. The need for counselling, using a well designed protocol, and the importance of focusing on suicide risk of participants in predictive testing programs is emphasized.

Introduction

Most investigators of Huntington's disease (HD) would agree that George Huntington's (1872) description of the disease was correct when he drew attention to the three peculiarities in this disease; "(1) its hereditary nature; (2) tendency to insanity and suicide; (3) its manifesting itself as a grave disease only in adult life". In this original description George Huntington refers to HD as "sometimes that form of insanity which leads to suicide", suggesting that mental illness and possible suicide is secondary to the disease. Indeed, the potential for suicide in HD has been recognised for more than a century as a serious consequence of HD (Di Maio, et al., 1993; Farrer, 1986; Hayden, Ehrlich, Parker, & Ferera, 1980; Huntington, 1872; Paulsen, Nehl, et al., 2005; Reed & Chandler, 1958; Robins Wahlin, 2007; Schoenfeld, Myers, Cupples, et al., 1984; Sørensen & Fenger, 1992).

HD is an autosomal-dominant neuropsychiatric disorder characterized by irreversible physical and mental deterioration, personality change, and increased susceptibility to mental disorder (Paulsen, Ready, Hamilton, Mega, & Cummings, 2001). Predictive testing for HD has been available from the middle of the 1980's, initially as linkage analyses, carried out first in USA (Gusella, et al., 1983; Meissen, et al., 1988) and Canada (Fox, Bloch, Fahy, & Hayden, 1989) and soon afterwards in many centres around the world (World Federation of Neurology Research Group on Huntington's Disease, 1993). Ten years after the discovery of markers for the HD gene (Gusella, et al., 1983), a CAG trinucleotide repeat showing expansion on the HD chromosome was identified (Huntington's Disease Collaborative Research Group, 1993), and this permitted direct testing of the mutation associated with HD. Following the first descriptions of a pilot project in British Columbia on genetic testing (Bloch, Fahy, Fox, Hayden, & James, 1989; Fox, et al., 1989), several testing centres from over the world have developed predictive testing programs for HD. Guidelines for predictive testing (Huntington's Disease Society of America, 1989) have provided an additional supplement to psychosocial, ethical, and legal considerations.

The preliminary guidelines for predictive testing protocols were developed to exclude adverse reactions and catastrophic events which may arise during the testing process. These protocols were designed to determine the psychiatric, psychosocial, and social consequences of predictive testing, whether pre-test characteristics of individuals can predict negative consequences of testing, and whether pre-test education and counselling, and post-test clinical follow-up, can prevent negative outcomes such as catastrophic reaction and suicides (Quaid, 1991). These guidelines have often been modified due to different health care systems and have thus presented a framework for testing programs around the world. However, although some countries have several testing centres, no unified test program has been established.

HD is a late-onset disorder characterized by increasing and irreversible physical and mental deterioration. It typically involves a very long course of disease before death. There is little that HD persons who find out their carrier status can do, although if they come early for genetic testing, they can make reproductive choices. If the diagnosis or the predictive information have little therapeutic value to the patient, especially if there is no family, one possible evasive action is available: planning suicide (Davis, 1999).

Suicidal Behaviour

Suicidal behaviour in HD is well recognized (Farrer, 1986; Hayden, et al., 1980; Huntington, 1872; Kessler, 1987a, 1987b; Ranen, 2002; Robins Wahlin, 2007; Schoenfeld, Myers, Cupples, et al., 1984). Higher suicide rates are reported for HD patients compared to the general population (Di Maio, et al., 1993; Hayden, et al., 1980; Robins Wahlin, et al., 2000; Schoenfeld, Myers, Cupples, et al., 1984; Sørensen & Fenger, 1992). It is also well known that persons with HD have an increased frequency of psychiatric dysfunction (Dewhurst, Oliver, & McKnight, 1970; Kessler, 1987b; Oliver, 1970; Robins Wahlin, et al., 2000; Wallace & Parker, 1973) and elevated rates of catastrophic social events (Kessler, Field, Worth, & Mosbarger, 1987). Furthermore, surveys of attitudes about likelihood of attempted suicide have indicated that 5-21% of at risk individuals would contemplate suicide if they received a result indicating increased risk (Kessler, 1987b; Mastromauro, Myers, & Berkman, 1987), and a later study by Robins Wahlin and co-workers (2000) indicated much higher figures. They reported that 29% of at risk persons would attempt suicide when developing symptoms of HD.

The prevalence of suicide and suicide attempts is difficult to estimate because in the majority of HD studies subjects are identified mainly from questionnaires and psychiatric hospital medical records. Another reason is that death certificates may give erroneous diagnoses of causes of death and therefore are subject to uncertainty. Sørensen and Fenger (1992) found the diagnosis of HD given in only 76% of the cases where they had evidence that the patient suffered from HD. Thirdly, data collected by questionnaire and genealogical reports from relatives may have been underreported. Under-identification of suicide attempts and suicides may result from reluctance of patients and families to admit the presence of a mental disorder and the need to seek medical attention. However, some estimates of the frequency of suicide indicate that this occurs between seven and 200 times more often in HD than in the general population (Farrer, 1986; Hayden, 1981; Hayden, et al., 1980; Wexler, 1979). Unfortunately, a variety of methodological problems in studies of patients makes calculation of suicidal behaviour virtually impossible (Stenager & Stenager, 1992).

Bates, Harper and Jones (2002) report that cases of suicide vary from 4% to 12.7% in HD population. A review from France reports a four times higher suicidal occurrence in manifest HD than in general population (Bindler, Travers, & Millet, 2009). This is probably a fair estimation as the observed frequencies of suicide in HD population were 5.6% in Denmark (Sørensen & Fenger, 1992), 3.4% in South Africa (Hayden, et al., 1980), 5.7% (Farrer, 1986), and 7.3% (Di Maio, et al., 1993) in the USA. Schoenfeld, Myers, Cupples, et al. (1984) found a rate of 4% in a sample of 506 Huntington cases also in the USA. However, suicides accounted for 12.7% of those cases (157) where a specific cause of death was reported (see Table 1). A somewhat lower frequency of 2% has been reported by Haines and Conealy (1986) also in the USA. Lanska, Lavine, Lanska, & Schoenberg, (1988) noted suicide as 6% cause of death. However, if death was attributed to suicide in all cases in which "accidents, poisonings, and violence" was reported as cause of death, this would account for 7.9% of the deaths. Chiu and Alexander (1982) observed a low frequency of 1.6% in Australia and a lower suicide rate has also been observed in Mexico (Alonso, et al., 2009). On the other hand an epidemiological study from Yugoslavia reported a normal prevalence of HD (4.46/100,000 population) but very high occurrence of suicide (Sepcic, Antonelli, Sepic-Grahovac, & Materljan, 1989). The authors traced ten families from Rijeka district and recorded nine deaths in the families, of which six were suicides. These variations in the suicide rates in different countries may be real, or accounted for by the variations in the methods used for identifying suicides, or both (Sørensen & Fenger, 1992). An overview of frequency of suicides and accidental deaths in HD patients and family members is provided in Table 1.

Table 1. Suicides and accidental deaths in HD patients and family members

Source, publication year (study period if available)	Country	No. of persons (deaths)	n (%) suicides - affected persons	n (%) suicides - family members	n (%) accidents affected persons	n (%) accidents - family members	Reported suicide - normal population	Comments
Reed and Chandler, 1958 (1940-1956)	USA	203	3	n/a	n/a	n/a	n/a	Group I: 3-4 times increased risk, Group II: no increased risk
Schoenfeld et al., 1984 (1956-1982)	USA	506 (157 known deaths)	20 (4%)	n/a	10 (6.4% of known deaths)	n/a	1%	20/506 = 4.0%. 20/157 = 12.7% (known deaths) = 4 times increased risk
Saugstad and Ødegård, 1986 (1916-1975)	Norway	199	1	n/a		n/a	n/a	Selected psychiatric population
Farrer, 1986 (n/a-1984)	USA	452	25 (5.7%)	n/a	14 (3.2%)	n/a	1.5%	4 times increased risk. Attempted suicide rate, HD 27.6%
Lanska et al., 1988 (1971, 1973-1978)	USA	3238	20 (0.6%)	n/a	257 (7.9%)	n/a	n/a	Suicide rarely reported, high rate of accidents, poisoning, and violence
Sepcic et al., 1989 (1946-1981)	Yugoslavia	87	6	n/a	n/a	n/a	n/a	Only 9 deaths were recorded and 6 of them were suicides
Sørensen & Fenger, 1992 (1943-n/a)	Denmark	677; 395 HD + 282 sibs	22 (5.6%)	15 (5.3%)	23 (5.8%)	20 (7.1%)	2.7%	2 times increased risk
Di Maio et al., 1993 (1815-1987)	USA	2793	114 (4.4%)	91 (3.5%)	n/a	n/a	1.0-1.3%	205/2793 = 7.3% = 7 times increased risk
Lipe et al., 1993 (1972-1992)	USA	300	9 (3.0%)	2	n/a	n/a	n/a	The 9 suicides were compared with 18 HD patients
Almqvist et al., 1999 (n/a)	International (Canada)	4527	5	n/a	n/a	n/a	n/a	21 attempted suicide and 18 hospitalizations were recorded
Baliko et al., 2004 (1920-1997)	Hungary	396 deaths	40 (10%)	27 (7%)	n/a	n/a	n/a	Increased suicide rates compared to the general Hungarian population.

n/a = not available

The National Huntington's Disease Research Roster at Indiana University, USA (Farrer, 1986), reported that 5.7% of deaths among affected persons resulted from suicide and 27.6% of patients attempted suicide at least once. Of these attempted suicides, 7.2% had two to three attempts. Suicide was the third most common primary cause of death in HD (Farrer, 1986). However, the frequency of suicide might be higher, because the cause of death was listed as

'Huntington's Disease' in 92 cases. Interestingly, they found that the female attempted-suicide rate was higher (22.0%) than the male rate (18.5%). This could be due to a higher success rate in males (who succeed after the first attempt rather than a long train of attempts). Farrer (1986) also noted that suicide in HD was four times greater than the reported rate 1.0% - 1.3 % for the general population in USA (Roy, 1995).

A later evaluation by Di Maio et al. (1993) studied 2793 people registered with the National Huntington's Disease Research Roster. Suicide was the reported cause of death in 205 cases, comprising 9.3% "affected," 21.5% "possibly-affected", 4.5% "at 50% risk", 2.6% "at 25% risk", 7.1% "children of possibly affected subjects", and 3.2% "spouses and family members with no risk". The authors recorded three times more male suicides than females. Although the exact suicide ratio between patients and family members and the US population were not calculated, their data show an increased frequency in all risk categories of HD family members including patients' spouses.

Schoenfeld, Myers, Cupples, et al. (1984) examined the proportions of deaths attributed to suicide among 506 deceased individuals with suspected or diagnosed HD from New England, USA. There were 20 documented suicides, nine in diagnosed HD and 11 in suspected HD individuals. These suicides accounted for 4% of all HD deaths, which was four times higher than the suicide rate in the Massachusetts general population. These suicides occurred during a period of 26 years. Sixteen of the 20 persons committing suicide were male. Methods of suicide included drowning, gunshot, hanging, drug overdose, fire, jumping from a high place, and asphyxiation by poisonous gas. The mean age of the persons was 47.7 years. This suggests that most of the suicides occurred early in the disease. Suicide was the third most common cause of death in HD patients after pneumonia and heart disease.

A Danish study examined causes of death for 395 HD subjects and 282 unaffected siblings and compared the causes of deaths with the general Danish population (Sørensen & Fenger, 1992). The proportion of deaths from suicide was 5.6% among the HD subjects, 5.3% among the sibs, both of which are higher than in the general Danish population (2.7%). Once again suicide was the third most common cause of death in HD subjects. Accidents as a primary cause of death occurred with a frequency of 3.8% in the affected and 7.1% among the siblings. The authors did not report the methods of suicides, but reported causes of accidents in HD and at risk subjects. In 13 cases of 23, patients aspiration was the cause of the "accidental" death and five "accidental falls" were reported.

Methods used to commit suicide by HD subjects are often drowning, jumping from a high place, hanging and poisonous gas (Baliko, Csala, & Czopf, 2004; Schoenfeld, Myers, Cupples, et al., 1984). Ready availability of firearms is associated with an increased risk of suicide in the home (Baliko, et al., 2004; Kellermann, Rivara, Somes, & Reay, 1992) and handguns are responsible for some of the suicides in HD families, mainly in USA. Lipe, Schultz, & Bird (1993) reported 11 suicides: four by drowning, three by handguns and one using a single vehicle auto crash. All of these 8 suicides were males. The three female cases were overdose, carbon monoxide poisoning and cutting of wrists. Sørensen & Fenger (1992) listed "accidents" among 23 HD patients and 20 at risk subjects. Thirteen patients were classified as dying from "aspiration", as mentioned before, and 11 at risk subjects had traffic accidents. Six of these accidents were a crash with a vehicle in the opposite lane and all were male single drivers. The attempted suicide methods reported by Robins Wahlin et al. (2000) were drug overdoses and cutting of wrists, all in females. This seems to indicate that females prefer less dramatic methods than men, who seem to choose methods which succeed better.

Lipe et al. (1993) also noted geographic influences in suicide methods, reporting three persons who jumped from bridges, reflecting the availability of water.

If the HD diagnosis is made when subtle signs are apparent, it gives an opportunity for patients to discuss feelings about the disease and problems that have brought them to the physician. However, in many cases patients suspect that the disease has taken its first steps, but they are not willing to seek help. Many of them are familiar with the disease and its symptoms and want to cope alone. Schoenfeld, Myers, Cupples, et al. (1984) reported that among 157 HD patients with an established cause of death, 18.4% of observed suicides occurred in the age group 10-49 years and more than half of the suicides occurred in individuals who showed early signs of the illness but who had not been diagnosed. Lipe et al. (1993) found that 5 of the 9 HD subjects committed suicide within the first year of symptom onset. This implies that people in early stages of HD are particularly prone to suicide. Indeed, there appears to be a consensus that those who are early in the disease course, and those who have not yet been diagnosed with the disease, are at greatest risk of suicide (Cina, Smith, Collins, & Conradi, 1996; Di Maio, et al., 1993; Farrer, 1986; Hayden, et al., 1980; Huntington, 1872; Lanska, et al., 1988; Lipe, et al., 1993; Paulsen, Hoth, Nehl, & Stierman, 2005; Reed & Chandler, 1958). Hence, suicide among HD patients is seen in the third through fifth decades.

An increased risk of suicide is not limited to the person with HD. Several studies suggest that family members, including spouses and children of the afflicted, also appear to be at higher risk (Cina, et al., 1996; Di Maio, et al., 1993; Kessler, 1987b; Lipe, et al., 1993; Robins Wahlin, et al., 2000). Huntington (1872) presents …"two married men, whose wives are living, and who are constantly making love to some young lady not seeming to be aware that there is any impropriety in it." These frontal symptoms of HD can lead to unhappy circumstances in the family, and personality changes are not rare in the early course of the disease. The tragedies of the HD families are difficult for an outsider to understand. Lipe et al. (1993) report suicides of an unaffected spouse with HD husband and two affected children and one unaffected at risk daughter of an affected mother. Kessler et al. (1987) found that 37.3% at risk subjects reported at least one attempted or completed suicide among relatives, and about 34.3% reported that a close relative had been hospitalised for psychiatric reasons. Two of 67 subjects (3%) had attempted suicide themselves (Kessler, 1987b). Similarly, Robins Wahlin et al. (2000) revealed that 69.2% of carriers and 85.7% of noncarriers had a close relative with psychiatric disorder related to HD. Furthermore, 30.8% of the carriers reported suicide or suicide attempts in the family, whereas the corresponding figure for noncarriers was 14.3%.

Sørensen & Fenger (1992) found that accidents as a primary cause of death occurred among the sibs with a frequency of 7.1% which is even higher than their suicide frequency (5.3%). These accidents were striking in their character, three were poisoned with carbon monoxide and six of the 23 "accidents" of at risk persons were car crashes with a truck in the opposite lane. All were male single drivers. This indicates clearly that some accidents leading to death may have been hidden suicides. Farrer (1986) questions also the possibility that several at risk sibs reported to have died accidentally may actually have been suicide victims. He remarks that only 1.8% of HD sibs were reported to have died by suicide and 5.5% known death cases were reported as accidents.

The increased occurrence of suicide among persons with HD may not represent a disproportionate increase over patients with other serious disorders (Schoenfeld, Myers,

Cupples, et al., 1984). Because of the late onset of HD, asymptomatic at risk persons who die at an early age may have been carriers. An association between suicidal potential and organic disease has been suggested among other diseases (Chapdelaine, 1993) such as Acquired Immunodeficiency Syndrome (AIDS, Komiti, et al., 2001), cancer (Louhivuori & Hakama, 1979), diabetes (Macgregor, 1977), renal failure (Washer, Schroter, Starzl, & Weil III, 1983) and psychiatric disorders (Tsuang, 1978). A better basis for computing the relative risk of suicide in persons with HD could be by comparing them with similar serious disease groups rather than the general population. Any disease leading to physical and mental disorder followed by death may predispose to higher risk of suicide. Declining cognitive function and eventual dementia involve barriers to social life, isolation and depression and may engender suicide attempts (Haw, Harwood, & Hawton, 2009). Some suffering, especially with physical pain, can be addressed by medication, but it is difficult to compensate for other forms of suffering such as that brought about by loneliness and indignity in HD. Davis (1999) argues that persons at risk of genetic disease resulting in long periods of mental incompetence have unique reasons for serious consideration of suicide.

A quarter of all HD patients attempt suicide at least once (Farrer, 1986). There is a well established association between suicide and psychiatric disorders (Tsuang, 1978). The propensity of suicidal behaviour could therefore be a direct consequence of familial neuropathology. However, Farrer (1986) did not find any hereditary patterns for suicide among the affected members of the families, although the recorded suicide attempts were 27.6% (148 known suicide attempts and 54 "unsure" suicide attempts in 727 individuals with HD). Di Maio et al. (1993) noted the occurrence of further cases of suicide in the same family after a first suicide, suggesting a possible genetic effect, but this phenomenon also occurs in families without HD (Roy, 1983). Furthermore, it should not be overlooked that psychosocial elements, such as witnessing gradual cognitive decline and dementia of the parent or relatives, learning behaviour patterns of the family, turbulent social background and years of constant symptom-seeking might aid the onset of the suicidal tendencies.

For several reasons, the studies of suicide among HD patients may underestimate its frequency. Suicide may be suspected in accidents leading to death but may not be recorded officially. As mentioned before, Sørensen & Fenger (1992) observed "accidents" among patients which were due to aspiration and to falls. Schoenfeld, Myers, Cupples, et al. (1984) excluded ten accidents (6.4%) from the suicide category because of lack of verification. Further, these authors classified only those patients for whom suicide was definitely reported as the cause of death by relatives, medical records and death certificates. This would exclude cases of which relatives may not wish to record the death as a suicide if there was some doubt. Farrer (1986) found 14 accidental deaths that accounted for 3.2% in their sample of 452 HD individuals. Religious reasons, the desire to protect the rest of the family, and other factors such as insurance benefits, may also play a role in underreporting of suicidal attempts and suicides (Kessler, 1987a). Similarly, inaccurate suicide reporting is also known in the general population (Brown, 1975).

Psychological Affect of Predictive Testing

The recognition of the risk of suicide in the at risk population is important for several reasons. As it is possible to accurately assess an individual's risk of developing HD years before symptoms begin, the fear of a positive test result may have devastating psychological implications. The fear of acquiring carrier status which will cut short one's career and family plans can be a heavy burden. The realization that a dementia disorder will lead to premature death could certainly predispose to suicide. The forthcoming diagnosis, the family history of suicidal behaviour and a depression may provide support for choosing this manner of death. Although in most cases predictive testing, especially with a negative screening test may relieve anxiety, in others it may place the person and family members at risk of suicide.

Knowledge about an individual's risk of HD can result in social stigma, employment discrimination, loss of medical and insurance benefits and give great psychological suffering. HD is a familial disease in every sense of the word (Wexler, 1985). Those affected face a progressive deterioration of their mind and body, while the rest of the family have to watch in the knowledge that they too may suffer the same fate (Farrer, 1987). For some, this can lead to strategies for evasive action (Davis, 1999). Although only a minority of those facing HD have suicidal ideation and choose suicide as an option, the genetic counsellor or physician cannot ignore this possibility.

Before the introduction of predictive testing serious concerns were raised as to whether it was ethical to offer predictive testing for HD when no cure was available (Marsden, 1981). It was feared that the predictive test results, especially positive (not favourable) ones, could precipitate anxiety, depression, suicide, stress in family and relatives, and survivor's guilt in family members testing negative (Wexler, 1985). Social implications such as stigmatisation and discrimination in employment and insurance were discussed (Kenen & Schmidt, 1978).

Prior to the implementation of predictive testing in Sweden, an interview study concerning attitudes towards testing indicated that 60% of persons who were at risk of developing HD had a generally positive attitude to the predictive test (Mattsson & Almqvist, 1991). Similarly, high interest in the test has been expressed by about two thirds of at risk persons interviewed in other studies (Evers-Kiebooms, Cassiman, & Van den Berghe, 1987; Kessler, et al., 1987; Meissen & Berchek, 1987; Schoenfeld, Myers, Berkman, & Clark, 1984; Stern & Eldridge, 1975). Barette & Marsden (1979) reported that around 75% of the persons at risk of HD would take advantage of accurate presymptomatic testing. Of particular interest to professionals is that Kessler et al. (1987) reported that, in an early study of predictive testing, all participants unanimously believed that predictive testing should be available even in the absence of a cure. However, the actual number of persons who have undergone predictive testing has not reached this high level. Only 3-5% of the population at risk have gone through predictive testing in Sweden during the first six years the test was available (Almqvist, 1996). Similarly, the Canadian collaborative study of predictive testing for HD (Bloch, et al., 1989) reported that about 13% enrolled in their program in its first 16 months. A retrospective study examined pre-natal and diagnostic testing in Canada from 1987 to 2000 and reported the uptake for predictive testing to be approximately 18%, ranging from 12.5 in Maritime to 20.7% in British Colombia (Creighton, et al., 2003). However, British Colombia is a major centre of HD research and has also taken an initiative in establishing a coordinated nationwide predictive testing programme. Further, British Colombia has a higher frequency of HD, finding an incidence of 6.9 per million per year (Almqvist, Elterman, MacLeod, & Hayden, 2001). The UK Huntington's Prediction Consortium observed that only around 15% of those at high risk undergo predictive testing in England (Harper, Lim, & Craufurd, 2000).

In summary, around 3 to 21% of at risk persons enter predictive testing programs now that DNA testing is available worldwide. Bird (1999) estimated that overall 25% of at risk persons are enrolled in genetic testing programs. However, in most countries the initiative for genetic testing stems from the individual and the uptake of predictive testing in most countries seems to be much lower than 20% (Evers-Kiebooms, Decruyenaere, Fryns, & Demyttenaere, 1997). So if the uptake is low, what characteristics compel the majority of HD population not to seek genetic testing?

Tibben, Niermeijer et al. (1992) sought information on the motives of persons who were not enrolled in the program although aware of its existence. They concluded that non-participants tended to overemphasize the negative consequences of the genetic test, such as depression, fear of HD, inability to cope with the results, if positive outcome, and depression, being banned from the family, and guilt feelings if the test is negative. 'Feeling of not having courage to get carrier status', was stated as a reason for not taking the predictive testing for HD in 45% of at risk persons in a small sample at Stockholm. Another frequent answer in an open-ended questionnaire by Robins Wahlin et al. (2000) was that it is easier to bear uncertainty than certainty (Robins Wahlin, unpublished data).

Individuals who enter a predictive testing program are faced with a variety of complex personal and ethical questions. It may be the most important decision in their entire life and is complicated by the fact that there are no absolutely right or wrong choices. What may be right for one person may not be right for another. It is widely agreed that the decision to proceed with testing should be entirely up to the individual (Robins Wahlin, et al., 1997).

At present, there is no cure or adequate treatment for HD. Therefore, are there any benefits of predictive testing for healthy individuals? At the beginning of the genetic testing era Wexler (1985) reviewed this complex issue and asked the poignant question: Would this new genetic knowledge be life enriching or destructive? In an invited editorial about the risk of suicide in genetic testing for Huntington's disease Bird (1999) asks about the emotional toll of such testing. The answer to these questions is always very complex and personal, but it has to be answered by the candidate who is considering entering a predictive testing program. Potential candidates from HD families face a twenty-first century dilemma in the deepest possible way. To paraphrase Hamlet (Shakespeare, 1992): To know or not to know, "That is the question" (Robins Wahlin, 2007).

Codori and Brandt (1994) pointed out that predictive testing may lead to both negative and positive psychological consequences, regardless of the outcome of the test. The major benefit they reported was relief from the uncertainty of not knowing. In addition, noncarriers were relieved by the fact that their children were spared. Negative effects of predictive testing for noncarriers included psychological distress and feelings of guilt. In addition, carriers may feel guilty for having transferred risk status to their children, whereas noncarriers may feel guilty because they were spared from HD (Kessler, 1994). The latter, paradoxical, reaction has been referred to as survivor's guilt (Lifton, 1979; Tibben, et al., 1990; Tibben, Vegter-van der Vlis, et al., 1992).

Although prior research has been informative regarding the psychosocial consequences of genetic testing, Wiggins et al. (1992) emphasized the need of continued longitudinal assessment of predictive testing groups to examine psychological and social effects of genetic testing. Several authors have argued that programs for predictive testing can potentially be harmful (Kessler, 1987b; Kessler, et al., 1987). However, at the same time these programs may enhance the quality of life (Wiggins, et al., 1992). The Canadian Collaborative Study of Predictive Testing for Huntington disease reported a slight improvement compared with baseline in well-being and lower depression scores, even for carriers, at the 12-month follow-up (Wiggins, et al., 1992). Predictive testing has also been noted to be beneficial in reducing overall ill-health symptoms and increasing well-being for those expressing concerns about HD (Larsson, Luszcz, Bui, & Robins Wahlin, 2006). However, Codori and Brandt (1994) noted that there was attrition in the high-risk group, leaving open the possibility that the most distressed persons dropped out during the one-year follow-up period. They also made the point that persons who were tested were self-selected, and questioned whether the favourable reactions to HD testing would continue in the future. Further, some at risk persons surveyed have predicted that they would experience anxiety and contemplate suicide if the test results indicated carrier status (Kessler, 1987b; Robins Wahlin, et al., 2000).

Baum, Friedman, and Zakowski (1997) suggested that, by reducing uncertainty and providing outcome information, some of the anxiety and stress associated with being at risk of HD may be reduced. Mutation analysis provides the individual with news of either being a carrier, which is bad news, or a noncarrier which ought to be perceived as good news. However, research suggests that this is not always the case (Bloch, Adam, Wiggins, Huggins, & Hayden, 1992; Codori & Brandt, 1994; Robins Wahlin, et al., 1997; Tibben, et al., 1993). Some carriers have the capacity to anticipate and arrange their future and manage to cope with a carrier status, whereas others experience the new status as a threat and feel distress facing the future disease (Bloch, et al., 1992; Codori & Brandt, 1994; Huggins, et al., 1992; Robins Wahlin, et al., 1997; Tibben, et al., 1993). Also a noncarrier status may be associated with unexpected costs or coping problems (Codori & Brandt, 1994; Huggins, et al., 1992; Tibben, et al., 1993). How people manage to cope with the new status varies with the coping strategy, how well they are prepared, what capacity they have in terms of active coping resources, and personal factors such as optimism, beliefs, and social support (Baum et al., 1997). Thus the coping outcome from participating in predictive testing is dependent on individual differences in various characteristics before entering the program (Larsson, et al., 2006).

When planning predictive testing programs, there are three obstacles which complicate the establishment of such programs. First, clinicians, psychologists, geneticists, and researchers do not commonly share their knowledge of benefits of programs that can be used in clinical work. Second, a relatively large proportion of the population in many countries lives in rural areas, and community-based programs may meet with economic obstacles. Third, cautious attitudes among people who want to participate and people who can offer genetic testing counteract potential benefits of the genetic services (Robins Wahlin, et al., 2000).

Baum et al. (1997) noted that HD society's technological capacities may have outpaced its understanding of psychological consequences. In a study by Kessler and colleagues (1987), all of the 69 persons taking part in the survey stated that the predictive testing should be available even though no cure for HD exists. Kessler (1987b) and Copley, et al. (1995) reported that persons taking part in predictive testing programs value the counselling highly. Further, Copley et al. (1995) found that 96% of their participants were satisfied with the testing program and less than 3% of those persons whose risk status was changed felt that the results had diminished their quality of life. Hence, the wheel of genetic testing progress rolls further in most developed countries.

Suicidal Ideation Before Predictive Testing

Are the persons who pursue genetic testing more inclined to catastrophic events? Have they higher suicidal ideation? Do carriers have higher suicidal ideation compared to noncarriers? What will happen when those family members who have not dared to test begin to pursue genetic testing? Will this coming population, those 80% that have not entered the predictive testing programs at present, have a greater risk of catastrophic events?

It may be hypothesized that HD carriers are also carriers of psychological vulnerability to psychiatric affection, given the inevitable stress that HD in families yields. To test this hypothesis, a Swedish study (Robins Wahlin, et al., 2000) compared 13 carriers and 21 noncarriers on demographic characteristics, pre-test attitudes and expectations, life style and life satisfaction, estimated sense of well-being and health, and thoughts about suicidal ideation and previous suicidal behaviour. Authors also used in-depth interviews indexing different aspects of psychological and social dysfunction in the case of catastrophic events.

Table 2. Number and Percentage of Participants with Suicidal Thoughts or Tendencies*

Thought / tendency	Carriers**** n = 13 n %	Noncarriers n = 21 n %	Total n = 34 n %
I have never thought or attempted to kill myself.	5 38.5	7 33.3	12 35.3
I have thought about suicide before I discovered that I were at risk for HD.	3 23.1	8 38.1	11 32.3
I have thought about suicide since I have known that I am at risk for HD.	5 38.5	8 38.1	13 38.2
I have thought about suicide within the past year.	6 46.1**	3 14.3	9 26.5
Suicide has sometimes been a passing thought and/or I have briefly considered it.	5 38.5	7 33.3	12 35.3
I have had a plan for killing myself and seriously considered carrying it out.	2 15.4	2 9.5	4 11.8
I attempted to kill myself.	1 7.7	5 23.8	6 17.6
It is likely that I will attempt suicide when I begin developing symptoms of HD.	1 7.7**	9 42.8	10 29.4
It is likely that I will attempt suicide when I get middle/advanced stages of HD.	3 23.1***	11 52.4	14 41.2

* Reproduced with permission from Acta Neurologica Scandinavica, Blackwell Publishing, Robins Wahlin et al, 2000.
** $p \leq .05$, *** $p \leq .10$
**** One carrier committed homicide-suicide, father and son, after the data collection.

This study did not reveal any significant differences in social and demographic characteristics of the carriers and noncarriers and the groups did not differ in any of the assessed pre-test attitudes, expectations, general well-being, life satisfaction or style, need for support, estimated sense of well-being, or degree of health. Although the study indicated few differences between the two groups in suicidal ideation and behaviour, carriers exhibited a significantly higher rate of suicidal thoughts within the past year than noncarriers, despite the fact that the carriers reported being fairly satisfied with their own health and indicated even marginally better health satisfaction than did the noncarriers in the Life Satisfaction Index (LSI). Importantly, six of the 34 participants (17.6%) had attempted suicide before entering the predictive testing program and five of them were given noncarrier status and only one of them was a carrier. One noncarrier had been hospitalised in a psychiatric ward for over a year in connection with three serious suicide attempts and another noncarrier had taken drug overdose on nine occasions. The remaining noncarriers and one carrier had tried to kill themselves either by overdose or by cutting their wrists. Of interest, method of suicide reflects country and social background. None of the suicide attempts were made by guns (which is the most common method of suicide in USA, Kellermann, et al., 1992). Four participants (two carriers and two noncarriers) were judged to be at an immediate suicidal risk, but these participants had not attempted suicide before entering the program. Robins Wahlin et al's. (2000) results suggest that psychosocial determinants rather than genetic factors may influence at risk persons' patterns of suicidal behaviour. An overview of the participants' suicidal thoughts or tendencies is provided in Table 2.

Further, both at risk groups in this study, showed considerably higher tendencies for suicidal behaviour and psychiatric dysfunction than the normal Swedish population (Center for Suicide Research and Prevention, 1994). A survey of a normal population aged 20-67 years ($n = 8800$) in the county of Stockholm, revealed that 19.7% had thought about suicide, and that 7% had thought about suicide within the last 12 months (Center for Suicide Research and Prevention, 1994). In the Robins Wahlin et al. (2000) sample which ranged in age between 23-57 years and came from the same area, 35% had considered suicide and 26% had done so within the last year (see Table 2). The attempted suicide rate of 17.6% was five times higher than the 3.5% documented for the general population between 20-67 years in Stockholm during the same period. This is in line with the estimated and completed frequency of suicide in HD population as compared to that of the general population in USA (Kessler, 1987a, 1987b; Schoenfeld, Myers, Cupples, et al., 1984; Wexler, 1979) confirming markedly elevated suicide rates for the HD population. In a survey by Kessler et al. (1987) 5% indicated that they would commit suicide if the predictive test results were positive and 11% of the subjects would consider suicide as a potential response. Mastromauro et al. (1987) found that about 21% at risk persons said that they might commit suicide if they did get positive results. The finding that 41% considered suicide if-or-when they reached middle/advances stages in the disease is in line also with a survey by Wexler (1979), in which around half of the 35 persons at risk for HD indicated that they would commit suicide if they became ill. Bloch et al. (1989) reported that 33% considered suicide a possibility in the future if the results indicated an increased risk. Counsellors and other professionals should take suicidal thoughts very seriously, if and when reported. The author has encountered one case where a carrier who had expressed suicidal thoughts subsequently killed not only himself but also his carrier son.

Farrer (1987) pointed out that HD is a slowly progressing disease and years of searching for signs of HD may aid in forming suicidal tendencies. A general population study in Sweden (Dahlgren, 1977) observed that about 11% of attempted suicides later resulted in completed suicides, suggesting that at risk persons in the Robins Wahlin sample would have a high possibility of repeating their self-injurious behaviour. The estimated casualty rate would be about 5%, a figure in line with a study of Beck et al. (Beck, Steer, Beck, & Newman, 1993), which followed 207 patients hospitalised because of suicidal ideation. Of all the data collected at the time of the hospitalisation, by Beck et al. only the pessimism item of the Beck Depression Inventory (BDI) predicted suicide rate. In Robins Wahlin's study (2000), 76.9% of the carriers and 52.4% of the noncarriers received a value of 1 or 2 on the BDI pessimism item. Thus, the observation that 61.8% of the total group felt discouraged about the future provides further support of high suicidal ideation in this sample. However, it is noteworthy that life satisfaction of at risk persons was average despite the high frequency of self-destructive behaviour. Robins Wahlin et al. (2000) noted that the Life Satisfaction Index (Campbell, Converse, & Rodgers, 1976) for eight domains of life experiences indicated for both at risk groups average or higher-than-average life satisfaction. Interestingly, only in the domain of economy were carriers less satisfied than the average hypothesised mean. In all eight domains of life experience the participants rated economy and work as least satisfying and marriage as most satisfying.

Previous documentation (Folstein, Folstein, & McHugh, 1979; Kessler, 1987b; Kessler, et al., 1987) indicates that a turbulent history predisposes the individuals at risk for psychiatric dysfunction and/or suicide attempts. Hence, should a person with a history of suicide attempts be excluded automatically from genetic testing? Lam et al. (1988) recommends that, along with personal and family history of suicidal behaviour, other factors need to be assessed. These would include suicidal ideation, personal resources, and social support system. The authors recommend that a history of serious suicidal behaviour should necessitate follow-up psychiatric evaluation and professional support after presentation of results of the genetic test (Lam, et al., 1988). In an early study by Kessler et al. (1987) 39% at risk subjects believed that even if a person was suicidal or psychologically unstable, that person had the right to be tested.

As at risk persons approach predictive testing an adequate support system is very important to assist them in dealing with potentially adverse outcomes (Kessler, 1987b). The Swedish sample (Robins Wahlin, et al., 2000) reported that their social support net involved an average of 4 persons and that they were fairly satisfied with their social network. They had better external sources of support than at risk groups surveyed in northern California, USA, where 5 to 11% reported that they had no external source of support (Kessler, 1987b; Kessler, et al., 1987). Moreover, an open attitude toward utilizing professional help is necessary to prevent catastrophic events. Robins Wahlin et al. (2000) reported that about 85% considered seeing a psychologist or social worker, and a support group was desired by 62%. The fears and stresses imposed by bad outcomes may erode even the best social support systems. Also, some at risk persons may not accept therapy or counselling when in crisis. In the study by Kessler et al. (1987) 19% of participants said that, if they were found to be carriers, they would not seek counselling to help them cope. A lack of social skills, personality factors such as avoidant or schizoid tendencies, or habitual pattern of social interaction, may affect the individual's personal responses. Some at risk individuals may be unprepared to deal adequately with the destructive feelings that a positive test results evoke (Kessler, 1987b). These 'marked' persons may develop suicidal tendencies and kill themselves prematurely.

Suicidal Behaviour After Predictive Testing

Already two decades ago, it was suggested that living with knowledge of carrier status might lead to major social catastrophic events, such as divorce, premature loss of employment and insurance benefits (Kessler, 1987b). Some of the psychological problems anticipated in the post-test period included depression, anxiety and panic disorders, other psychiatric disorders and suicide. The Canadian Collaborative Study of Predictive Testing studied the frequency of adverse reactions after the predictive test (Lawson, et al., 1996). Adverse psychological reactions were defined as suicidal ideation, psychiatric hospitalisation, depression, relationship breakdown, substance abuse, or psychological distress. In the first year after predictive testing for Huntington's disease such an event occurred in 16% (15/95) of the tested individuals. However, there was no significant difference in frequency between the increased-risk and decreased-risk groups (Lawson, et al., 1996). An earlier study by Quaid (1993) reported 2.1% (4/189) hospitalisation after predictive testing results (three received increased risk results and one a decreased risk result). However, adverse reactions without hospitalisation were not reported.

Catastrophic events have been studied in a worldwide assessment of the frequency of suicide, suicide attempts, or psychiatric hospitalisation after predictive testing for Huntington's Disease (Almqvist, Bloch, Brinkman, Craufurd, & Hayden, 1999). The authors surveyed catastrophic events after predictive testing results, through questionnaires administered in 100 centres in 21 countries from five continents. They did not include symptomatic participants suspected of having HD (diagnostic tests). A total of 4527 participants had received a predictive testing result, 16.4% via linkage analyses and the remaining 83.6% via direct DNA testing. Some (5.3%) were tested by both methods and included in mutation analyses. All these persons were followed up after receiving the predictive testing results. An additional 441 tested participants, who were not followed-up, were not included in the study. About forty per cent of participants received increased-risk results or expanded CAG repeats (>35) and 57.5% of participants received a decreased risk result or noncarrier status. Additionally, 0.7% received an uninformative result from linkage analyses and 1.5% had an intermediate allele. This sample consisted of 59.3% females and 40.7% males. Furthermore, previous psychiatric history indicated that 38.5% of participants with a catastrophic event had a psychiatric history prior to the predictive testing.

The results of this large, international study can be briefly summarized as follows. Forty-four persons (0.97%) of the total 4527 participants reported catastrophic events. There were five completed suicides, all women, and 21 (13 women and eight men) attempted suicides. Eighteen persons (13 women and five men) were hospitalised for psychiatric dysfunction. All persons that committed suicide were symptomatic. The catastrophic event rate for increased risk group was 2% and for decreased risk group 0.3%. Twenty-four of 37 (62.1%) of those catastrophic events with an increased risk were symptomatic. The increased risk group experienced the majority of the catastrophic events, 62.1% during the first year, in contrast with those who received a decreased risk results when the catastrophic events were experienced more than the one year after receiving the results (57.1%). During the first two years, there were three suicides, 18 suicides attempts, and 14 hospitalisations. The rest of the catastrophic events happened during the follow-up period, 2-4 years. The authors reported no difference in frequency of catastrophic events between persons receiving results through linkage analyses and those receiving definite results after the mutation analyses.

How should we interpret these results? Is the genetic testing for HD a success story as Bundey (1997) stated? Are suicide rates and attempted suicide rates as low as they appear? Bird (1999), in an invited editorial, calculated that five suicides in 1,817 (increased risk) individuals equates to about 275 suicides per 100,000 persons. Because most suicides occurred in the first two years after genetic testing, an overall suicide rate would be 138/100,000 persons per year. The reported rate for suicide among the general population in USA is 1.0% - 1.3% which equates 10-13/100,000 persons (Roy, 1995). Hence, the suicide rate of 13.8% reported by Almqvist et al. (1999) is 10 times higher than the general suicide rate in USA. However, as pointed out by Bird (1999), suicide rates vary among countries, and they are affected by factors such as socio-economic status, age, and sex (Roy, 1995). Bird quotes suicide rates of <10/100,000 persons in Ireland, >30/100,000 persons in Hungary and 21/100,000 persons in Sweden. In fact, the rate of suicide for the year 1999 in Sweden was 14/100,000 persons and for the year 2008 13/100.000 persons (Official Statistics of Sweden, 2008), indicating that suicide rate varies even from one year to another. If we take the suicide rate of 14/100,000 for Sweden, then the suicide rate reported by Almqvist et al. (1999) is approximately 10 times the average rate.

How does predictive testing affect suicide rate in the HD population? Bates et al. (2002) reported that cases of suicide in the general HD population vary from 4% to 12.7%, based on the study of Schoenfeld, Myers, Cupples, et al. (1984). Farrer (1986) reported 25 cases of suicide out of 440 affected HD persons, which extrapolate to a 5.7% suicide rate. Hayden et al. (1980) calculated the suicide rate for the HD population in South Africa to be 3.3%. These figures imply that the rate of suicide after predictive testing, extrapolated as 13.8% (Almqvist, et al., 1999) is even higher than in the general symptomatic HD population.

Another important piece of information provided by Almqvist et al (1999) is the 2% catastrophic event rate for the increased risk group within two years of receiving the results of genetic testing. This indicates that the majority of individuals receiving carrier status appear to cope well with this new information. On the other hand, a 2% catastrophic event rate gives us a message that, in predictive testing programs a follow-up is necessary to reduce the catastrophic event rate. Overall, the Almqvist et al. (1999) study indicates that predictive testing for HD has serious risks.

Almqvist et al (1999) suggests that women who are unemployed, who have a history of psychiatric illness, and who develop early symptoms of HD, are at greatest risk of catastrophic events. However, Bird (1999) notes that it is not a helpful profile. Generally more women than men choose to be tested, the majority of people who have catastrophic events have no psychiatric history, and furthermore, in a study of Lipe et. al. (1993), all the completed suicides were men. Similarly, Schoenfeld, Myers, Cupples, et al. (1984) found that approximately four times as many male patients committed suicide as females and Farrer (1986) reported twice as many suicides among men (7.7%) as among women (3.2%) with HD. This highlights the necessity of careful evaluation of all persons undergoing HD predictive testing (Bird, 1999) in order to be able to screen the 2% of at risk population for catastrophic reactions.

Almqvist et al (1999) indicated that 40.1% of participants received increased-risk results. This suggests that counsellors and physicians should encourage all at risk persons to think of the possibility of a positive result. However, the strong hope for good news could emotionally blind the at risk person. The at risk individuals should face the thought of how they might react to bad news, and how this change would affect their lives (Bird, 1999). Many predictive programs offer detailed printed information of predictive testing (Robins Wahlin, et al., 2000) but the people seeking genetic information are not always psychologically mature enough to change their genetic status. The genetic counsellors should not promote nor advocate presymptomatic DNA-testing as many at risk persons do not feel strong enough to go through the emotional changes that may follow a positive result. Given everything known about responses to medical diagnosis of HD, the likelihood of catastrophic reactions to positive test outcomes among at risk individuals is high (Kessler, 1987a).

On the contrary, a two year follow-up study of depression and suicidal ideation after predictive testing, reported much lower suicidal ideation than expected in Sweden (Larsson, et al., 2006). The authors reported a follow-up of depression, self injurious behavior, life satisfaction, general health, well-being, and lifestyle of 35 carriers and 58 noncarriers before the predictive test and 24 months afterwards. Both carriers and noncarriers showed high suicidal ideation before the genetic testing. However, depression scores and frequency of suicidal thoughts increased more for carriers, compared to noncarriers, over the restrictive time period. Interestingly, no differences were found regarding life satisfaction or life style between carriers and noncarriers (Larsson, et al., 2006).

Sometimes a person may enter the predictive testing program thinking that the outcome concerns if-and-when the disease will develop in the future. However, the early signs of the disease may already have started and the news changes the character of the counselling. These persons receive a double dose of bad news and are extremely vulnerable. Patients with unsuspected early disease, who would not otherwise have learned of HD presence if predictive testing were not offered, are not prepared to deal with the unexpected news and are most likely to suffer severe catastrophic reaction (Lam, et al., 1988). Although the importance of a sensitive approach to conveying bad news to the patient, the possibility of depressive reactions and suicidal behaviour is not always seen as a likely outcome (Draper, Peisah, Snowdon, & Brodaty, 2010). Post-test casualties due to suicidal ideation and suicide, need an optimal management of these adverse reactions and may be an important determinant of the number of catastrophic events.

Depression in Huntington's Disease

Depression is an early manifestation of HD (Paulsen, Nehl, et al., 2005). Prevalence of major depression one year after predictive testing for HD has been reported for mutation carriers to 6% and noncarriers to 3% (Codori, Slavney, Rosenblatt, & Brandt, 2004). However, the same authors report 20% for carriers and 13% for noncarriers of HD experiencing clinically significant depressive symptom (Codori, et al., 2004). Severe depressive disorders are documented (Folstein, et al., 1979) and these can occur before significant cognitive decline or motor difficulties (Paulsen, Hoth, et al., 2005). Anguish, dysphoria, agitation, irritability, apathy, anxiety, and denial are very common in both the patients and family members of HD (Alonso, et al., 2002; Paulsen, et al., 2001). In the initial phase of shock and disbelief when the diagnosis of HD is delivered, the patient may feel extremely low and contemplate suicide. Indeed, the literature of psychiatric symptoms in HD suggests that depression, apathy, and aggression are common and suicide rate are well over those of the general population (Di Maio, et al., 1993; Paulsen, et al., 2001). Although patients early in the clinical course of the disease are at the greatest risk of suicide, the depression may occur in any phase of HD (Paulsen, Nehl, et al., 2005). If a major depression with dementia occurs in the late course of the disease, it may impair the patient's ability to commit suicide.

In the general population suicidal behaviour has many contributing factors: it may be based on the belief of being trapped in an impossible situation; it may be a symptom of mental disorder or specific disease; it may be the result of hostility turned against the self; or it may be " a cry for help" (for an overview, see Beck, Kovacs, & Weissman, 1975). In short, depression is often characterized by feelings of hopelessness (Folstein, 1989).

Depression is commonly discussed in terms of feelings of hopelessness (Farrer, 1986; Kessler, 1987a). Beck et al. (1975) argued that, irrespective of the diagnosis, the negative expectations or hopelessness appear to be a primary feature in suicidal intent. Beck and Kovacs (1979) found that 51% of females and 67% of males, in a group of 90 hospitalised for suicide had received a diagnosis of depressive disorder. Furthermore, Beck et al. (1993) found that higher levels of suicidal ideation were associated with increasing severity of self-reported depression, and negative expectancies about the future.

Affective disorders are the most commonly described psychiatric syndrome (excluding dementia) associated with HD (Hayden, 1981). Stern and Eldridge (1975) found that patients with early HD and their families were often concerned about mental changes. Folstein, et al. (1979) noted that HD patients who were not referred for psychiatric reasons nevertheless had a high frequency of psychotic disorders. Specifically, 45% of the patients met the criteria for manic-depressive disorder and 18 % were diagnosed with an auditory hallucinatory state. Moreover the HD patients were reactively depressed in stressful situations brought on by the disease, such as learning of the diagnosis, losing their jobs, being isolated at home, and losing manual skills. These kinds of complications are familiar to HD families and may also be brought on by a change in genetic status.

Roy (1983) found that there was a marked association between suicidal history, attempted suicide and the presence of depression and/ or recurrent affective disorder within family. These patients may be more vulnerable than usual to major depression (Hayden, 1981). The author (1981) implied that patients in early stages of HD who are minimally demented but severely depressed run the highest risk of suicide. Later studies report evidence that brain changes precede the traditional clinical diagnosis and psychiatric manifestations may predate neurological signs (Gutekunst, et al., 1999; Paulsen, et al., 2001). The comparison of onset age of HD and suicide age shows that at risk subjects' suicide may occur at the first appearance of symptoms (Di Maio, et al., 1993). The fact that depressive episodes can appear before the neurological symptoms of HD, suggests a direct neurobiological link between depression and the disease (Lam, et al., 1988). The degenerative disease process could directly result in depression, rather than being a simple psychological reaction to illness.

Paulsen et al. (2001) described 69.2% in their sample of 52 patients as having dysphoria and that it was the most common symptom (along with agitation) in HD. The authors note that most of their patients were prescribed antidepressant medication and that it is likely that rates of depression are even higher than that reported in literature. Depression has been associated with the orbito-frontal circuit that projects from the anterior and lateral orbito-frontal cortex via the ventro-medial caudate nucleus to the ventral anterior and medial dorsal nuclei of the thalamus (Paulsen, et al., 2001). This is consistent with the view that HD is characterized as a frontostriatal dementia, in which cognitive deficits may result from pathological changes at multiple sites in the frontostriatal circuitry (Bäckman, Robins Wahlin, Lundin, Ginovart, & Farde, 1997; Robins Wahlin, Larsson, & Luszcz, 2010; Robins Wahlin, Lundin, & Dear, 2007). Furthermore, most research and clinical reports have found that, in virtually all cases, the manifestation of depression is either simultaneous with or subsequent to the motor onset of HD (Di Maio, et al., 1993; Hayden, 1981).

Lipe et al. (1993) found in a retrospective study of nine instances of suicides in HD families that 50% of suicides and 39% of controls with HD but without suicide attempts, had a history of depression. Although many of HD subjects had a history of alcoholism and violence, it did not distinguish the suicide subjects from other HD patients. These behaviours may reflect the use of alternate coping methods (Lipe, et al., 1993).

Nevertheless, the stress of learning the diagnosis of HD, which results eventually in losing one's job, being isolated at home, and losing cognitive and manual skills could also contribute to a reactive depression. The need to carefully explore these issues with at risk persons prior to predictive testing may be an important factor in clarifying thinking and helping them to plan appropriately for the post-test period (Kessler, 1987a). Taking into account that many at risk subjects seek predictive testing when they are near the critical age, it is of importance to consider carefully the possibility of an increased suicide risk of these persons.

Folstein et al. (1979) stressed that the demoralizing feelings usually subsided following counselling, and when patients accepted their situation. They also observed that the clinical state of persons at risk varies widely. Hence, they recommended quantification and documentation of psychiatric symptoms. This would permit classification of clinical symptoms and assist with treatment and research. Further, Lester, Beck and Mitchell (1979) concluded that specific scores on depression inventories of persons who attempt suicide might be important in predicting suicidal deaths. The authors extrapolated two monotonic trends from attempted and completed suicides; the more serious attempters had higher scores both on depression (Beck, Ward, Mendelson, Mock, & Erbaugh, 1961) and on a hopelessness scale (Beck, Weissman, Lester, & Trexler, 1974). Beck, Herman, & Schuyler (1974) devised a test to measure the cognitive component of depression, which they labelled "hopelessness". They found that "hopelessness" correlates more highly than does depression with suicidal intent in attempted suicides and with suicidal ideation. Therefore, depression, hopelessness, anxiety, emotional distress, suicidal tendencies, and social dysfunction grading should be important in predictive testing of HD. The psychological well being of predictive testing program participants' should be documented by questionnaires and/ or in depth interviews, which should index different aspect of psychological and social dysfunction in the case of catastrophic events, such as suicide attempts. Farrer (1987, in a letter to editor) urged that professional involvement should become mandatory in the setting up of predictive testing.

Low mood or depressive thoughts are often reported in counselling for HD predictive testing. To evaluate the degree of depression of participants and General Health Symptoms before predictive testing for Huntington's disease, Robins Wahlin et al. (2000) employed BDI (Beck, et al., 1961) and the General Health Questionnaire (GHQ; Goldberg & Williams, 1988). These results are depicted in the Table 3. Robins Wahlin et al. (2000) observed that 26% of the at risk participants reported mild to severe depression as measured by BDI, and

47% reported mild to severe degrees of general health symptoms as indicated by GHQ. 46% of the carriers and 48% of noncarriers scored above the normal range in the GHQ-30. This is a pure state measure of general mental health. Hence, a high score should be taken as an indication for the counsellor to focus the interview, to follow up the person carefully, and to prepare him or her for possible adverse effects of the genetic result. Depressive and self-destructive attitudes, thoughts, tendencies, and wishes, are often manifest and identifiable in persons at risk of HD.

Table 3. The Degree of Depression of Participants and General Health Symptoms As Measured by The Beck Depression Inventory and The General Health Questionnaire before Predictive Testing for Huntington Disease*

Beck Depression Inventory (BDI)	Normal range [a]	Mild - Severe [b]
Carriers (n = 13)	11 (84.6%)	2 (15.4%)
Noncarriers (n = 21)	14 (66.7%)	7 (33.3%)
Total (n = 34)	25 (73.5%)	9 (26.5%)
General Health Questionnaire (GHQ)	Normal range [c]	Mild - Severe [d]
Carriers (n = 13)	7 (53.8%)	6 (46.1%)
Noncarriers (n = 21)	11 (52.4%)	10 (47.6%)
Total (n = 34)	18 (52.9%)	16 (47.0%)

* Reproduced with permission from Acta Neurologica Scandinavica, Blackwell Publishing, Robins Wahlin et al, 2000.
a BDI normal range = 0-9, b BDI mild to severe ≥ 10
c GHQ normal range = 0-4, d GHQ mild to severe = 5-30

Huggins et al. (1992) observed that concerns about predictive testing centred around the counsellor's ability to identify persons who would be unable to cope with the consequences of a carrier status. They concluded that approximately 10% of noncarriers might also be vulnerable to adverse effects of the test results, and require professional intervention and ongoing counselling. As pointed out by Codori and Brandt (1994) and Robins Wahlin et al. (2000; 1997), at risk persons should be cautioned that testing could create, rather than alleviate, problems. A high frequency of depression and psychiatric affection, even for noncarriers (Larsson, et al., 2006; Robins Wahlin, et al., 2000), suggests that a careful follow-up is justified not only for carriers but also for noncarriers (Tibben, et al., 1993; Tibben, et al., 1990).

For those people for whom autonomy, mental competence and independence are prime values, predictive testing for HD presents an opportunity for acting according to those values before it is too late. These life tragedies take radically different forms, with different levels of burden and adaptation, and some of the newly diagnosed HD persons might choose to end their life. Hence, a planned suicide can also be a response to diagnosis of future dementia and thus be a rational and defensible strategy (Davis, 1999). Suicide may not always be an 'adverse reaction' to bad news. The candidate may view suicide as an option, i.e. rational suicide. It can be viewed in the light that "a diagnosis of impending dementia is a warning that one is about to be invaded by an enemy army that will always win. It is entirely sensible to burn down the fort and refuse it a home" (Davis, 1999).

Kessler and Bloch (1989) ascribe system factors that may account for suicidal ideation and behaviour. They observe that professional pessimism about the lack of cure and treatment of HD often affects the perceptions of patients and their relatives, and may contribute to the avoidance of further professional contact. Suicidal ideation can be fostered by social attitudes towards the chronically ill, their long-term care and social costs. Further, Kessler and Bloch (1989) claim that "the prospect of financial drain unsupported by societal willingness to provide for such catastrophic situations provides the backdrop for the shame, neglect, depression, and suicidal behaviours among affected individuals and their relatives." They describe, by using case examples, that events leading to the suicidal act reflect a system pressure promoting behaviour that leaves little choice but suicide. The belief to be a burden to the rest of the family, and specifically when depression narrows the scope of social interactions and support, may cause the affected to try the suicidal outcome.

It is understandable that some members of the HD population may see suicide as an appropriate response to impending dementia. However, biological, psychological, social, and even situational factors may interact and end up in catastrophic reactions and eventual suicide. Counsellors and other professionals face a painful ethical dilemma in dealing with issues of suicidal ideation and suicide in HD. Kessler (1987a) makes an important point that if suicide is viewed as a rational and dignified act to an increasingly deteriorating life, the professional course of action may differ towards the affected and family from if the professional believes that suicide is a demented act. Kessler (1987a) asks, should professionals act in good faith that the affected has impaired judgment, and protect the patient against the suicidal act? Furthermore, should the professionals act against the family members who out of anguish, despair, lack of will or resources may encourage suicide, even in a passive manner or covertly? Potential for suicidal ideation and suicidal behaviour exists particularly for those testing positive for HD and the affected.

New technologies, such as positron emission tomography (PET), have increased the possibility of early diagnosis of HD (Paulsen, 2009). Research using PET suggests that underlying brain abnormalities such as dopamine neurotransmission parameters for caudate and putamen as well as the volumetric measurements of these structures may show changes and precede overt symptoms (Aylward, et al., 2004; Bäckman, et al., 1997). Early diagnosis without adequate treatment and cure of HD, may have traumatic effects on the affected individual and their family. Taken together, degenerative, rapid course of HD, cognitive decline, and absence of cure, suicidal behaviour may provide an outlet for feelings of hopelessness, shame, depression, and endless suffocation by negative affect.

Looking at the milestones of the progress in HD research huge steps have been taken. Gusella et al. made predictive testing possible by linkage analyses in 1983 and direct testing became available in 1993 after the discovery of the mutation associated with HD (Huntington's Disease Collaborative Research Group, 1993). However, central to these issues are effective treatment and eventually cure for HD. In the 1980's and afterwards many persons believed that a cure for HD would probably be found in their own or their child's lifetime (Kessler, et al., 1987). Now in the twenty-first century, clinical trials are underway to slow the progression of HD in clinically affected patients and there are also plans to treat pre-clinical patients. This may give new hope to the at risk population and might even decrease the incidence of suicide in HD and their relatives. At the present, the problems associated with suicidal ideation and suicide in HD populations, worldwide, are much the same as two decades ago. Literature published about suicide and suicide attempts in HD populations is sparse. More research with larger samples is needed in order to be able to focus on catastrophic events and suicidal behaviour in at risk persons, carriers, noncarriers and HD families alike.

Acknowledgments

The research was supported by grants from Stockholm County Council, PickUp, grant No 20070033 and No 20050940; Alzheimer's Australia Research; and the Swedish Association of Neurologically disabled (NHR) to Tarja-Brita Robins Wahlin. I am indebted to A. Frick, M. Cook, and A. Holmes for valuable comments on the manuscript.

References

Almqvist, E. W. (1996). *Molecular genetic studies on Huntington disease* Unpublished Doctoral dissertation, Karolinska Institute, Stockholm.

Almqvist, E. W., Bloch, M., Brinkman, R., Craufurd, D., & Hayden, M. R. (1999). A worldwide assessment of the frequency of suicide, suicide attempts, or psychiatric hospitalization after predictive testing for Huntington disease. *American Journal of Human Genetics, 64*(5), 1293-1304.

Almqvist, E. W., Elterman, D. S., MacLeod, P. M., & Hayden, M. R. (2001). High incidence rate and absent family histories in one quarter of patients newly diagnosed with Huntington disease in British Columbia. *Clinical Genetics, 60*(3), 198-205.

Alonso, M. E., Ochoa, A., Boll, M. C., Sosa, A. L., Yescas, P., Lopez, M., et al. (2009). Clinical and genetic characteristics of Mexican Huntington's disease patients. *Movement Disorders, 24*(13), 2012-2015.

Alonso, M. E., Yescas, P., Rasmussen, A., Ochoa, A., Macias, R., Ruiz, I., et al. (2002). Homozygosity in Huntington's disease: new ethical dilemma caused by molecular diagnosis. *Clinical Genetics, 61*(6), 437-442.

Aylward, E. H., Sparks, B. F., Field, K. M., Yallapragada, V., Shpritz, B. D., Rosenblatt, A., et al. (2004). Onset and rate of striatal atrophy in preclinical Huntington disease. *Neurology, 63*(1), 66-72.

Bäckman, L., Robins Wahlin, T.-B., Lundin, A., Ginovart, N., & Farde, L. (1997). Cognitive deficits in Huntington's disease are predicted by dopaminergic PET markers and brain volumes. *Brain, 120*, 2207-2217.

Baliko, L., Csala, B., & Czopf, J. (2004). Suicide in Hungarian Huntington's disease patients. *Neuroepidemiology, 23*(5), 258-260.

Barette, J., & Marsden, C. (1979). Attitudes of families to some aspects of Huntington's chorea. *Psychological Medicine, 9*(02), 327-336.

Bates, G., Harper, P. S., & Jones, L. (2002). *Huntington's disease* (Third ed.). Oxford: Oxford University Press.

Baum, A., Friedman, A., & Zakowski, S. (1997). Stress and genetic testing for disease risk. *Health Psychology, 16*(1), 8-19.

Beck, A. T., Kovacs, M., & Weissman, A. (1975). Hopelessness and suicidal behavior. An overview. *Journal of the American Medical Association, 234*(11), 1146-1149.

Beck, A. T., Kovacs, M., & Weissman, A. (1979). Assessment of suicidal intention: the Scale for Suicide Ideation. *Journal of Consulting and Clinical Psychology, 47*(2), 343-352.

Beck, A. T., Schuyler, D., & Herman, I. (1974). Development of suicidal intent scales. In A. Beck, H. Resnick & D. Lettieri (Eds.), *The prediction of suicidal behaviour* (pp. 45–56). Bowie, MD: Charles Press.

Beck, A. T., Steer, R. A., Beck, J. S., & Newman, C. F. (1993). Hopelessness, depression, suicidal ideation, and clinical diagnosis of depression. *Suicide and Life-Threatening Behavior, 23*(2), 139-145.

Beck, A. T., Ward, C. H., Mendelson, M., Mock, J., & Erbaugh, J. (1961). An inventory for measuring depression. *Archives of General Psychiatry, 4*, 561-571.

Beck, A. T., Weissman, A., Lester, D., & Trexler, L. (1974). The measurement of pessimism: the hopelessness scale. *Journal of Consulting and Clinical Psychology, 42*(6), 861-865.

Bindler, L., Travers, D., & Millet, B. (2009). [Suicide in Huntington's disease: a review]. *Revue Medicale Suisse, 5*(195), 646-648.

Bird, T. D. (1999). Outrageous fortune: the risk of suicide in genetic testing for Huntington disease. *American Journal of Human Genetics, 64*(5), 1289-1292.

Bloch, M., Adam, S., Wiggins, S., Huggins, M., & Hayden, M. R. (1992). Predictive testing for Huntington disease in Canada: the experience of those receiving an increased risk. *American Journal of Medical Genetics, 42*(4), 499-507.

Bloch, M., Fahy, M., Fox, S., Hayden, M. R., & James, F. R. (1989). Predictive testing for Huntington disease: II. Demographic characteristics, life-style patterns, attitudes, and psychosocial assessments of the first fifty-one test candidates. *American Journal of Medical Genetics, 32*(2), 217-224.

Brown, J. H. (1975). Reporting of suicide: Canadian statistics. *Suicide, 5*(1), 21-28.

Bundey, S. (1997). Few psychological consequences of presymptomatic testing for Huntington disease. *Lancet, 349*(9044), 4.

Campbell, A., Converse, P., & Rodgers, W. (1976). *The quality of American life*: Russell Sage Foundation New York.

Center for Suicide Research and Prevention, S. C. C. (1994).

Chapdelaine, P. A., Jr. (1993). Terminal illness and risk of suicide. *Journal of the American Medical Association, 269*(22), 2847-2848.

Chiu, E., & Alexander, L. (1982). Causes of death in Huntington's disease. *Medical Journal of Australia, 1*, 153.

Cina, S. J., Smith, M. T., Collins, K. A., & Conradi, S. E. (1996). Dyadic deaths involving Huntington's disease: a case report. *American Journal of Forensic Medicine and Pathology, 17*(1), 49-52.

Codori, A. M., & Brandt, J. (1994). Psychological costs and benefits of predictive testing for Huntington's disease. *American Journal of Medical Genetics, 54*(3), 174-184.

Codori, A. M., Slavney, P. R., Rosenblatt, A., & Brandt, J. (2004). Prevalence of major depression one year after predictive testing for Huntington's disease. *Genetic Testing, 8*(2), 114-119.

Copley, T., Wiggins, S., Dufrasne, S., Bloch, M., Adam, S., McKellin, W., et al. (1995). Are we all of one mind? Clinicians' and patients' opinions regarding the development of a service protocol for predictive testing for Huntington disease. *American Journal of Medical Genetics, 58*(1), 59-69.

Creighton, S., Almqvist, E. W., MacGregor, D., Fernandez, B., Hogg, H., Beis, J., et al. (2003). Predictive, pre-natal and diagnostic genetic testing for Huntington's disease: the experience in Canada from 1987 to 2000. *Clinical Genetics, 63*(6), 462-475.

Dahlgren, K. (1977). Attempted suicides-35 years afterward. *Suicide & life-threatening behavior, 7*(2), 75.

Davis, D. S. (1999). Rational suicide and predictive genetic testing. *Journal of Clinical Ethics, 10*(4), 316-323.

Dewhurst, K., Oliver, J., & McKnight, A. (1970). Socio-psychiatric consequences of Huntington's disease. *The British Journal of Psychiatry, 116*(532), 255.

Di Maio, L., Squitieri, F., Napolitano, G., Campanella, G., Trofatter, J. A., & Conneally, P. M. (1993). Suicide risk in Huntington's disease. *Journal of Medical Genetics, 30*(4), 293-295.

Draper, B., Peisah, C., Snowdon, J., & Brodaty, H. (2010). Early dementia diagnosis and the risk of suicide and euthanasia. *Alzheimer's and Dementia, 6*(1), 75-82.

Evers-Kiebooms, G., Cassiman, J., & Van den Berghe, H. (1987). Attitudes towards predictive testing in Huntington's disease: a recent survey in Belgium. *British Medical Journal, 24*(5), 275.

Evers-Kiebooms, G., Decruyenaere, M., Fryns, J. P., & Demyttenaere, K. (1997). Psychological consequences of presymptomatic testing for Huntington's disease. *Lancet, 349*(9054), 808.

Farrer, L. A. (1986). Suicide and attempted suicide in Huntington disease: Implications for predictive testing of persons at risk. *American Journal of Medical Genetics, 24*, 305-311.

Farrer, L. A. (1987). Letter to the Editor: Response to Kessler: Suicide and presymptomatic testing in Huntingtons disease. *American Journal of Medical Genetics, 26*, 319-320.

Folstein, S. E. (1989). *Huntington's disease: a disorder of families*: Johns Hopkins University Press Baltimore.

Folstein, S. E., Folstein, M. F., & McHugh, P. R. (1979). Psychiatric Syndromes in Huntington's Disease. In N. S. Wexler & A. Barbeau (Eds.), *Advances in Neurology* (Vol. 23, pp. 281-289). New York: Raven Press.

Fox, S., Bloch, M., Fahy, M., & Hayden, M. R. (1989). Predictive testing for Huntington disease: I. Description of a pilot project in British Columbia. *American Journal of Medical Genetics, 32*(2), 211-216.

Goldberg, D., & Williams, P. (1988). *A user's guide to the General Health Questionnaire*: Nfer-Nelson Windsor.

Gusella, J. F., Wexler, N. S., Conneally, P. M., Naylor, S. L., Anderson, M. A., Tanzi, R. E., et al. (1983). A polymorphic DNA marker genetically linked to Huntington's disease. *Nature, 306*(5940), 234-238.

Gutekunst, C., Li, S., Yi, H., Mulroy, J., Kuemmerle, S., Jones, R., et al. (1999). Nuclear and neuropil aggregates in Huntington's disease: relationship to neuropathology. *Journal of Neuroscience, 19*(7), 2522.

Haines, J. L., & Conneally, P. M. (1986). Causes of death in Huntington disease as reported on death certificates. *Genetic Epidemiology, 3*(6), 417-423.

Harper, P. S., Lim, C., & Craufurd, D. (2000). Ten years of presymptomatic testing for Huntington's disease: the experience of the UK Huntington's Disease Prediction Consortium. *Journal of Medical Genetics, 37*(8), 567-571.

Haw, C., Harwood, D., & Hawton, K. (2009). Dementia and suicidal behavior: a review of the literature. *International Psychogeriatrics, 21*(03), 440-453.

Hayden, M. R. (1981). *Huntington's chorea*: Springer-Verlag New York.

Hayden, M. R., Ehrlich, R., Parker, H., & Ferera, S. J. (1980). Social perspectives in Huntington's chorea. *South African Medical Journal, 58*(5), 201-203.

Huggins, M., Bloch, M., Wiggins, S., Adam, S., Suchowersky, O., Trew, M., et al. (1992). Predictive testing for Huntington Disease in Canada: Adverse effects and unexpected results in those receiving a decreased risk. *American Journal of Medical Genetics, 42*, 508-515.

Huntington's Disease Society of America (1989). *Guidelines for Predictive Testing for Huntington's Disease*. New York: Huntington's Disease Society of America, Inc.

Huntington, G. (1872). On chorea. George Huntington, M.D. *Journal of Neuropsychiatry Clinical Neuroscience, 15*(1), 109-112.

Huntington's Disease Collaborative Research Group (1993). A novel gene containing a trinucleotide repeat that is expanded and unstable on Huntington's disease chromosomes *Cell, 72*, 971-983.

Kellermann, A., Rivara, F., Somes, G., & Reay, D. (1992). Suicide in the home in relation to gun ownership. *New England Journal of Medicine, 327*(7), 467-472.

Kenen, R., & Schmidt, R. (1978). Stigmatization of carrier status: social implications of heterozygote genetic screening programs. *American Journal of Public Health, 68*(11), 1116.

Kessler, S. (1987a). The dilemma of suicide and Huntington disease. *American Journal of Medical Genetics, 26*(2), 315-317.

Kessler, S. (1987b). Psychiatric implications of presymptomatic testing for Huntington's disease. *American Journal of Orthopsychiatry, 57*(2), 212-219.

Kessler, S. (1994). Predictive testing for Huntington Disease: A psychologist's view. (Invited Editorial). *American Journal of Medical Genetics B Neuropsychiatric Genetics, 54*, 161-166.

Kessler, S., & Bloch, M. (1989). Social system responses to Huntington disease. *Family Process, 28*(1), 59-68.

Kessler, S., Field, T., Worth, L., & Mosbarger, H. (1987). Attitudes of persons at risk for Huntington disease toward predictive testing. *American Journal of Medical Genetics, 26*(2), 259-270.

Komiti, A., Judd, F., Grech, P., Mijch, A., Hoy, J., Lloyd, J. H., et al. (2001). Suicidal behaviour in people with HIV/AIDS: a review. *Australian and New Zealand Journal of Psychiatry, 35*(6), 747-757.

Lam, R. W., Bloch, M., Jones, B. D., Marcus, A. M., Fox, S., Amman, W., et al. (1988). Psychiatric morbidity associated with early clinical diagnosis of Huntington disease in a predictive testing program. *Journal of Clinical Psychiatry, 49*(11), 444-447.

Lanska, D. J., Lavine, L., Lanska, M. J., & Schoenberg, B. S. (1988). Huntington's disease mortality in the United States. *Neurology, 38*(5), 769-772.

Larsson, M. U., Luszcz, M. A., Bui, T. H., & Robins Wahlin, T.-B. (2006). Depression and suicidal ideation after predictive testing for Huntington's disease: a two-year follow-up study. *Journal of Genetic Counseling, 15*(5), 361-374.

Lawson, K., Wiggins, S., Green, T., Adam, S., Bloch, M., & Hayden, M. R. (1996). Adverse psychological events occurring in the first year after predictive testing for Huntington's disease. The Canadian Collaborative Study Predictive Testing. *Journal of Medical Genetics, 33*(10), 856-862.

Lester, D., Beck, A. T., & Mitchell, B. (1979). Extrapolation from attempted suicides to completed suicides: a test. *Journal of Abnormal Psychology, 88*(1), 78-80.

Lifton, R. (1979). *The broken connection: On death and the continuity of life.* New York: Simon and Schuster.

Lipe, H., Schultz, A., & Bird, T. D. (1993). Risk factors for suicide in Huntingtons disease: a retrospective case controlled study. *American Journal of Medical Genetics, 48*(4), 231-233.

Louhivuori, K., & Hakama, M. (1979). Risk of suicide among cancer patients. *American Journal of Epidemiology, 109*(1), 59.

Macgregor, M. (1977). Juvenile diabetics growing up. *The Lancet, 309*(8018), 944-945.

Marsden, C. (1981). Prediction of Huntington's disease. *Annals of Neurology, 10,* 203-204.

Mastromauro, C., Myers, R. H., & Berkman, B. (1987). Attitudes toward presymptomatic testing in Huntington disease. *American Journal of Medical Genetics, 26*(2), 271-282.

Mattsson, B., & Almqvist, E. W. (1991). Attitudes towards predictive testing in Huntington's disease--a deep interview study in Sweden. *Family Practice, 8*(1), 23-27.

Meissen, G. J., & Berchek, R. L. (1987). Intended use of predictive testing by those at risk for Huntington's disease. *American Journa of Medical Genetics, 26,* 283-293.

Meissen, G. J., Myers, R. H., Mastromauro, C. A., Koroshetz, W. J., Klinger, K. W., Farrer, L. A., et al. (1988). Predictive testing for Huntington's disease with use of a linked DNA marker. *New England Journal of Medicine, 318*(9), 535-542.

Official Statistics of Sweden (2008). Statistics-Health and Diseases, Causes of death 2000. The National Board of Health and Welfare, Centre for Epidemiology. Stockholm, Sweden.

Oliver, J. (1970). Huntington's chorea in Northamptonshire. *The British Journal of Psychiatry, 116*(532), 241.

Paulsen, J. S. (2009). Functional imaging in Huntington's disease. *Experimental Neurology, 216*(2), 272-277.

Paulsen, J. S., Hoth, K. F., Nehl, C., & Stierman, L. (2005). Critical periods of suicide risk in Huntington's disease. *American Journal of Psychiatry, 162*(4), 725-731.

Paulsen, J. S., Nehl, C., Hoth, K. F., Kanz, J. E., Benjamin, M., Conybeare, R., et al. (2005). Depression and stages of Huntington's disease. *Journal of Neuropsychiatry and Clinical Neurosciences, 17*(4), 496-502.

Paulsen, J. S., Ready, R. E., Hamilton, J. M., Mega, M. S., & Cummings, J. L. (2001). Neuropsychiatric aspects of Huntington's disease. *Journal of Neurology, Neurosurgery, and Psychiatry, 71*(3), 310-314.

Quaid, K. A. (1991). Predictive testing for HD: maximizing patient autonomy. *Journal of Clinical Ethics, 2*(4), 238-240.

Quaid, K. A. (1993). Presymptomatic testing for Huntington disease in the United States. *American Journal of Human Genetics, 53*(3), 785-787.

Ranen, N. G. (2002). Psychiatric management of Huntington's disease. *Psychiatric Annals, 32*(2), 105-110.

Reed, T., & Chandler, J. (1958). Huntington's chorea in Michigan. I. Demography and genetics. *American Journal of Human Genetics, 10*(2), 201.

Robins Wahlin, T.-B. (2007). To know or not to know: a review of behaviour and suicidal ideation in preclinical Huntington's disease. *Patient Education and Counseling, 65*(3), 279-287.

Robins Wahlin, T.-B., Bäckman, L., Lundin, A., Haegermark, A., Winblad, B., & Anvret, M. (2000). High suicidal ideation in persons testing for Huntington's disease. *Acta Neurologica Scandinavica, 102*, 150-161.

Robins Wahlin, T.-B., Larsson, M., & Luszcz, M. (2010). WAIS-R features of preclinical Huntington's disease: Implications for early detection. *Dementia and Geriatric Cognitive Disorders, 29*(4), 342-350.

Robins Wahlin, T.-B., Lundin, A., Backman, L., Almqvist, E., Haegermark, A., Winblad, B., et al. (1997). Reactions to predictive testing in Huntington disease: case reports of coping with a new genetic status. *American Journal of Medical Genetics, 73*(3), 356-365.

Robins Wahlin, T.-B., Lundin, A., & Dear, K. (2007). Early cognitive deficits in Swedish gene carriers of Huntington's disease. *Neuropsychology, 21*(1), 31-44.

Roy, A. (1983). Family history of suicide. *Archives of General Psychiatry, 40*(9), 971-974.

Roy, A. (1995). Suicide. In H. Kaplan & B. Sadock (Eds.), *Comprehensive textbook of psychiatry/VI*: Williams & Wilkins, Baltimore, MD.

Saugstad, L., & Odegard, O. (1986). Huntington's chorea in Norway. *Psychological Medicine, 16*(1), 39-48.

Schoenfeld, M., Myers, R. H., Berkman, B., & Clark, E. (1984). Potential impact of a predictive test on the gene frequency of Huntington disease. *American Journal of Medical Genetics, 18*(3), 423-429.

Schoenfeld, M., Myers, R. H., Cupples, L. A., Berkman, B., Sax, D. S., & Clark, E. (1984). Increased rate of suicide among patients with Huntington's disease. *Journal of Neurology, Neurosurgery, and Psychiatry, 47*(12), 1283-1287.

Sepcic, J., Antonelli, L., Sepic-Grahovac, D., & Materljan, E. (1989). Epidemiology of Huntington's disease in Rijeka district, Yugoslavia. *Neuroepidemiology, 8*(2), 105-108.

Shakespeare, W. (1992). *Hamlet (Original work published 1602)*. Ware, Hertfordshire: Wordsworth.

Sørensen, S. A., & Fenger, K. (1992). Causes of death in patients with Huntington's disease and in unaffected first degree relatives. *Journal of Medical Genetics, 29*(12), 911-914.

Stenager, E. N., & Stenager, E. (1992). Suicide and patients with neurologic diseases. Methodologic problems. *Archives of Neurology, 49*(12), 1296-1303.

Stern, R., & Eldridge, R. (1975). Attitudes of patients and their relatives to Huntington's disease. *Journal of Medical Genetics, 12*(3), 217-223.

Tibben, A., Duivenvoorden, H. J., Vegter-van der Vlis, M., Niermeijer, M. F., Frets, P. G., van de Kamp, J. J., et al. (1993). Presymptomatic DNA testing for Huntington disease: identifying the need for psychological intervention. *American Journal of Medical Genetics, 48*(3), 137-144.

Tibben, A., Niermeijer, M. F., Roos, R. A., Vegter van de Vlis, M., Frets, P. G., van Ommen, G. J., et al. (1992). Understanding the low uptake of presymptomatic DNA testing for Huntington's disease. *Lancet, 340*(8832), 1416.

Tibben, A., Vegter-van der Vlis, M., Roos, R. A., van de Kamp, J. J., Frets, P. G., & Verhage, F. (1990). [Presymptomatic DNA diagnosis in Huntington's chorea: reactions to the certainty of not being a genetic carrier]. *Nederlands Tijdschrift voor Geneeskunde, 134*(14), 701-704.

Tibben, A., Vegter-van der Vlis, M., Skraastad, M. I., Frets, P. G., van der Kamp, J. J., Niermeijer, M. F., et al. (1992). DNA-testing for Huntington's disease in The Netherlands: a retrospective study on psychosocial effects. *American Journal of Medical Genetics, 44*(1), 94-99.

Tsuang, M. (1978). Suicide in schizophrenics, manics, depressives, and surgical controls: A comparison with general population suicide mortality. *Archives of General Psychiatry, 35*(2), 153-155.

Wallace, D., & Parker, N. (1973). Huntington's chorea in Queensland: The most recent story. In A. Barbeau, T. N. Chase & G. W. Paulson (Eds.), *Advances in Neurology* (Vol. 1, pp. 223-236). New York: Raven Press.

Washer, G., Schroter, G., Starzl, T., & Weil III, R. (1983). Causes of death after kidney transplantation. *Journal of the American Medical Association, 250*(1), 49.

Wexler, N. S. (1979). Genetic "Russian Roulette": The experience of being at risk for Huntington's disease. In S. Kessler (Ed.), *Genetic Counseling: Psychological Dimensions* (pp. 190-220). New York: Academic Press.

Wexler, N. S. (1985). Genetic jeopardy and the new clairvoyance. In A. Bearn, A. Motulsky & B. Childs (Eds.), *Progress in Medical Genetics* (Vol. VI, pp. 277-304). New York: Praeger.

Wiggins, S., Whyte, P., Huggins, M., Adam, S., Theilmann, J., Bloch, M., et al. (1992). The psychological consequences of predictive testing for Huntington's disease. Canadian Collaborative Study of Predictive Testing. *New England Journal of Medicine, 327*(20), 1401-1405.

World Federation of Neurology Research Group on Huntington's Disease (1993). The presymptomatic testing for Huntington's disease: worldwide survey. *Journal of Medical Genetics, 30*, 1020-1022.

In: Encyclopedia of Neuroscience Research
Editors: Eileen J. Sampson and Donald R. Glevins
ISBN 978-1-61324-861-4
© 2012 Nova Science Publishers, Inc.

Chapter XXXVIII

Making Reproductive Decisions in the Face of a Late-Onset Genetic Disorder, Huntington's Disease: An Evaluation of Naturalistic Decision-Making Initiatives[*]

Claudia Downing[1]
Research Fellow, Centre for Family Research,
University of Cambridge, Free School Lane, Cambridge, England CB2 3RF

Abstract

This chapter makes two contributions to the psychology of decision making. It draws on empirical work to document how family members make and live with reproductive decisions as they become aware of their risk for a serious late-onset genetic disorder, Huntington's disease (HD). Decision-making involves negotiating two dimensions of reproductive risk – that for any child which might be born and the uncertainty that arises about the at-risk parent's ability to sustain a parenting role should he or she become symptomatic. A detailed account is given of how the model of responsibility was generated from their narratives. This model encapsulates what families find important when making reproductive decisions. It demonstrates that how people make decisions can become as important as what they decide. Examples are given to show how the model provides a framework for comparing how people deal with each dimension of risk, to compare different people's decision-making and to illuminate decision-making in the face of change. This includes how they address new options generated by recent developments in molecular genetics which can resolve uncertainty about, or avoid these risks. A detailed account is also given of the process of their decision-making. These findings are used to evaluate claims being made by naturalistic decision-making (NDM) initiatives to account for decision-making in the real world. An outline of this initiative prefaces the research findings. Beach's (1990) image theory, used to illustrate the NDM

---
[*] A version of this chapter was also published in *Psychology of Decision Making in Medicine and Health Care*, edited by Thomas E. Lynch published by Nova Science Publishers, Inc. It was submitted for appropriate modifications in an effort to encourage wider dissemination of research.
[1] Correspondence to: Centre for Family Research, University of Cambridge, Free School Lane, Cambridge, England CB2 3RF cd10008@cam.ac.uk

approach, emphasises the role that values play in forming, negotiating, implementing and living with decisions that arise in everyday lives. Participants' accounts support this claim - revealing how values such as responsibility become established and contribute to decision-making. This and other findings lend support to the NDM claims. The chapter concludes with suggestions about how the model might be used more generally to further our understanding of the psychology of decision-making in the face of risk.

Introduction

This chapter is about decision-making in the face of genetic risk. It presents a model of responsibility that encapsulates what families find important when making reproductive decisions in the growing awareness of their risk for a late-onset genetic disorder, Huntington's disease (HD). It will be argued that this model facilitates our understanding of how those living with genetic risk plan their lives in the age of the new genetics. These comprise advances in molecular genetics making it possible to offer genetic testing – both prenatal and predictive – to those identified as at-risk from their family history for an increasing number of disorders. Testing establishes whether mutations, which are likely to result in these disorders, are present in genes. It differs from other forms of medical testing in that any body tissue can be used and it can be performed at any stage of life from conception. Results may be sought to inform personal decision making rather than medical care.

Concerns about one value, responsibility, have been consistently cited as motivating the new decisions that people faced about predictive testing (Cox, 2002, 2003; Kenen, 1994, 1996; Kay, E. & Kingston H.,2002; Hallowell, 1999; Robertson, 2000; Smith, Stephenson and Quarrel, 1999; Taylor, 2004). Decision-making is instigated and constrained by social and biological connections to others such as biological family members who share similar risks, future generations, and partners (Hallowell, 1999; Burgess and d'Agincourt-Canning, 2001; Downing, 2002, 2005). Responsibility is not enacted from a single script: those at-risk for HD report aiding their adult children's reproductive decision-making as an important consideration both when choosing to have or not have predictive testing (Smith, Stephenson and Quarrel, 1999). This finding suggests how new decisions become part of ones people have always faced in their everyday lives - such as those around reproduction. It also shows that these decisions can continue to be made without recourse to clinical settings and the additional information that genetic testing provides.

Reproductive decision-making in the face of genetic risk has been an under-researched area. A review of 547 studies published between 1986 and 1996 categorised as being about informed decision-making yielded only seven that dealt with genetic factors, and none which addressed late onset disorders or examined how reproductive decision-making is experienced as new options to clarify risk status have become available (Bekker et al, 1999:32). The in-depth qualitative study of reproductive decision-making in the face of late-onset genetic risk whose findings are reported in this chapter was undertaken to address this lacunae.

The chapter also addresses suggestions that adopting a decision theoretic approach could enhance our understanding of how people make reproductive decisions under these conditions (Huys, Evers-Kiebooms and d'Yedwalle, 1992; Shiloh, 1996; Broadstock and Michie, 2000). The introduction notes problems inherent in traditional approaches used to study decision-making in the face of risk and suggests why the naturalistic decision-making (NDM) initiative offers to become a more informative resource. The study findings will be used to test their claim to account for personal decision-making. In preparation for this evaluation the introduction presents key tenets of NDM, reasons for focusing on one NDM model, Beach's (1990) image theory, what this comprises, what was previously known about perception of genetic risk, reproductive decision-making in the face of genetic risk and what is different about late-onset risks.

Naturalistic Decision Making

The NDM initiative sets out to address limitations inherent in the approaches, such as health psychology and normative decision theory, more traditionally deployed to study decision-making in the face of risk. These approaches were initially constrained by an underlying assumption that people lived their lives in a planned and rational fashion, and that, once informed about risks would engage in preventative behaviour. This assumption continues to drive research agendas, as can be seen from the following comment concluding a systematic review of 547 studies published between 1986 and 1996 categorised as being about informed decision-making (Bekker et al, 1999: 22). The authors' key recommendation was that:

> Genetics and prenatal diagnosis ... is a good place to begin [further studies] because the nature of the information base, with relatively simple clearly-defined options with known risks corresponds closely to the terms that decision theories use

Another limitation has been these approaches' tendency to focus on the individual. Individual factors *do* shape decision-making: holding internal or external attributions of causality (Wallston, 1992) being risk averse or seeking (Lopes, 1987) and decision averse or confronting (Beattie et al, 1994) contributes much to how people assess problems confronting them, constrains their ability to deal with them, and to negotiate with others. But, as Kahneman (1991: 145), in a personal reflection on the field of the psychological study of judgement and decision-making points out, it is essential to acknowledge that 'significant decisions are made in a social and emotional context'. The value of a multiplicity of studies over the years has been to document social constraints in which risk is actually experienced and acted on. Health psychology has identified other factors, such as perceived social support and efficacy, which can mitigate these constraints. Efforts have been made to incorporate these factors in subsequent health psychology models such as the protective motivation theory (Rogers, 1994) and self regulation theory (Leventhal, Nerenz and Steele, 1994).

Considerable overlaps occur between evolving health psychology models and decision theories. Theoretical overlaps occur in concepts, such as risk perception, perceived benefits of actions, beliefs about self-efficacy and response-efficacy and perceived control, used to explain behaviour in the face of risk. Correspondences exist in findings about how people perceive and react to the presentation of risk information. For example, laboratory based normative decision theory studies have yielded a taxonomy of heuristics and biases that shape the way people process risk information when making decisions (Kahneman and Tversky, 1984) which have been replicated in health settings and used to inform clinical practice. Both approaches have come to depict decision-making as a process, with a preliminary period of cognitive restructuring preceding choice. The health belief model now postulates that taking preventative measures is a function of considering perceived susceptibility to the risk, benefits of taking actions such as genetic testing and barriers to taking this action whilst the protection motivation theory emphasises the role played by beliefs about magnitude and likelihood of threat, and the efficacy of envisaged responses.

Three characteristics distinguish NDM initiatives from previous decision-theoretic approaches: their focus on decision-making in real life contexts; their recognition of the competence that people achieve when making decisions under difficult conditions; and, being open to a range of disciplines and perspectives thought to contribute to a richer understanding of decision-making. These characteristics are considered in more detail below.

Decision-Making in Real Life Contexts

NDM initiatives have concentrated on two aspects of real life: decision-making in field settings which may result in disaster situations,[2] and personal decisions that people see as important in their private lives including divorce (Willen and Montgomery, 2000), and childbearing (Willen, 1994). Personal decisions, which can entail negotiating risks, are characterised by:

> ... deep and long-term significance to the subject. By making the decision in question the person is attempting to make his or her life turn out differently (Willen, 1994: 24:1).

Decision-making frequently involves, or has to consider, significant others, precisely because people are embedded in social networks. Far less empirical work has been conducted into personal decision-making. Montgomery (1999) remarked on the absence of studies which in addition to risk explore 'hot' psychological factors such as "social influences ... emotions, interests and values".

NDM's initial remit was to identify what decision-making encompasses (Cannon-Bowers, Salas and Pruitt, 1996). The value of resulting working models has been in delineating and broadening the parameters within which decision-making actually occurs rather than in specifying *exactly* how each decision-maker will operate (Lipshitz, 1993).[3] Considerable agreement exists between models about the nature of these parameters. NDM models can be categorised into two groups. The first group, *typology* models, identify contrasting cognitive processing styles, such as intuitive and analytic, which may be employed at various stages of decision-making. The second group, *process* models, suggests what these stages might be. How these approaches coalesce is outlined below.

Process models distinguish three broad stages. These are: (i) the preliminary work needed to come to a decision, (ii) implementation of choices, and (iii) living with decisions. Preliminary work involves perception of the situation, which may include evaluation of risks, formulation of goals, and assessment of options open and strategies available to realise these

---

[2] For detailed accounts of NDM approaches see Montgomery and Svenson (1989); Klein (1999); Klein, Orasanu, Calderwood and Zsambok, (1993); Svenson and Maule (1993); Svenson, (1996), Klein (1997); Zsambok and Klein (1997).

[3] For further details of these models see Klein et al (1993) and in particular Klein's diagrammatic representation of their communalities in figure 22.1, page 391.

goals. Models differ in how they conceptualise this work: as screening out unacceptable options according to values held (Beach, 1990); as 'the search for a dominant option' (Montgomery, 1993); or, as 'differentiation and consolidation' of chosen options (Svenson, 1991). It is probable that people act in all of these ways, but by limiting their focus, models remain manageable. Implementation is characterized by mental simulation of chosen strategies, realising them, and modifying them in the light of progress decisions about their ability to achieve goals. Progression through these stages may be both linear and recursive. For example, living with decisions involves post decision evaluation, which can then contribute to preliminary work for subsequent decisions. It may also include drawing on decisions made by others.

If decisions are perceived as clear, acceptable, feasible and familiar, passage through these stages becomes unproblematic. Little overt consideration is required, and the processing style employed is described as intuitive. It will be suggested that this frequently characterises reproductive decision-making in general populations.

A different picture of decision-making emerges when decision-makers perceive themselves as confronting unfamiliar decisions or uncertain, ambiguous, redundant or incomplete information - such as genetic risk. Under these circumstances decision-making may necessitate a more analytical processing style. Action feedback loops (which may include previous decisions) and mental simulations may be employed in order to structure and implement decisions and reflect on them. Feedback loops and mental simulations blur distinctions between the stages. As Orasanu and Connolly (1993:19) explain:

> Instead of analyzing all facets of a situation, making a decision, and then acting, it appears that in complex realistic situations people think a little, act a little, and then evaluate the outcomes and think and act some more. ... Development of this knowledge [is] ... an integral part of decision making.

Competing and shifting goals may emerge which have to be resolved within a temporal emotional framework of anticipation, hindsight and regret (Kahneman, 1991). Conflict, which may lead to procrastination, is seen as 'the price one pays for the freedom to choose' (Tversky and Shafir, 1992: 358).

Recognition of Competence

NDM models emphasize the competence people achieve whilst operating within the constraints inherent in naturalistic settings. As Klein (1999: 1-2) explains:

> Instead of trying to show how people do not measure up to ideal strategies for performing tasks, we have been motivated by curiosity about how people do well under difficult conditions.

Competence is conceptualized as feedback loops showing the ability to benefit from past experience. This can take several forms: shaping values (Beach, 1990); recognition of ability to adapt to changing situations and priorities (Payne, Bettman and Johnson, 1993), which may involve the creation of new pathways and perspectives (Willen and Montgomery, 2000); and, developing the expertise to utilise appropriate and disparate sources of power, such as intuition, mental simulation, metaphor and storytelling. Klein (1999: 3) elaborates on how these strategies coalesce:

> The power of intuition enables us to size up a situation quickly. The power of mental simulation lets us imagine how a course of action might be carried out. The power of metaphor lets us draw on our experience by suggesting parallels between the current situation and something else we have come across. The power of story telling helps us consolidate our experiences to make them available in the future, either to ourselves or to others.

Competence is also associated with exercising control, in that decisions represent a desire to:

> ... act upon the world, to make sure that the future does not look like the past. Decision makers go to great lengths to insure that they have the ability to control future events, (Beach and Lipshitz, 1993: 26).

Where decisions involve risks, which can be defined as uncertainty about negative or adverse outcomes (Teigen and Brun, 1997), it may not be possible to realise this objective. Under these circumstances competence may become making decisions that it is possible to live with.

Transtheoretical Aspects

The third distinguishing aspect of NDM is that it is open to a variety of theoretical perspectives thought to provide additional insights into decision-making. These include phenomenology (Karlsson, 1987), identity formation (Sloan, 1986; Nelson and Nelson, 1995), social approaches to risk (Wartofsky, 1982; Krimsky and Golding, 1992; Palmlund, 1992) and narrative accounts (Beach, 1997; Cox, 2002, 2003; Klein; 1999). What these perspectives comprise and can contribute is outlined below.

Phenomenology captures how people experience exercising their will, develop a sense of competence, reflect on what they should do and structure decisions as realisable projects with others (Karlsson, 1987). Decision-making shapes identity in that it contributes to the way we are perceived, by others and ourselves. As Nelson and Nelson (1995: 136) explain:

> Choices made ... in a very important way determine who we are, who we have been, and who we will be as moral people; this gives the choices their proper weight, and defends us from the kind of self-deception that is sometimes a part of having to decide in morally difficult circumstances.

Social approaches to risk elaborate how identity is shaped. Conceptualising decision-making as social drama, consisting of competing stories involving justifications of actions and attributions of blame, identifies a range of roles that are played out when making decisions in relation to risk (Palmlund, 1992). These include risk bearers, risk generators, risk informers and risk arbiters. Questions that can be explored in empirical work include how these roles come to be allocated and assumed in families, the expectations they arouse, and how these expectations can constrain or facilitate decision-making (Downing, 2005).

Answering these questions involves formulating 'looser' definitions which draw on narrative approaches. For example, Beach (personal communication, August 1997) describes decision-making as 'allowing people to tell stories to construct alternative futures and find ways of making the stories they like happen'. This definition connects us to studies such as those of Cox (2002, 2003) which enhance our understanding of the different ways in which people 'progress' from intentions to implementation of predictive genetic testing through examination of this as a 'narrative moment'. Her participants' stories of 'evolving towards it', 'having to know' and 'taking the decision' bear little similarity to conventional notions of decision-making as a rational planned process of decision and implementation. In this way, stories become 'natural experiments, linking a network of causes to effects' (Klein, 1999: 196). Exploration of narratives reveals and challenges prevailing cultural stories, as shown in Sandelowski and Jones' (1996) qualitative study of women's constructions of choice after the detection of fetal abnormalities.

Image Theory as an Example of NDM Approaches

Underlying similarities in how NDM models conceptualize decision-making enable their principles to be illustrated in more detail by one particular model, image theory (Beach, 1990). An additional reason for focusing on this model was that it evolved from a normative study of reproductive decision-making in a general population and thus can demonstrate how NDM approaches offer a far richer understanding of decision-makers' perspectives and processes.

The original research involved construction and employment of a decision analysis tool, the Optional Parenthood Questionnaire, to help couples think more rationally about the costs and benefits of having a/another child (Beach, Townes, Campbell and Keating, 1976). Partners were asked to rate a list of pre-determined values - gleaned from birth planning literature - which children could fulfil for parents. Values were narrowly defined as utilitarian benefits and costs. They indicated considerations such as caring for children with health problems which people might take into account when making reproductive decisions. Questions were weighted equally making it impossible to assess, even under hypothetical circumstances, the relative importance people placed on each aspect. These constraints, inherent in a traditional decision-analysis approach, were subsequently removed when values were re-conceptualised more loosely as 'images' in image theory.

Processes Identified in Image Theory

Decision-making is depicted as comprising two different but connected processes: screening and choice. Findings that people do not necessarily return to previously rejected options when those in the choice set become unavailable demonstrate that screening is more than a means of reducing cognitive overload, as had previously been suggested by Simon's (1955) model of bounded rationality. According to Simon people reduce problems to a limited number of variables in order to decrease cognitive overload, enabling them to choose options that seem reasonable, but which are not necessarily optimal. In contrast, Beach (1990) showed that screening limits choice to options that meet decision makers' minimal standards - or, produces an empty choice set, thereby prompting a search for better options. Choice only becomes possible when options exist that are not discordant with the decision makers' morals, beliefs, attitudes and values - henceforth referred to by Beach as *images*.

Role Played by Images

Images are defined as schema with cognitive and affective components that organize decision-makers' values and knowledge, guiding their decisions and creating the frames within which decision-making occurs. Three types of images form, and have to be reconciled, in order for decision-making to proceed. These are:

value or principles, relating to how people feel they should behave;
trajectory or goals, relating to what they want their future to be like (which may be different);
strategic or plans, encompassing ways of securing that image.

Images continue to be drawn on to monitor progress, demonstrating that they are influential at *all* stages of decision-making.

Empirical support exists for Beach's (1990) claim that these considerations shape decision-making. A meta-analysis of involvement studies in relation to attitude formation showed that the form of involvement differed according to whether values, impressions or outcomes were prioritised (Johnson and Eagly, 1989). High value-relevant involvement was associated with selection and bias, whilst high outcome-involvement was associated with accuracy driven processing, and impression-relevant involvement with a more flexible approach, which allowed decision-makers to respond to social interaction.

Verplanken and Svenson (1997) drew on Svenson's (1991) NDM differentiation and consolidation model to demonstrate that the form of involvement affected how decisions were made. Pre-choice cognitive processing is depicted as differentiating between options in order for one to emerge as preferable and become chosen, and post decision processing as consolidation of this option by accentuating its positive aspects whilst stressing negative aspects of rejected ones. Consolidation corresponds closely to the process of cognitive dissonance (Festinger, 1957) but differs from it in that it commences before decisions are implemented.

Further insights into the relationship between emotional engagement and decision-making are provided by Willen and Montgomery (2000). They show that people in troubled marital relationships can only move towards a decision situation after they have changed their form of involvement, termed 'perspective'. Reconfiguring partners' previously positively regarded qualities as negative produces psychologically more 'distant' (or less emotionally involved) perspectives on these relationships. This shift enables them to contemplate divorce. As they distance themselves from their relationship, decision-makers become increasingly emotionally involved with future perspectives. These incorporate possible outcomes of divorce, such as how it will affect them, their life and those they care about and how these concerns can be reconciled with values important to them.

Contribution of Narrative Approaches to Image Theory

Beach does not elaborate how images arise, change or become shared. Incorporating narrative approaches to family therapy (Dallos, 1997) and family aspects of health (Nelson and Nelson, 1995) makes it possible to see how this happens. Dallos amalgamates attribution theory (Heider, 1958), personal construct theory (Kelly, 1955) as applied to families (Proctor,

1981). Dallos's (1997: 96) starting point is the personal constructs (reflecting individual values) that people develop to make sense of the world. Personal constructs affect how disclosures and sharing of life stories are presented. They signal what is important (but not necessarily agreed) for family members about their decision-making:

> Remembering in families can be seen as a joint, collaborative activity in which partners may remind each other of the facts - what happened, when it occurred, who was there, what was said - but also the meaning of these events, in particular what people's intentions were, how they felt at the time. There may also be a reflexive view about what has been learned from the event, how perceptions of the event differ now, and how in the same circumstances they might now act differently.

The resulting narratives identify stakeholders in decisions; encompass alternative representations of events, predictions or anticipations of other's actions. They are shaped by internal and external attributions (Heider, 1946) including justifications and blame. They contain themes such as responsibility; maintain family myths and link family members together over time. These become shared social realities, indicating 'family' constructs or values, which are more than a collection of individual constructs.

Nelson and Nelson (1995: 130) suggest that narratives from families of birth provide working models to be drawn on and adjusted in subsequent decision-making in families by marriage:

> In early adulthood we begin to plot out our life story, using the first chapter our parents wrote with us as a point of departure. That first chapter can now take on meaning as a guide to action, if we integrate the present moment of decision into the overall narrative of our lives, either by ratifying our existing course or by deliberately charting a new one. It is open to us to make our present choices in such a way as to weave our past into our future, consciously and thoughtfully constructing a meaningful moral autobiography.

In this way, narratives come to provide an ethical basis for decision-making:

> The narrative ... asserts that the quality of the decision matters as well as who is making it. (Nelson and Nelson, 1995: 107).

Support is therefore shown for claims that additional insights into the decision-making process will be provided by links made to other domains, such as narrative approaches (Beach, 1997). Testing these claims involves deploying qualitative methods that provide access to decision-makers' narratives.

Accounting for Shared Decision-Making

Work conducted in corporate contexts led Beach (1993: 91) to state that individual decision-making precedes shared decision-making:

> Ultimately every decision is made in the isolation of a single human mind. Even when working in a group, each individual has to make his or her own decision. Through negotiation, the group's decision is derived from its members' private decisions.

He postulates that a process of shared screening, termed negotiation, subsequently occurs. Negotiation only becomes possible when an option exists that is acceptable to at least one individual. Different patterns of negotiation observed by Beach (1996) were:

yielding, defined as passive acceptance of decisions formulated by others;
compromise, defined as the abandonment of an individual's own decision but not requiring wholesale acceptance of someone else's;
contending, defined as each person stating what they think the decision should be, then attempting to persuade others of this until other options are dropped or a majority consensus is formed;
problem solving, defined as looking for a further option that might suit all participants.

His taxonomy provides a starting point to compare joint decision-making in other contexts, such as the family, where individuals may be connected biologically as well as having shared a past and be anticipating a shared future, and in relation to other types of decisions, such as the reproductive ones which form the focus of the next sections and this chapter

Reproductive Decision-Making in General Populations

General populations comprise those who do not have to contend with reproductive problems such as infertility or the known risk of a genetic disorder. Understanding their experiences helps clarify how reproductive decision-making differs when people face considerations about genetic risk. This section summarises what demographic and exploratory studies reveal about their decision-making.

Demographic studies report outcomes of reproductive decision-making, such as family size. They provide valuable contextual information by documenting how this changes over time for different cohorts, but cannot access the decision-making that culminates in these outcomes. As Wexler (1980: 322) - herself at-risk for HD - states, decision-making, being essentially a private matter, requires different approaches. :

> We know so pitifully little about the myriad of complex motivations that propel people to propagate their kind, particularly now that birth control methods are giving people such freedom of choice. Counting facts retained and babies born is not enough. What goes into that crucial decision?

Answering this question involves qualitative approaches that seek out and prioritize perspectives of those making decisions. Exploratory interview studies reveal decision-making to be a process, characterised by a multiplicity of decision points rather than a single 'choice' (Busfield and Paddon, 1977; Payne, 1978; Richards, 1978; Dowrick and Grundberg, 1980; Fox, 1982; Currie, 1988; Willen, 1994). As well as deciding whether to have a child, these include with whom, whether to have further children, when to do this, under what conditions, and what action should be taken to achieve these goals. These considerations face everyone including those contending with difficulties, such as infertility or genetic risk even if as will be shown, they come to conceptualize them differently.

These studies also show how people construct and resolve these decision points. Establishing shared social realities forms a significant preliminary step. For example, many of

Willen's (1994) respondents framed reproductive decision-making primarily as a relationship question, focusing not so much on *whether* to have children as *with whom*. Reproduction becomes part of the 'natural' pattern of acceptable relationships and, as such, not requiring overt reflection or negotiation, except about what might be the 'right time' to implement previously unexpressed and unquestioned intentions. This is revealed to be a subjective judgment of circumstances rather than a chronological point (Currie, 1988). These observations are consistent with other findings, such as those from an earlier study (Mansfield and Collard, 1988) which found that newly-wed couples did not discuss reproduction, and were evocative of the commonly used expression, 'falling for a baby' which suggest little in the way of conscious planning.

Adopting a NDM perspective enables us to construct a coherent account of these findings. This confirms that decision-making *does* occur but in a more intuitive form, consistent with the values reproduction represents for those concerned. As Beck and Beck-Gernsheim (1995:196) postulate:

> A child represents ... the 'natural' side of life ... Motherhood seems to offer the woman an alternative refuge from the working world, where it is imperative to behave responsibly, and soberly, and emotions are generally considered a nuisance. Committing yourself to a child means contradicting the cognitive side of life, and finding a living counterweight to all that soul-destroying routine.

Taking a 'natural' stance can thus be interpreted as conforming to existing norms that children will 'happen' within relationships and do not need to be consciously 'decided' or planned. The existence of these norms is corroborated by the experiences of the involuntarily childless who cannot fulfil them (Pfeffer and Woollett, 1983; Mason, 1993; Read, 1995), and the voluntary childless who must resist them (Campbell, 1985; Kiernan, 1989; Bartlett, 1994; Morell, 1994; McAllister and Clarke, 1996).

Reproductive Decision-Making in the Face of Genetic Risk

Questions arise about whether the seemingly unquestioning acceptance of norms to have children and associated intuitive processing can be sustained (and how) once decision-makers become aware of genetic risk. In contrast with the 'normal' experience outlined above, reproduction in the face of genetic risk is depicted as involving the recognition of uncertainty about potentially severe and possibly irreversible consequences, and that decision-making has significant implications for the decision-makers, other family members and future generations. Factors thought to influence reproductive decision-making under these circumstances are: the value placed on having children; the perception of genetic risk; and personal experience of the condition (Shiloh, 1996). How these factors are thought to operate is considered below.

The Value Placed on Having Children

The experiences of the childless illuminate how the value placed on having children is construed by those for whom reproduction is problematic. Examination of the contrasting standpoints of the infertile and the voluntarily childless reveal how having children becomes

Involuntary Childlessness

Being confronted with infertility creates feelings of abnormality, leading to loss of self-esteem and self-identity, a sense of not being in control of one's life, and an existential anxiety about the purpose of life (Read, 1995). These losses may be experienced differently by men (Mason, 1993) and women (Pfeffer and Woollett, 1983), and hence between partners (Mason, 1993). Problems relating to one partner can lead couples to feel disconnected, or 'out of step' with each other. The supposition that those facing genetic risks experience similar losses and disconnections is borne out in a qualitative study of women carriers of X-linked genetic disorders (Kay and Kingston, 2002). Loss of normal expectations became one of the women's principal concerns.

Hesitations were also noted in various studies about employing strategies, such as artificial insemination by donor (AID), to redress these problems. This phenomenon indicates another dimension: the considerable, and sometimes overriding, value people place on being genetically connected to their children. Not wishing to lose this connection can prevail over the envisaged loss of childlessness.

Voluntary Childlessness

Accounts from the voluntarily childless stress what they see themselves as having *gained* by not having children: for example, self-fulfilment, and alternative ways to establish identity (Morell, 1994) and bond with partners - such as working on shared projects (Willen, 1994). In contrast with the infertile, they construe reproduction as leading to the loss of their identity. McAllister and Clarke (1996) found that only a third of women self-categorised as voluntarily childless had initially made a firm decision never to have children and ten percent were still ambivalent. Many of their respondents had regular, and valued, contact with other people's children. These observations suggested that factors other than values held about children were crucial in implementing women's decisions. It emerged that a key aspect was holding negative images of parenthood: they viewed parenthood as sacrificing hard won social lives, financial independence and jobs. Not wanting to experience these losses made them reluctant to become mothers.

Perception and Experience of Genetic Risk

The psychometric work conducted by Slovic, Lichstenstein, and Fischhoff, (1984) and Slovic (1987) provides a useful starting point for learning how people perceive risk. Analysis of people's ratings of a variety of risks on a number of given dimensions generated three underlying factors. These are:

(i) *dread* risk, which relates to perceptions of the uncontrollability of exposure to risk, dread associated with its impact and whether distribution is judged as inequitable;

(ii) *unknown* risk which reflects the extent of personal experience and expert knowledge that exists of a risk, how observable it is, and whether effects are considered to be immediate or delayed. The impact of experience of risk on decision-making is likely to be complex. It has been suggested that risks which are known are more likely to be perceived as less threatening (Slovic, 1987; 1992);

(iii)   the *number of people* likely to be exposed to or affected by a risk.

Their findings (despite not drawing specifically on genetic risk) *are* consistent with prevailing discourses about genetics. For example, Rothman (1998: 233) explains how genetic determinism views an individual's genetic makeup as an uncontrollable accident of fate which has a great impact in that it holds the key to identity, and because it is readily observable to those with expert knowledge:

> A story is written into our bones and skin and hair, there to be read by anyone who knows the code.

Reproductive decision-making involves transcending this reductionist stance, in that:

> gametes are not private possessions. Biologically they are shared values, realised in sexual union. They have to be combined and merged with the genetic values of others, (Fletcher, 1980: 134).

Awareness of potential and actual genetic connections changes an independent concept of self into an 'interdependent' social one, heightening one's awareness of genetic responsibilities towards others who may also be affected by how one acts on this information (Kenen, 1994, 1996). They include biological family members who share similar risks, future generations, and partners. Their stories portray risk as experienced subjectively and emotionally (Gifford, 1986; Hallowell, 2006), as "threats to people and things they value" (Kates and Kasperson, 1983: 7029).

Thus what risk means and what people are prepared to *do* about it is likely to vary according to their underlying value systems. If, as has been claimed earlier in this chapter, values are formed in social contexts, decision-making about genetic risk will also be shaped by the wider social and cultural framework in which decision-makers live (Douglas, 1985). As Rothman (1998: 39-40) remarks, risks become enmeshed with:

> our own personal troubles, our daily lives, our intimate concerns, and the world in which we live, the issues that face us collectively. We live a biography, a personal tale, but we live it in a moment of history, in a collective time.

Rothman's comment reminds us of the need to acknowledge that reproductive decision-making differs from other decisions about genetic risk in that it is embedded in a particular history. In addition to their biological inheritance, families have had to contend with a 'social' legacy of opposition to their existence and reproduction. Social legacies, as well as current attitudes, vary between cultures. Within living memory, members of HD families were murdered in Nazi Germany. In other western countries sterilization was advocated, and sometimes enforced in some American states (Davenport and Muncey, 1916; Wexler, 1980; Harper, 1996a; Thom and Jennings, 1996). Reducing the incidence of at-risk births remained as an objective of genetic counselling in Britain for many years (Carter and Evans, 1979; Harper et al, 1979; Thom and Jennings, 1996). These eugenic measures were justified as being the responsible way to promote the wellbeing of future generations. Perception of this legacy is thought to have caused many of those at-risk who wished to have children to remain "doctor shy" (Wexler, 1979: 200) long after more liberal attitudes came to prevail.

In many countries it has subsequently become unacceptable to suggest that those at-risk should refrain from reproduction. It is now generally assumed that people want to feel in charge of their own destinies, are able to exercise freedom of choice in ways that benefit their wellbeing and wish to take an active role in managing their own affairs, including negotiating genetic risk (Conrad and Gabe, 1999). What has *not* changed is an underlying assumption on the part of health care professionals and the general public that, given a choice, people would prefer not to pass on known risks.

This assumption is borne out by prenatal studies in general populations which identify a moral and a 'technological imperative' to make use of risk avoiding possibilities (Tymstra, 1989). It explains why some women terminating an abnormal pregnancy identified through routinely offered prenatal testing report having had "no choice" - when clearly they did - because the alternatives were rejected as unacceptable, or too difficult to justify (Green, Statham and Snowdon, 1992). Pregnancy is changed even for those who decide not to make use of fetal diagnosis. As Green (1990: 38) explains:

> mothers who decline to take part ... will always know that they could have had that knowledge and could have acted on it .

Nelson and Nelson (1995: 69-76) argue that expectations arise that *both* parents will act in their child's interests:

> "Parents' duties to children, ... arise from their direct responsibility for bringing the children into existence, whether they meant to or not - and for putting them into the world in a condition of extreme vulnerability ... Nor can one parent release the other from this responsibility, as it is the child, and not the other parent, to whom the debt is owed.

Parents are likely to be viewed as irresponsible if they act in a way that is judged by others not to be in their fetus' best interests (Lippman, 1992).

Potential quality of life, becomes a paramount consideration when making decisions about testing and termination for fetuses at-risk for, or affected by, disorders that manifest at birth or in infancy (Kitcher, 1996). The more a disorder is perceived to negatively affect quality of life the more willing people are to consider termination (Evers-Kiebooms et al, 1993; Drake, Reid and Marteau, 1996).

Little is known about how people view these considerations in relation to a life that might be 'normal' until adulthood, but then be affected by a genetic disorder. HD offers a particularly rich context to explore them, in that the clinical picture presents families with two decision-making dilemmas. The first is whether to have children who, in later life, may develop the disease. The second is whether it is justifiable to have children - given that uncertainty arises about the at-risk parent's ability to sustain a parenting role. HD's late-onset makes it likely that people will have had had to make decisions about childbearing, and know of their risk, before developing any overt symptoms. Affecting both men and women these comprises cognitive changes, loss of control over voluntary movements, psychiatric symptoms and marked physical decline over a period of ten to fifteen years leading to inevitable death both men and cognitive changes, loss of control over voluntary movements, psychiatric symptoms and marked physical decline over a period of ten to fifteen years leading to inevitable death (Harper, 1996b). The mutation is highly penetrant, meaning that, providing they live long enough, most people who inherit it will develop HD. If they were to develop the disorder, the current lack of any effective treatment or cure would then threaten their ability to mother or father their children.

Few have utilized genetic testing since its introduction in the late 1980s to resolve these dilemmas (Harper, Lim and Craufurd, 2000; Hayden, 2000; Simpson and Harding, 1993). Low uptake of predictive testing has been attributed to reluctance to make contact with clinical genetics services, reluctance to lose hope, uncertainty about ability to cope given the continuing lack of treatments, and threats to identity. Van der Steenstraten et al (1994) found that non-requesters of predictive testing were more likely to have learnt of their risk before adulthood than requesters. This led them to hypothesise that non-requesters were more likely to have incorporated HD into their identity and have less need to resolve their uncertainty than requesters who perceived their risk as threatening their previously established identity.

Making decisions about prenatal testing concerns both general populations who undergo routine antenatal screening as well as those aware of genetic risk. Decisions about whether or not to employ prenatal and subsequent decisions about terminations may need to be made under severe time constraints. Factors found to be associated with use of prenatal testing for early onset disorders include perceived accuracy of the test, willingness to consider termination, and other family members' attitudes towards termination (Shiloh, 1996).

Values held about abortion can be either absolute or relative, in that people might accept abortion for disorders that affect children's quality of life but not see it as appropriate for HD. If, as has been suggested above, those growing up in affected families see HD as part of their identity, they can be reluctant to consider termination, seeing it as rejection of themselves (Hayes, 1992; Wexler, 1992; Richards, 1993; van de Steenstraten et al, 1994). Those unable to agree (for whatever reason) to termination of high risk fetuses are unlikely even to be offered prenatal testing. However, documented cases of women discharging themselves from hospital at the penultimate moment rather than undergoing a termination (Tolmie et al, 1995) and of continued high risk pregnancies (Bloch and Hayden, 1990) reveal that people cannot always predict how they will react as the decision-making process unfolds.

Learning more about clinical encounters can shed some light on what they can contribute to the decision-making process. For those uncertain about their reproductive decisions - one third of the participants in one study (Wertz, Sorenson and Heeren, 1984) - counselling can be helpful at an early stage of decision-making, as part of (Kessler 1989: 350):

> an ongoing process, over time, of evaluation, weighing of options, and of responses to personal and interpersonal factors all of which contribute to reproductive decision making

Jungermann's (1997) advice giving and taking (AGT) model offers an explanation of what happens when counselling occurs at this pre-choice stage of decision-making. He focuses on interactions, such as those which arise between consultants prepared to give advice and clients prepared to consider advice. Advice is defined as the preferences of another person. Four components thought to comprise this process are: description of the problem followed by identification of appropriate options; justification of particular options; particular options being offered as advice by consultants; and clients' evaluation and consideration of this advice culminating in either acceptance or rejection. Willingness to accept advice is linked with factors such as perceived credibility and confidence felt about the person giving advice. Jungermann (1997) suggests that uptake of advice is limited because decision-making is frequently guided by values, about which clients are the experts, rather than facts, such as the probability of a negative genetic outcome, which are the provenance of the consultant.

Other research reveals Jungermann's dichotomy as too simplistic, in that it is not possible for facts to be provided in a neutral way. Comparisons made between professional groups have shown them to evaluate risks differently (Marteau, Drake and Bobrow, 1994) which can then influence how they convey information. It is also necessary to acknowledge the taxonomy of cognitive heuristics and biases that shape the way people process information presented to them (Kahneman and Tversky, 1979; 1984). These findings have implications for clinical practice. As Wexler (1992: 219) comments:

> It is so important to explain genetic information both in terms of gain and loss. Telling clients they have one chance in four of having an affected child conveys one psychological message, saying that they have three chances in four to have a normal baby conveys a different one - even though the statistics are the same.

It is likely that consultants and clients prioritize different forms of involvement (Verplanken and Svenson, 1997) which affect how they present and perceive risk and responses to it.

It has also been shown that when and where people are offered these tests influence their uptake. Women are less likely to question them if they are presented in antenatal clinics as 'routine' prenatal care (Richards and Green, 1993; Green, Statham and Snowdon, 1994). Testing for rare late-onset disorders is less likely to be presented in this way, as it is felt that more detailed information will be provided by genetic counselling clinics. This factor may partially explain why uptake of prenatal testing is very low for HD (Hayden, 2000) and practically non-existent for other late-onset disorders such as inherited breast, ovarian and bowel cancers (McAllister, 1999). Very little interest has been expressed in prenatal testing for HD (Mastromauro, Myers and Berkman, 1987) and even fewer have actually used it after it became available (Hayden, 2000).

Genetic counselling has a different impact when it coincides with a later stage of decision-making. A review of prospective studies of genetic counselling concluded that counselling then *confirms or reinforces* decisions already taken (Kessler, 1989). Counselling can also *reduce anxiety* about implementing decisions already taken. This supposition would account for an observed post counselling increase in births for those who perceived themselves to be at high risk.

Clinical studies, some of which are cited above, report only the views of a small and unrepresentative proportion of the total estimated HD at-risk population willing to make their family history known. Clinical studies' focus on the 'patient' also means they fail to address the social aspects of decision-making. Aggregation of existing data makes it impossible to learn how couples come to see HD as an issue for their reproductive decision-making and decide to contact clinicians.

Studies adopting a family systems approach produce useful insights into how risk about HD is experienced, perceived and communicated. Focusing on interactions within families, they reveal that crucial cognitions and behaviour are shaped long before reproduction is contemplated. Members may collude to 'preselect' one of their members, sometimes on the basis of lay beliefs about gender vulnerability or physical resemblance, as the future sufferer (Kessler, 1988). It produces an illusion of control and alleviates uncertainty about other family members' risk. This provides welcome reassurance if, as Korer and Fitzsimmons (1985) found, the statement that HD 'does not skip generations' is sometimes interpreted as meaning that someone from each generation *will* definitely be affected. Interpreting the same nominal risk quite differently for, and by, each family member, generates contrasting expectations about what comprises responsible decision-making (Downing, 2005).

Knowledge of HD also influences how family members relate with outsiders. Families can become either "HD orientated" (defined as becoming excessively preoccupied with HD) resulting in them leading socially isolated lives, or "independently orientated", with HD being given a less prominent role and families remaining open to contact with others (Korer and Fitzsimmons, 1985). These different perspectives affect how partners are informed about HD. Existing family members may collude to keep information from them. The nature and extent of information divulged, when this occurs, and the impact that information has on decision-making can vary considerably (Barette and Marsden, 1979; Shakespeare, 1992). Wives of HD patients felt vital information had been kept from them until *after* they had had children, and that they would have made different reproductive decisions if they had been fully informed (Hans and Koeppen, 1980). This could explain why Barette and Marsden (1979) found that over sixty percent of their respondents were prepared to tell their *children* not to procreate.

In order to comprehend the minutiae of how HD is experienced it is necessary to access personal accounts. Their evidence, though compelling, is often criticised as being of an anecdotal nature. The onus is placed on the researcher to impose rigor on how they are gathered and analyzed in order to decipher underlying commonalities yet preserve their diversity. The following aspects of the grounded theory approach adopted in the study help meet these requirements. These include grounding the analysis in the data, making analytic journeys explicit and moving *beyond* description of themes to develop an explanatory framework, or model, around the core category to facilitate subsequent evaluation (Hamburg et al, 1994; Boulton and Fitzpatrick, 1997; Creswell, 1998; Reid and Armstrong, 1998; McAllister, 2001). Spelling out the substantive area, or context, allows others to decide whether the framework is relevant in other situations, leading to transferability and the building up of general theories, (Glaser and Strauss, 1966; Seale, 1999). The next sections detail how grounded theory principles were applied to generate the model from the personal accounts gathered in the study.

Ethical Approval

The Local Clinical Research Ethics Committee, the Huntington's Disease Consortium and the Cambridge University Psychology Research Ethics Committee granted ethical approval for the study. All names and some details have been changed to protect confidentiality.

Methods

Recruitment and Characteristics of Participants

Theoretical sampling (to find those encompassing a range of relevant experiences and address the gaps noted earlier in this chapter) involved seeking out people at various stages of the reproductive process; of different ages and gender; with varied experiences of HD, Huntington's Disease Association (HDA) support groups and clinical genetics services; making a range of choices such as accepting or avoiding passing on the risk; and, of doing this in several ways, including having predictive, prenatal testing and undergoing termination of high-risk pregnancies. Seventy-six participants, comprising two generations, lateral kin and their partners within 17 kinships, were recruited to the study through the HDA, a Regional Clinical Genetics Centre and 'snowballing', or seeking chain referrals from initial respondents.

Recruiting through family members offered a broader range of people the opportunity to opt in to the study than a purely direct approach by the researcher would have done. This technique provided access to people *not* in contact with either the clinical genetics centres or the HDA, and who would not have otherwise known about the study.

Characteristics of participants are shown in Table 1 below.

Table 1. Sample Characteristics at Time of Interview
Total N = 76 (generation 1 = 26, generation 2 = 50)

Category	Details	Gen 1	Gen 2
Gender	Male	9	21
	Female	17	29
Age	range	46-70 yrs	20-49 yrs
relationship	Married	23	38
	Single	0	8
	Separated/ Divorced	1	3
	Widowed	2	1
Reproduction	No children	0	25
	One child	4	13
	More than one child	22	12
	Intending to have child/ more children	0	23
Category	**Details**	**Gen 1**	**Gen 2**
Risk	At risk	4	20
	Partner at risk	1	14
	Affected	2	0
	Partner affected	12	0
Genetic testing	Tested positive	0	2
	Partner tested positive	0	2
	Tested negative	1	6
	Partner tested negative	1	5
	Used prenatal testing	0	4 couples
Contact	No contact with Clinical Genetics	15	18
	No contact with HDA	11	10

The older generation had completed their reproductive decision-making before genetic testing was introduced. They could still play an active part in the younger generation's decision-making in the following ways: ascertaining their status through predictive or diagnostic testing; providing samples for linkage or prenatal testing; or by being supportive or critical of their children's decision-making. The younger generation, who formed the main focus for the study, were those making reproductive decisions prior to and at the time of interview. Some of them had made decisions before as well as after the introduction of genetic testing. Including siblings and cousins who face similar risks enabled the study to explore whether lateral kin provide guidelines for each other and how different responses are tolerated

Data Collection and Analysis

Preparatory work

Qualitative studies often start off with broad questions, such as in this study about how reproductive decision-making is experienced. They are also initially imprecise about sample size and type as well as the form results might take. This means that traditional 'pilot studies' are unlikely to be useful. Rather what is needed is a period of careful preparation in order to approach sensitive issues of interest (Lee, 1993). Preparatory work involves familiarizing oneself with the relevant literature. A multidisciplinary approach resulted in that the thinking and methodology of psychology was augmented with that of related disciplines, such as family therapy, sociology, social anthropology, medicine and genetics. Adopting a social anthropological approach helped accessing populations and becoming sensitized to their ways of thinking. As Yow (1994: 156) points out, it is helpful to first locate "the places where people (of interest) go to talk":

> For the hospital history, I found it was the employee cafeteria. For the history of the women's cooperative art gallery, hanging around after gallery meetings and going to shows enabled me to learn a lot.

Accordingly, in addition to reading about HD, contact was made with people at the following locations: a regional clinical genetics centre, HDA support groups and organisations such as Sue Ryder homes and the Crossroads service providing care for those affected. Another reason for contacting the regional clinical genetics department was to establish how many HD families they knew to exist in the vicinity. Concerns had been expressed that it might be difficult to recruit enough families. It was reassuring (from a research point of view) to establish that, despite HD being thought of as a rare disorder, approximately 250 families had been seen there over the last twenty years. Contact with families had varied and information about family structure was sometimes incomplete. Being given permission by clinicians to examine these records gave me an invaluable understanding of the work of clinical genetics departments in pre- and post-testing eras. I learnt that it was not unusual for clinicians to present an optimistic picture to those at the beginning of their reproductive life, setting up expectations that genetic testing would become available one day. This had lead them to encourage the use of temporary rather than permanent methods of birth control. It also became apparent that people had faced considerable practical difficulties in realising certain options, such as prenatal testing, which were first introduced in only one or two centres. Being in the department and sitting in on consultations informed me of current work, including preparation for the predictive testing programme. Contact with a second regional clinical genetics department refined these ideas.

Visits to Sue Ryder care homes and spending time with the local Crossroads family support service enabled me to learn what happens when families can no longer cope unaided. Regular attendance at one HDA support group's meeting, and joining in social activities with families over a period of a year and a half - which overlapped with the commencement of data collection - enabled me to 'hear the talk' outside medical settings. These meetings varied from informal chat to structured discussion of particular issues, such as reproduction. With the group's permission, I tape recorded some of these more focused sessions in order to reflect on them.

Key points emerging from preparatory work

The preparatory work enabled me to show the families that I had made some effort to understand their perspectives before we met for the interviews. It facilitated drawing up a list of core questions and phrasing them in terms reflecting participants' language. Hearing of the difficulties people experienced when talking about HD within families indicated that it would be advisable to interview each person separately. Stories told of affected parenting led me to realise that what are normally regarded as non-threatening questions such as early childhood memories (Yow, 1994), and which are therefore often asked first in interviews, might need to be approached later in the interview.

During the group discussions I first heard people prefacing their stories with comments about knowing or not knowing about HD, or of knowing but not really understanding what it could mean. I was to return to these observations when I subsequently identified 'awareness' as the causal condition for my model.

Subsequent data collection and analysis

Taking part in the study involved completing a short questionnaire to gather background information and enable participants to state what they felt the study should address. They were subsequently individually interviewed at a place of their choice, normally their home. The semi-structured in-depth interviews were tape-recorded with participants' consent. A check list ensured that participants had covered key areas such as their reproductive decision-making and future plans, their own childhood, how they had become aware of HD, attitudes to and experiences of genetic testing and termination, what they felt the future held for those at-risk and decision-making by other family members. Interviews took between one and a half to six hours and were transcribed in full. Approximately six months later participants completed a follow up questionnaire or spoke with the researcher to report any developments and comment on their experience of taking part in the study. Management of the data and analysis was aided by NUD*IST, qualitative analysis software (Richards and Richards, 1994).

Decision-making has traditionally been plotted as flow charts, or decision trees which show how people apply decision criteria to choice alternatives and come to decision outcomes. This study was guided by a combination of ethnographic decision tree modelling (Gladwin, 1989) and mental model constructions (Bostrom, Fischhoff and Granger Morgan, 1992) instead. These approaches were thought more able to encompass the variety of ways that decision-making is actually structured and to meet grounded theory guidelines.

Ethnographic decision tree modelling employs qualitative methods to learn how decision makers perceive decisions and structure the decision-making process. Questions asked by the interviewer vary for each participant, but the claim is that it is possible to incorporate these into decision trees which encompass individual variations, interrelated sequential decisions and feedback loops that the decision-making process involves. Content analysis of participants' narratives distinguished key concepts. The method of constant comparison then revealed the different meanings these concepts hold for people. Decisions were then represented by creating simple mental models linking these concepts, using an approach originally developed to compare lay and expert perceptions of environmental risk (Bostrom, Fischhoff and Granger Morgan, 1992). Models were then compared with those from other individuals, such as partners, involved in the decision-making process to search for overlaps, contradictions and omissions. These were refined into the model of responsibility presented below which encapsulates what families facing a late-onset genetic disorder, HD, find important when make reproductive decisions in the face of changing options being presented by developments in genetics and reproductive technology. The detailed analysis that generated the model will be presented next as part of the Results section.

Results

This section provides an overview of the model, discusses the analysis that generated it, and then outlines how it provides a framework to account for variations in what people do. A subsequent section discusses what emerged about the process of decision-making.

Overview of the Model of Responsibility

The grounded theory analysis was facilitated by applying Strauss and Corbin's (1998) paradigm model, which it was felt could accommodate the process of reproductive decision-making. In the following summary the components of their paradigm are placed in parenthesis and the form these take in the study model are italicised. The 'causal condition' is one that provides the context in which the phenomena of interest occur. Comparisons of aware and unaware decisions identify this to be *awareness* of the risks for HD. A key aspect of the model is that it encapsulates the two dimensions of reproductive risk identified above: that which may be passed on to any children born, and that of not being able to sustain a parenting role. These are the *child-centred* and the *parenting* risks. The 'consequence' of becoming aware of these risks is that *reproductive decision-making becomes defined as problematic*. This prompts people to *redefine* certain elements of their situation - such as the nature of the risks, themselves, fertility, and relationships. The form this 'strategy' takes reflects 'modifying factors' such as the values that people hold, their concept of the future and their perception of social support. The 'consequence' of redefining is to enable people to tell a different story when subsequently *accepting, modifying* or *avoiding* one or both of these risks. Concerns about one value, responsibility, dominate these stories - identifying *responsibility* as the 'core concept' for the model. A key aspect of the model is that it demonstrates that *how* people are perceived to make decisions can become as important as *what* they decide.

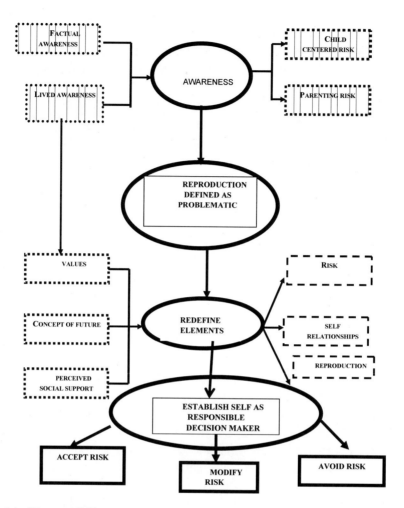

Figure 1. Model of Responsibility

Concepts Associated with the Model of Responsibility

Key aspects of the analysis underlying the model are:

- responsibility
- the nature of awareness
- the impact on decision-making of growing awareness of the risks that HD presents for reproduction;
- coping strategies such as redefining
- the role and nature of modifying factors such as individuals' values, perceptions of the future and social support

Responsibility

Responsibility is shown to be both a private and a public concern. Fundamentally people want and need to be seen as responsible decision-makers, by themselves and others. What this involves becomes apparent from exploring how people account for their decision-making in the face of growing awareness of HD.

Awareness

Awareness refers to the subjective perception of the risks that HD presents for reproduction, which both reflects and influences how people engage with them. Comparing new examples with previously coded extracts generated sub-codes of *factual* and *experienced*, or lived, dimensions. Factual awareness of HD corresponds to textbook or medical models, incorporating (in participants' words) just the "blatant black and white facts" such as diagnosis, prognosis and how genetic mechanisms work. In contrast, lived awareness feels "sort of arbitrary", encompassing "anything that happened" as well as emotional reactions. Becoming fully aware necessitates knowing in both factual and lived dimensions and integrating these perspectives.

Factors Facilitating or Limiting Awareness

Awareness is a *temporal* process in that it accumulates over time. It can also arise retrospectively, as hindsight. The ongoing analysis clarified links between responsibility and increasing awareness. One woman's attempt to reassess her mother-in-law's response to her first pregnancy shows how hindsight shifts the focus of responsible decision-making to others and introduces a contrasting facet, notions of blame.

> When you think back, how, how sort of gob-smacked she WAS when I told her I was expecting. We'd been married four years, so, you know, a different reaction altogether to what my mother was - my mother was over the moon. ... Looking back now, did she know something? And, was she in a quandary whether or say, or not?

Her reflections identify a second factor, *communication*, which impinges on awareness. All of the study families - including those who could communicate freely about most subjects - reported difficulty in talking about HD, supporting the initial fieldwork memos.

Further analysis revealed that barriers to communication took different forms in each generation. The older generation spoke of those affected being hidden away by their families. Their reluctance to make the condition known was not altogether surprising in view of the 'social' legacy of opposition to their existence and reproduction noted above. Another explanation to emerge was there was less incentive for the previous generation to consider the outcome of their decisions because they held different expectations about life.

> Then, if you got to fifty you were lucky ... You might have died of pneumonia, or TB, or half of the children that had Huntington's could have been stillborn anyway. ... It's only NOW that you expect to have longer quality of life that it's become such a problem. That you feel you're entitled, you know, you sort of say, "Oh, I'm not even going to get fifty years."

Older generation participants who *had* known about HD reported information being linked in uncompromising terms to reproduction. They were told either not to have children or to go away and "get on with things". In the latter case this could be interpreted as not questioning widely held expectations that marriage would lead to having children.

The younger generation reported that their parents had waited divulge relevant but limited information until their teenage years or when they had formed serious relationships. Now that HD is cited on the school science curriculum families report feeling obliged to discuss it earlier and in more depth than they might otherwise have done.

Finally, individual variations in *information accessing and processing skills* affected the speed with which awareness occurred. This included the prior establishment of schemas - for example, from professional experience, or from contact with sufferers of other illnesses - with which to make sense of information as it emerged. Such schemas could both help those at-risk to explain what HD involves and partners to make sense of the information.

Cognitive Processing

Distinctions participants made between decision-making before and after becoming aware of HD revealed *how* it changed with growing awareness. People commented that before becoming aware "there had been no decision to make" that was any different to other people's reproductive decisions, i.e. taking into account desires, financial and travel plans and waiting for the right time (Currie, 1988; Willen, 1994). Accounts of aware decision-making drew on physical metaphors of burden, such as weight and size. These metaphors conveyed the effort required to weigh, balance, or set aside the child-centred and parenting risks associated with HD. As had been surmised, like those facing infertility cited in the introduction, references were made to a sense of loss - the loss of normal expectations for reproduction. It was having to consciously address these considerations that turned reproductive decisions into *difficult* ones. Difficulty was compounded by complex and multiple analytical concerns, such as those around termination and determining risk status. Decision-making also felt onerous because as awareness increased people felt obliged to account - to themselves and others - for their intentions and actions. This observation provided an early indication that responsibility would become the core concept for the model.

Clarifying Links between Awareness, Responsibility and Blame

The process of becoming aware comprised five stages. Who could be held accountable, and how, varied at each stage. In the first *unaware* stage there was no conscious consideration of potential problems in the family. Information may have been available - as can be seen from comments made by other family members - but not divulged. Other explanations were that it was not consciously thought about or that people were not able to put it into words. Consequently, there was no recalled need for explicit consideration of risk in reproductive decision-making. If people had had children at this stage they could feel happy in retrospect that they had 'not really known' and could not therefore themselves be held accountable. Guilt could subsequently be felt about having unwittingly passed on the risk or not having pursued enquiries. However, as the example cited above showed, not knowing could also permit guilt to be displaced as blame on to other people who it was felt had been in possession of information and had not passed it on.

In the second stage, *pre-awareness*, concerns began to be expressed about whether problems existing within the family might have implications for reproduction. This could occur when questions were asked about family illnesses in prenatal clinics. It was not unusual, especially in the older generation, for HD to be attributed to non-genetic causes, such as the menopause or shellshock, which had no implications for other family members. If this had been the case, reassurances from the medical profession enabled reproduction to resume on a 'normal' track. People could subsequently blame those responsible for misdiagnoses or themselves for unquestioningly accepting them.

The third stage, *limited awareness*, arose when HD had been diagnosed in a family member but there was limited factual understanding of the implications this had for reproduction and for other family members. As Daniel explained:

> You know, Huntington's disease didn't sort of mean anything. It's just - a word.

As in the previous stage concerns arose in connection with reproduction. How health care professionals reacted to the mention of HD then became the crucial factor in determining how seriously these concerns were taken. Naomi held her doctor accountable for having had a child at-risk in that she felt he had not ensured she and her at-risk husband had understood the magnitude of the risks:

> He SHOULD'VE warned us, properly, and MADE us understand, how serious, it all is. And if he had the SLIGHTEST doubt that we didn't understand, he should've SAID it again, and again and again, until that, he was fully satisfied that we, understood FULLY what we were about to enter into.

The fourth stage, *detailed awareness*, encompassed comprehensive understanding of the HD risks. Engaging with the risks associated with HD became a central part of reproductive decision-making. Being aware made it much more difficult to delegate or deny responsibility for decisions that were being made. There was a need to show that one was unable rather than unwilling to act in what might be perceived as a conventional responsible way (Finch and Mason, 1993). Intervening factors (see below) then became crucial in construing responsibility. Many accepted the risks, supporting previous suppositions that although most people in Huntington's families have, or intend to have children, few choose to employ strategies such as prenatal testing that could ensure that these children would not be at-risk. Modifying strategies comprised timing of reproduction and determining family size. People could even have *more* children than they might otherwise have done, in the hope that at least one had not inherited the mutation and could care for any affected siblings. Avoidance strategies included not having children to seeking gamete donation, clarifying risk by testing and termination and seeking (unsuccessfully) to adopt or foster.

Some participants experienced a fifth stage, *refined awareness*, when risks became certainties as a result of genetic testing or diagnosis. A few chose a non-disclosing form of prenatal testing, prenatal exclusion testing (PNET) which did not clarify the parent's status but enabled them to have children known to be at low risk. This form of testing is complicated to understand. It requires parents to terminate any 'high risk' fetus known to be at the same fifty percent risk of having inherited the HD mutation as the parent. This means that there is a fifty percent chance that it does *not* have the mutation. It works by comparing markers on the gene from three generations, to see if the fetus has inherited the at-risk parent's copy from the affected or non-affected grandparent. The idea becomes to only proceed with pregnancies shown to be at minimal risk in that they have almost certainly not inherited a copy from the affected person. For a detailed account of how couples live with their decisions under these circumstances see Downing (2005). Donating samples to inform testing enabled the older generation to retrospectively define themselves as responsible decision-makers.

Growing awareness also involves acknowledging how reproduction becomes embedded in a 'HD life cycle'. This consists of the following events and stages: being born at-risk, being parented by an affected parent, discovering the risk, developing ways of living with risk, deciding whether to tell prospective partners about HD as a preliminary to reproductive decision-making, undergoing changes in risk status arising from diagnosis and predictive testing, becoming a suffer or carer for parents, one's children or other relatives, and coping with death and loss. People do not necessarily experience all of these events, the order in which they occur can vary, and events may coincide. For example, some people only learn that they were born at-risk when an affected relative died. Learning how HD events are experienced, resolved and map onto the 'normal' life cycle (Erikson, 1963) can elucidate the varied responses that people show to the reproductive risks for HD. As intimated above, how and when people discover their risk seems to affect how HD is taken into account when subsequently making reproductive decisions. Being brought up with the knowledge that HD is in their family, enabled some people to assimilate having at-risk children with a 'normal' life trajectory. Others, appraised of their risk at a later age, found it more difficult to accommodate reproductive risk with normal expectations for reproduction.

Redefining as a Response to Awareness

Redefining denotes a strategy of restructuring elements of people's situation, either cognitively or through actions, in order to have a different story to tell. New narratives incorporate changes in their perceptions of risk, fertility, themselves or their partners, and reproduction thereby enabling them to account for having accepted, modified or avoided either or both of the risks in a responsible way.

Deconstructing participants' references to *denial* - a powerful form of redefining - revealed three ways this could manifest. The first corresponded to the accepted clinical psychological definition as unconscious repression of painful information, such as learning of the risk. The second, 'affected' denial, was associated with the medical onset of HD. The third and most common way was a conscious strategy of 'distancing' oneself from threatening information. In contrast with clinical denial people could acknowledge this was happening and unwanted information remained accessible. Distancing was evidenced by HD only being spoken of in connection with more distant kin, such as cousins, or in relation to physical distancing, or severing contact between affected and non-affected kin.

Resorting to distancing was a function of the degree of fear engendered by the risk, control felt over it, and individual differences in the amount of information that could be tolerated about HD. If distancing was widespread in families, a *closed awareness* context resulted - similar to that identified by Glaser and Strauss (1966) when studying dying patients. As Angela explained, family members knew about HD but colluded in not openly acknowledging or discussing it.

> We used to live a life of pretending it isn't happening. .... When we were at home, we used to react as if nothing had happened, and it became like a family trait. If dad choked on some food, you'd sort it out, organize it, and then carry on with the meal as if nothing had happened.

A closed awareness context had considerable and contrasting implications for reproductive decision-making. It provided the freedom for people to take risks without fear of voiced criticism. But, as Frances explained, it also made it hard for those like her marrying into the family to find out what they felt they needed to know in order to make an informed responsible decision.

> Every time I've - I tried to talk about Huntington's, to the family, I'm more or less told, "Yeah okay, all right, you know, forget about it, shut up." They don't - they just - just don't want to know.

It also became difficult to implement choices, such as prenatal or linkage testing, which depended on other family members acknowledging the need for their assistance. As Tara said:
> We had to plead with him to give his blood, for the comparisons, for mine and my brother's test, because he didn't want to know, and he didn't want us to know.

As well as denying risks, narratives could redefine them in ways that appeared to lessen them or clarify them through genetic testing. Those at-risk, or their partners, whose values allowed them to accept a "semi-biological" child could redefine themselves as temporarily infertile in order to gain access to new reproductive technologies and bypass their genetic risk. Reproduction could be redefined as a challenge to the disease, or as replacing an affected person who had died. Others redefined partners as potential single parents so as to counteract the uncertainties surrounding their own ability to sustain a parenting role. Parenting could also be redefined, for example, through becoming an uncle or by caring for children in a professional capacity which would enable a person to relinquish responsibility on becoming symptomatic.

The versatility of redefining is illustrated in the following examples relating to unplanned pregnancies. Redefining them as chance events made it possible to abdicate responsibility. Maria's comment was typical of this not uncommon response:

> Well, if I did just happen to get pregnant ... the child can't turn round to me and say, "You shouldn't have done this to me," because I'd say, "Well - it just happened.

Choosing not to proceed with an at-risk pregnancy enabled one to be redefined by others as acting responsibly. Sabina explained what else she had gained from her unplanned, and terminated, pregnancy. Making decisions about another life allowed her to redefine *herself* as a responsible caring mother, through having:

> the experience of GENUINELY meddling in another person's life and shaping it irrevocably so that it can never be as it would have been without my interventions ... knowing that you have taken upon yourself that kind of assumption of responsibility which you have, as a parent, I haven't missed. The, the knowledge of being wholly responsible for life and death of another individual umn, I had with deciding to have an abortion. .... I talked to Sam - Sam's the guy, I named him / her, and sort of explained what I was doing, and all the rest of it, and explained why it was, this was the best life for that person to have. So, I did all of that.

Redefining thus illustrates how people make creative and adaptive responses to difficult dilemmas, consistent with naturalistic decision making (NDM) accounts of personal decision-making (Payne, Bettman and Johnson, 1993).

The Role Played by Modifying Factors

Modifying and interacting factors which determined *what* and *how* people redefined aspects of their situation comprised: values held about a range of issues, how the future was conceptualized, how social support was perceived and factors subsumed under the heading of individual differences.

Values

It became crucial for those seeking to justify accepting either or both of the risks to demonstrate that they had been *unable*, rather than unwilling, to have acted otherwise. This can be seen from the references to limited awareness that prefaced many of the accounts of acceptance of genetic risk. After becoming aware of the risks, acceptance began to be justified in terms of the values that the participants held - for example, it was accepted that those known to be strongly opposed to abortion were unlikely to consider prenatal testing.

The grounded theory analysis identified three clusters of interacting values which people found salient in their decision-making. Table 2 indicates how different aspects of these values, relating to reproduction, quality of life and decision-making, influenced reproductive decision-making.

Table 2. Values Shaping Reproductive Decision-Making in the Face of HD

Reproductive values, related to *feelings about:*	Quality of life values related to *perceptions & experiences of:*	Decision-making values related *to the importance placed on:*
knowing that you could conceive or make your partner pregnant	a life lived at-risk for HD	Making decisions
having a child	Living with definite knowledge of HD status	risk taking & avoiding
having an at-risk child	a life affected by HD	information seeking and avoiding
Having a biological child	being parented by parent an affected	Identity as decision-maker a responsible
abortion	the parenting that one might offer to a child	
Envisaged identity as a father or mother	Alternative opportunities for fulfillment of goals seen as related to reproduction	

As had been suggested in the introduction (Dallos, 1997; Nelson and Nelson,1995) values started to form long before any thought of reproduction. Their impact depended on how peripheral or central they were for people. This could vary at different points of time and with the introduction of each person involved in decision making. Those who saw reproduction as central to their existence found it difficult to think of a life without children and easier to downplay other considerations.

Flora explained:

> I wanted children MORE than I thought the risk would be to her.

Conversely, men and women who placed less value on having a child found it convenient to draw on HD as a responsible reason for remaining childless. Matthew's comment was typical of this standpoint:

> Plenty of other people are busily producing babies. And if you are unsuitable to provide them, why worry?

More frequently decision-making entailed resolving central but conflicting values, such as wanting to have biological children, not wanting to pass on the risk, valuing an at-risk life, and being opposed to abortion. Thus certain strategies such as prenatal testing or utilizing donated gametes, could be screened out by the values that people held. For a detailed analysis of uptake of various risk limiting and avoiding strategies according to the values that they involved see Downing (2002).

Vision of the Future

How people viewed the future affected whether or not they were prepared to accept the risks. It also illuminated how some people (who would have been prepared to accept termination of pregnancy) justified rejecting genetic testing. They construed availability of testing as an indication that treatments for HD would follow in the near future. Envisaging the future in terms of *promise* made it easier to redefine the risk as acceptable and reject testing or other risk avoiding strategies. Hearing about on going research into possible treatments fed into this optimism.

Others, more cautious about the future, were more likely to take steps to avoid passing on the risk. Whilst they were able to accept their own existence as a result of the previous generation's limited awareness they doubted that the next generation would necessarily be so tolerant. They saw the future as a time of *reckoning*, when they would become accountable for the decisions that they had made. Their priority was to be seen as responsible decision-makers by their future as well as existing families. As Rupert explained:

> I would really HATE to have the discussion with a child of mine which went: "Why did you let me live when you knew I had this disease?"

How the future was conceptualised reflected concepts of agency. Many participants holding a 'chance' perspective of the parenting risk saw their future as predetermined and their role as reactive. Others holding a 'choice' perspective saw themselves as shaping the future. They deployed options such as genetic testing to clarify and control their risk of becoming an affected parent, or had children earlier than they might otherwise have done in the hope that they would be grown up before the parent became symptomatic.[4] This dichotomy accorded with well established psychological concepts relating to internal and external locus of control (Rotter, 1966; Wallston, Wallston and DeVellis, 1978; Wallston, 1997). It clarified contrasting constructions of responsibility in connection with the parenting risk. Those holding

---

[4] These terms were coined by Marshall (1995) to account for contrasting attitudes to new reproductive technology.

chance perspectives took a reactive perspective of responsibility, seeing it as coping with whatever fate presented. Conversely those holding choice perspectives became proactive about responsibility, employing interventions such as predictive testing or gamete donation in order to inform and control outcomes. It became possible to explain why people might not act consistently in regard to the two risks, seeing the parent's risk as predetermined and that for any child as a matter of choice.

Perceived Social Support

This factor similarly started to shape decision-making long before any thoughts of reproduction. One interesting finding was that those growing up in families affected by HD were more willing to accept the uncertainty associated with their ability to sustain a parenting role *if* they had experienced a supportive sibling relationship. Conversely, only children, and participants whose siblings had not been perceived as supportive, reported feeling more vulnerable and less willing to take on this risk.

Support concurrent with decision-making ranged from endorsement of, or practical assistance with choices made to tolerance of differences. This was more likely to be forthcoming if people could demonstrate that they had acted as responsible decision-makers, or in a manner consistent with their central values. Those whose partners, families and friends were able to show non critical support for whatever path had been chosen acknowledged how helpful this had been. Case studies drawing on the model as a framework illuminated how support changed with repeated decision-making and could come to involve clinicians as well as families (Downing, 2005).

Individual Factors Contributing to Decision-Making

Individual factors that contribute to decision-making styles include differences in information processing approaches, corresponding to the monitor (information seeking) and blunter (information avoiding) categories suggested by Miller (1987). It has been suggested that these differences are indicative of general patterns of responding to risk (Lopes, 1987), and attitudes to decision-making (Beattie, Baron, Hershey and Spranca, 1994).

Maturational and physical characteristics, such as age, risk status, fertility and gender, also impacted on decision-making. Many participants reported that they had found it easier to distance themselves from the risk when younger. Being able to conceive easily encouraged those undergoing prenatal testing to continue along this path (Downing, 2005). How long a couple could wait for a risk-avoiding strategy to become available, such as genetic testing, was to some extent an interaction of age and gender, in that men could contemplate reproducing at a later age than women.

Impact of Genetic Testing on Reproductive Decision-Making

A key aim of the study was to explore the impact that the introduction of genetic testing has had on reproductive decision-making. Support was shown for previous consistent findings concerning the reluctance of many facing HD to use predictive testing, seeing it as a *threat* to the hope that uncertainty conveys about their risk status.

The study augments existing work by revealing additional perceptions that people have of genetic testing and how these facilitate or constrain decision-making. Construing genetic testing as *reconciling previously incompatible goals*, such as having biological children and avoiding passing on the risk, facilitated decision-making. Similarly, perceiving testing as *an indication of technological and medical progress,* implying that treatments for HD would become available in the near future, made decision-making feel less constrained. Participants holding this view could justify *not* employing testing and accepting the risk. Conversely, participants who perceived genetic testing primarily as a *technological imperative*, imposing a responsibility on them to employ it in order to make an informed decision, experienced their decision-making as more constrained. However, because using genetic testing was likely to be interpreted by others as *exercising responsibility*, support was shown by other family members and clinical staff which mitigated this feeling of constraint.[5] It was not always necessary to employ testing to be judged as acting responsibly. Considering testing was sometimes sufficient to ensure support (Downing, 2005). This observation contributed to the key finding that *how* decisions were made can be perceived as being as important as *what* is decided.

The study also makes a significant contribution to what is known about the ways in which deciding to undergo genetic testing changes family dynamics (Roos et al, 1990; Quaid and Wesson, 1995; Cox, 2002, 2003; Sobel and Cowan, 2000). It was shown, for example, that requests for samples to be used in linkage and prenatal testing of other family members enabled some people to view testing as providing an opportunity to *alleviate the guilt* they felt about having unknowingly passed on the risk, and to retrospectively redefine themselves as responsible decision-makers. Conversely, envisaging genetic testing as a *threat*, in that important information about oneself might be revealed to others, made people less likely to cooperate in providing samples and could result in the breakdown of relationships within families.

The impact of genetic testing needs to be seen as a function of which *form* of testing is being employed or considered. Linkage predictive testing, for example, had a limited impact because it was not available for all who considered it. Another constraint encountered in connection with predictive testing, both linkage and mutation, was that this was seen as a decision that could only be made by those at-risk. In contrast, pursuing prenatal testing was seen as a decision involving *both* partners. This observation was an important one as it challenged the appropriateness of the widespread tendency in the literature to refer solely to women when speaking of decision-makers.

Partners of those who did not want to know their status faced making uninformed decisions or having non-disclosing exclusion prenatal testing (PNET). Uptake of PNET has been very low worldwide (Adam et al, 1993; Maat-Kievet et al, 1999; Hayden, 2000, Simpson and Harper, 2001). Fewer than ten prenatal tests were done in Canada between 1990-1995, and the rate in the USA was similarly low. It was therefore a significant achievement for this study to have accessed ten couples who seriously considered employing it, and four who actually used it. Their experiences were very informative about the decision-making of those considering but not using prenatal testing, the problems associated with negotiation of its use, considerations arising with its repeated employment, and the impact of different outcomes.

Some of the ways that using prenatal testing made a difference - such as creating tentative pregnancies (Rothman, 1986) whilst waiting for results, setting up assumptions that people will act on the information received and terminate wanted but high risk pregnancies,

---

[5] Although the word 'mitigated' suggests that constraint has a negative effect on decision-making, this may not necessarily be the case. Constraint may help people arrive at a decision by reducing the alternatives.

or having to involve others in decision-making - were not unique to HD. Neither was the experience of two of the couples who had to cope with the subsequent miscarriage of pregnancies shown to be at low risk (Robinson et al, 1991). The predicament associated with PNET of terminating 'high risk' fetuses which may not be affected, was also not a new one. This dilemma has been previously documented in connection with Duchenne muscular dystrophy and other sex linked conditions. Prior to the introduction of genetic testing some people chose to terminate any male fetus (a proportion of which would not have been affected) in order to ensure that they only had a child that they knew would not be affected (Parsons and Clarke, 1993).

The low uptake of prenatal testing observed in this study appeared to be influenced by weighing these issues in relation to a risk that would almost certainly not manifest until adult life. What the study revealed was that participants were faced with considerations of the quality of a 'tentative life', that is, the life to be lived at-risk. This meant that those who *would* have been prepared to consider termination for early onset disorders could have difficulty in seeing it as appropriate for HD because they saw their, or their partner's, life as worthwhile.

Interesting findings emerged about repeated reproductive decision-making. Those who did not initially use prenatal testing, but then employed it for subsequent pregnancies, were able to accept that their decision-making would result in them having different information about the status of each child. However, couples who started to use a certain type of prenatal testing felt compelled to use it for subsequent pregnancies. This effect was apparent for individuals who had both high and low risk results. What this meant was that couples continued to employ PNET even after a more informative and predictive test, direct mutation prenatal testing, became available. Those who had previously terminated on the basis of a high risk result stated that the distress experienced was preferable to being confronted with the realization that, in the event of the parent having a negative predictive test, termination had not been necessary. This finding illustrated the limitations of classic decision theory models which fail to take account of the input from past decisions. It was consistent however, with studies that have showed what has been called the "sunk cost" phenomenon (Arkes and Blumer, 1985). This refers to a situation that arises favouring continuing with a pattern of behaviour because the investment made - which can be emotional, financial or practical - makes it difficult to assess subsequent decisions independently.

Participants saw genetic testing primarily as an extension of *existing* methods - such as sterilization and using donated gametes - offering control over risks. This was compatible with Strathern's (1995) observations of the different ways in which people make sense of new reproductive technology, seeing it as either innovative or linked to established ways of having children. It suggests that there may be underlying consistencies in how people make sense of a range of technological changes.

Conceptualizing testing as an extension of existing methods of control generates additional questions. First, to what extent do the various technological developments offering control over reproduction and risk - which includes improved contraception, early identification of pregnancy and prenatal testing techniques - enable people to actually *feel* more in control of reproductive risks? Perception and tolerance of residual uncertainty associated with testing varied. One of the most interesting findings to emerge from the study was that perceived lack of confidence in *low* risk PNET results formed a significant barrier to its use. Doubts about the reliability of low risk results could also arise later, after repeated high risk results, through actual experience of laboratory error, or from learning that testing for *other* conditions, such as HIV, was fallible. These findings are illustrative of cognitive biases, such as the availability bias (Tversky and Kahneman, 1973), which have been previously well documented in relation to a range of risks (Kahneman, Slovic, and Tversky, 1982). The key point to be made in relation to HD was that although experiences such as these could engender uncertainty about the reliability of testing, they did not necessarily dissuade people from employing it.

A related question arises about how much people *want* to feel in control of reproduction. It was interesting to note that unplanned pregnancies did occur to both those accepting and wanting to avoid passing on the risk. Defining a pregnancy as a chance unplanned event - for which one cannot therefore be held accountable - was shown to facilitate acceptance of the risks. Another indication of ambivalence relating to control was that those who felt strongly about not passing on the risk did not necessarily employ contraception, or, if they did, did not use the most reliable methods.

The few who had learnt their status felt that the increased information clarified their subsequent decision-making. Previous research had suggested that learning there is no longer any risk appears to facilitate implementing previously expressed reproductive intentions. Decruyenaere, Evers-Kiebooms, Boogaerts, Cassiman, et al (1996) et al (1996) showed that within a year of the partner testing negative all 13 couples had had a child, were pregnant or were intending to have a child. In contrast, 13 other couples who learnt that the partner had tested positive had decided not to have a child or more children, were contemplating reproduction only in conjunction with prenatal testing, or were still undecided. However, what people then *did* with this information varied considerably, adding further support to the claim that people do not necessarily prioritize exercising control over the risk in their decision-making. This finding supports that from another study which interviewed couples after a longer time period to show that reproductive behaviour of those who test positive is complex and does not always correspond to pre or post test intentions (Tibben, Frets, van de Kamp, Niermeijer et al,1993).

Applying the Model

The model provides a framework in which to re-examine the observed complexity of decision-making. This involves systematically comparing and accounting for:

- variations in responses shown between people, including partners and other family members
- variations in responses to each dimension of risk
- choices of risk avoiding or modifying strategies
- responses to change such as increasing awareness of HD, pre and post testing reproductive decision-making, decision-making with different partners
- making decisions before and after becoming parents
- decision-making at different stages of the reproductive decision-making process including how decision-making becomes shared.

A comprehensive account of these applications is given in Downing (2002). A few examples are presented here to indicate the versatility of the model. The first deconstructs why first and subsequent pregnancies presented different issues, making them feel like different decisions. First pregnancies involved framing decision-making as about *whether* to reproduce and the extent to which identity is shaped by becoming a father or a mother.

Families were more supportive of those at-risk having one child for this reason. Making decisions about subsequent children became more complicated - involved reconciling expectations voiced by others outside the family about 'having another one' with diminishing support from other family members, the projected needs of existing children and the at-risk parent, from the perspective of being a parent. Having additional children therefore included acknowledging these needs and presenting oneself as addressing them in a responsible way. Becoming a responsible decision-maker involved deciding which of these needs would be prioritized.

The second example, set out in Table 3, uses the model as a framework to compare different responses to the parenting risk.

The third example illuminates what is different about risk splitting and linking and provides insights into how couples' decision-making became shared. As Beach (1993; 1996) had suggested, individuals engaged in preliminary restructuring of their situation and options to reflect their own values before negotiating with partners. Negotiation involved checking whether their values were compatible. Difficulties could arise if partners were not in accord - as Ingrid's comments revealed:

> You know, I'm not prepared to let those children have what he's had.. .... He never agreed with me on that - he said, I should have children at-risk. He didn't see that it was that bad. He said, "I've had a good life. You know, I'm thirty seven. A lot of people don't get to that age, if you're over in Rwanda or something." But maybe, I wonder whether, if, he's in a different position to me, isn't he? Because he's < > if he agrees that you shouldn't have kids, maybe he feels that his life isn't, wouldn't have been worth anything. I mean he'd have been, you know, I don't know if he'd have wiped out, because I don't know the, what how his result would have been.

In the event of disagreement about prenatal testing and terminations the woman's views prevailed. As she explained:

> One of you's got to do it, you see, and I, I decided it was MY choice as far as whether to have the tests for the baby, and my choice whether to have a termination and you can't compromise on that. You either do it or you don't do it. So my husband didn't really feature in it, which was TERRIBLE.

For most couples decision-making about predictive testing was seen as the prerogative of the at-risk partner. Focusing on non-disclosing prenatal testing and termination enabled Ingrid to 'split' the risks and redefine both risks: that for the child as negligible and that for her partner as less stressful:

> I'm cutting down on my anxiety so it's quite selfish, in a way. ... I want to reduce my anxieties, see? And, at the moment, I'm only worrying about my husband. I don't want to worry about them as well.

In contrast, Martin and his at-risk partner agreed in their views that the risks were firmly 'linked'. He explained, that for them:

> The logical approach is Mum wants kids. Mum is at-risk. Mum has the test. If mum's got it, mum cuts the tubes. But only when you've cleared mum, then the whole thing -then, and then you're back to normality. ... UNTIL - until somebody comes up with some - is it cystic

fibrosis where they've now got something you can pop up people's noses to do wonderful things against the gene? Fine. If somebody can come up with, with some wonderful thing which takes all your, all your cells out, scrubs the genes over, shakes them into order and bungs them back in again.

His partner concurred:

If you're not prepared to test yourself, you don't - no, absolutely. ... If you test the child, and then you get Huntington's, that's out of order too - as far as I'm concerned. You DON'T put a child in a family, knowledgeably, within a Huntington's family. ... Part of the reason you're not having children is not only to pass it on, but not to inflict Huntington's on them. So therefore you should be testing yourself. And if you got a positive ... you don't have them. It's ABSOLUTELY simple.

Table 3. Using the Model Of Responsibility to Account for Accepting, Modifying or Avoiding the Parenting Risk

	ACCEPT	MODIFY	AVOID
VALUES	Identity as a parent, parenting as "natural"		Value child's needs over need to reproduce
FUTURE	optimistic, focus on present ability		cautious, focus on future limitations
SOCIAL SUPPORT	compensating, forthcoming from partner and previous generation	compensating, forthcoming from partner and previous generation	diminished, children and partners seen as having competing needs
REDEFINING	risk splitting: between child and parenting risk seeing parenting risk as predetermined	timing: redefine oneself as a responsible informed decision-maker by having children earlier to lessen possible impact of affected parenting.	risk linking: redefining self as responsible by drawing comparisons with others as selfish
STRATEGIES	self as single parent or dependant child		redefining at-risk parent as risk free through predictive testing.

The model has also been used to clarify clinical encounters with multiple decision-makers. It is being used as an on-line teaching tool in conjunction with case studies exploring decision-making about prenatal testing (Downing, 2005).

The Process of Reproductive Decision-Making

The study generated the first detailed account of the *process* of reproductive decision-making in the growing awareness of HD. This was shown to involve a multiplicity of decision points, requiring people to draw on a variety of resources to engage in private and shared considerations of a number of areas. These included how genetic risks are perceived, the responsibilities that genetic risks entail, how technological developments are assimilated and, as Richards (1993) and van der Steenstraten et al (1994) had suggested, how identity is constructed. This section provides an overview of what emerged about the four stages comprising the decision-making process: pre-contemplative; pre-implementation; implementation; post-decision evaluation.

Pre-contemplative stage

The qualitative approach adopted in the study showed the importance of acknowledging an additional, pre-contemplative, stage where no knowledge might be held of HD, nor any considerations entertained about procreation, but that shaped later reactions by forming individual attitudes and values. Support is therefore shown for NDM claims claim that additional insights into the decision-making process will be provided by links made to other domains, such as narrative approaches (Nelson and Nelson (1995) which provide access to how values are formed in families.

This stage could encompass being parented by a parent who was affected but not yet diagnosed. Connections were shown to exist between how affected parenting was experienced, intervening factors such as social support, and the subsequent value placed on avoiding this risk for future generations. Narratives from different family members confirmed the concept of preselection in connection with risk status (Kessler, 1988) and extended it to incorporate particular children being singled out for *abusive* treatment. Younger children in families were more likely to be exposed to affected parenting, but this could be mitigated by sibling support. Only children were the most vulnerable. This meant that affected parenting as well as risk could be *interpreted* quite differently by each family member leading to them placing different values on avoiding this risk for the next generation.

Pre-Implementation stage

The second, or pre-implementation, stage of decision-making was initiated by becoming aware of the risk and culminated in forming potential responses to it. Intermediary steps involved at this stage were:

- becoming aware of the risks that HD presented;
- assessing the impact of these risks on reproduction;
- defining reproductive decision-making as a difficult decision;
- redefining reproductive goals and evaluating them for compatibility with values held;
- evaluating strategies for their ability to realise these goals and their compatibility with values;
- mental simulation of implementation and outcomes;
- re-assessment.

Decision-making became shared at this stage. Comparison of different family members' accounts revealed how this happened. As the example presented in the previous section shows the findings supported Beach's (1993) claim that individual processing preceded negotiated, or social, decision-making. People considered the issues in relation to the values they held and these values shaped the way they subsequently presented them to others. I had speculated that partners, because they had not shared the same family experiences, might hold different values about passing on the risk, or ability to parent. This was shown not to be the case. Members of a couple did not necessarily hold the same values but acceptance, modification or avoidance responses were made to the risk by *both* those at-risk and their partners. Partners were able to draw on other experiences, such as having a parent die from cancer, to understand the dilemmas that HD presented.

Analysis of how the younger generation's accounts revealed three distinct forms that negotiation could take. These were 'partnership', characterised by collaboration, 'facilitating', where one person made a decision and the other facilitated it, and 'deadlock', where it was difficult to find common ground. These patterns were similar to those identified by Beach (1996) in a corporate context. Successful negotiation created shared values that became a resource to be drawn on in times of stress, for example when facing terminations. This suggested that once shared values were established they became more powerful and took precedence over individual ones.

These patterns of negotiation extended to others who were involved in the decision-making process. This included lateral kin such as siblings and cousins, the previous generation from both the affected and non-affected families, work colleagues, friends and health care professionals and sometimes even laboratory staff. Conflicting ideas about appropriate responses could cause temporary breakdowns in relationships between family members. Case studies presented elsewhere documented how the health care professionals and laboratory staff can come to be seen as a substitute family when this happened (Downing, 2005).

Certain areas emerged as not negotiable. One was the decision whether to have a predictive test - which was seen as the prerogative of the at-risk partner. Another was whether to have a termination - which was seen as the woman's decision. However, couples might still *discuss* these decisions and partners decide whether to *facilitate* the choices made.

Information seeking formed another aspect of social decision-making. Constraints could arise according to how HD was managed within a family. If the family coped by not talking this could result in frustration and bitterness if it was not compatible with partners' needs to know. Tensions could also arise if partners had more factual knowledge than the family, especially if it contradicted the myths, incompleteness and misinformation that could exist about HD.[6] For some, this stage included making contact with clinical genetics services.

Implementation

Passage through these stages could be rapid or slow. The retrospective design of the study made it possible to encompass this variation. For example, one couple waited twelve years for a new strategy - genetic testing - to become available to achieve their goal of a child free from risk. The limitations of studying personal decision-making through hypothetical situations were shown: what people actually did when in a situation could be quite different from what they thought they would do. Further difficulty could arise at this stage from having to confront additional choices, for example, around termination. Previous experience aided implementation of subsequent decisions by providing procedural expertise and feelings of increased confidence. However, it could prove to be disconcerting if new scenarios arose, such as the contrast between reproductive decision-making for first and subsequent children described above.

---

[6] This was especially the case for those who had come across HD in a professional capacity.

Post-decision evaluation

The final stage of decision-making was post-decision evaluation which could be positive, negative or mixed. Individuals could be reticent about divulging post decision distress in order to protect partners. They could also be reluctant to share their feelings if they sensed that their partner was at a different stage to them in working through post-decision reactions. This was especially noticeable in the accounts given of post termination distress. These observations suggest that decision-making is most likely to be shared in the pre-implementation and implementation stages.

People subsequently justified having children at-risk in diverse ways. Many felt they have been deprived enough - citing lack of nurturing in childhood and repeated bereavements - and questioned why they should have to give up having children as well. Reproduction enabled them to do something normal, reassuring affected relatives that at-risk lives were not constrained, compensating those they love and bringing them a measure of hope. Children continued the generations, replacing other family members who had died. They became symbols of their parent's continuing health, providing a notion of immortality to counteract fears of premature death. One of the commonest fears expressed by those at-risk is that of being abandoned if they develop HD. Having children, provided potential carers - what better carer than someone who has seen it all and seen others coping with it? Having more than one child provided potential carers for the carers should they in turn become affected.

Progression through stages

Although identification of these stages suggests a linear progression this was not in fact necessarily the case. For example, mental simulation could incorporate projected post-decision evaluation - in the form of anticipated decision regret - prior to implementation. Previous decisions could also feed into subsequent ones, or function as progress decisions to modify assessments of strategies or goals. Progression was aided by the development of competence as outlined by Klein (1999). Family case studies (Downing, 2005) showed this could encompass decision-making by other family members such as siblings.

Evaluating the retrospective approach taken to studying decision-making

There were clearly demonstrable advantages of taking a retrospective approach. As hoped, this provided the flexibility to encompass variations in time taken to make decisions, enabled contact to be made with people at various stages of decision-making, and encapsulated variations in the social contexts experienced by each generation, thus allowing considerable access to what Kemdal and Montgomery (1997: 75) have called the "full picture of all the factors that have been of importance for the decisions". Although the interviews gave a snapshot of one person's views and interpretations at that time, their stories revealed ways in which these had changed. Decisions could also be analyzed in the face of changes which occurred over time, such as moving from being at 25% risk to 50% risk after a parent was diagnosed, the introduction of genetic testing and subsequent altered risk status.

Another way of evaluating the retrospective approach taken was to consider whether the findings might be different, or what could be added, if the study had adopted a prospective design. In fact, some prospective data was generated in that participants were sent a short questionnaire approximately six months after being interviewed, which provided information about how taking part in the study had affected the way HD was discussed in families. Six months was possibly too short an interval to capture relevant developments in risk status, relationships and reproduction. If a longer interval had been scheduled would it have been possible to say something different linking HD status and reproductive decisions?

Some participants remained in contact following their interviews, providing some informal longer term prospective data, including results of predictive tests, which can be drawn on to reflect on this question. What these limited data suggested was that changes other than clarifying risk status influenced participants' responses. In particular, continued contact with participants revealed that decisions made when in one relationship do not necessarily predict those that will be made in the context of a different relationship. Two women who had been adamant that they would not have children had been prepared to do so after forming new relationships. Their behaviour was consistent with the theoretical model, in that it demonstrated how reproductive decision-making was mediated by perceived social support and shared values. These instances reinforced the claim made in this chapter of the need to consider couples, as well as individuals, as the unit of reproductive decision-making.

It is undeniable that retrospective data, such as that gathered in the interviews, is subject to biases of recall and omission. However, it became possible to see these gaps and biases as informative, in that they elucidated *how* people recalled events and structured them as stories in the evaluative stage of decision-making. Corroborations and contradictions between different accounts - from multiple family members - were especially useful in tracking down where these biases arose and in providing indications of what they might signify.

This can be illustrated by the following example, concerning a considerable disparity that occurred in accounts given by a sister and a brother about *when* they learnt of their parent's diagnosis. The sister placed it at a time before either of them had had children, the brother after they both had. Had this time gap actually occurred? Was the initial information not in fact passed on by the woman to her brother - which then tells us something about the time that such information may take to filter through families? Or, was it because her attempt to challenge the existing diagnosis of Parkinson's disease was met with extreme hostility from other family members, and, a firm refusal to accept that it might be HD until much later - when their remaining sibling was diagnosed with HD? Or, did it mean that she recalled the information because it was more salient to her than to her brother, as she was considering having children at that time, and he was not? It was likely that in this case the distress that she could envisage from not having children also helped to structure the vivid memory that she had of having to confront this diagnosis. Bruner (1986;1990), amongst others, has stressed how important emotion and affect are in structuring memories and how events become coloured by the emotional state people are in when they experience and recall them. This example reinforces the point made earlier in this chapter that 'knowing about' HD is not the event of a single moment, but is rather an unfolding awareness which was evidently happening at a different time or pace for these siblings.

Another interpretation of the contradictions in their accounts was that she could accept that she had knowingly had children at-risk whilst he could not. This last point suggests that people may have very good reasons for choosing to remember experiences in ways which may not reflect the actual sequence of events. Evidence existed within their narratives for *all* of these explanations and it was difficult to prioritize any one of them. However, by looking at a number of these instances it became possible to create a detailed picture of the complexity in which events occurred, were recalled and came to have significance for people. Similar conclusions are deduced by Song (1998: 115) in a study of siblings' experiences of working in family businesses. She too had found that "the result of combining and analysing two siblings' interviews was greater than the mere sum of the two interviews." This was not necessarily because of any additional information provided but because "multiple interviews helped reveal the complexities, contradictions, and tensions in people's accounts and in their daily lives."

Conclusion

The specific findings presented above are of particular importance as reproductive decision-making in the face of genetic risk remains an under-researched area. Bekker et al's review cited in the introduction of 547 studies published between 1986 and 1996 categorised as being about informed decision-making yielded only seven that dealt with genetic factors, and none which addressed late onset disorders. Their description of reproductive decision-making as facing "relatively simple clearly-defined options with known risks" does not accord with actual accounts of how decision-making is experienced in the face of genetic risk. The grounded theory analysis revealed that considerable complexities exist in the way genetic risks are perceived which need to be identified and acknowledged in order to understand the multiplicity of responses that people show to them.

From an early stage in the analytic process concepts emerging from the data were able to be integrated with and extend existing NDM theoretical frameworks. Overall similarities in the ways in which NDM models convey the process of personal decision-making enabled their principles to be illustrated by one particular model, image theory (Beach, 1990). Comparisons made between the model of responsibility generated from data and image theory revealed shared features. For example, both models stress that the values that people hold play a crucial part in shaping their decision-making. An important contribution made by the grounded theory analysis conducted as part of the study was to identify which values participants saw as salient in the context of reproductive decision-making and HD risks. It then became possible to account for varied responses shown to the child-centred and parenting risks by identifying which of these values decision-makers chose to prioritize, in combination with other contributing factors.

Another point of correspondence between image theory and the study model was the recognition of the importance of pre-choice processing, including coming to an understanding of the decision situation. Awareness was shown to be the prime causal condition in this process. Empirical support also was found for the claim forwarded in the introduction that decision-theoretic and health psychology form complementary rather than competing approaches. For example, a related finding that awareness included emotional and cognitive aspects is consistent with theoretical frameworks advanced in health psychology models such as the parallel response model (Leventhal et al, 1980, Leventhal et al, 1997). However, as McAllister (1999) also noted in relation to her model of engagement, the grounded theory analysis revealed that emotional and cognitive aspects interact dynamically rather than as parallel responses. Decision-making became shared in the way that Beach (1996) had suggested, in that individual decision-making preceded shared decision-making. Patterns of negotiation emerging from the data mapped onto those identified by Beach in other contexts. What the study additionally contributed was to reveal how negotiation can engender shared values that become a valuable resource facilitating implementing difficult decisions, such as proceeding with the termination of a wanted pregnancy. These correspondences suggest that NDM models have much to offer as frameworks in which to study decision-making in the face of genetic risk.

Further Applications of the Model

The introduction alluded to the fact that how risk is negotiated is shaped by being lived in a particular moment in history. Deconstructing how parents viewed reproductive responsibility in relation to genetic inheritance had previously been seen by Callahan (1979) as entering uncharted territory. Opportunities at that time for genetic information were extremely limited and motivations for reproduction had not normally been subjected to detailed scrutiny. However, as Callahan pointed out it was possible to view this question as part of existing work exploring locus of control and intervention in natural processes. Employing this framework yielded contrasting constructions of what could be expected in relation to acting responsibly. If control was perceived as external then parental responsibility could only be reactive, characterised as coping with whatever fate presented. If control was perceived as internal, parental responsibility became proactive, raising expectations that people would employ interventions such as predictive testing in order to inform and control outcomes. These constructions map onto the choice or chance perspectives referred to earlier in this chapter.

Other features of Callahan's analysis accord with the present study's findings. Callahan (1979: 235) distinguished between "exercising responsibility *as* a parent and responsibility in *becoming* a parent", which corresponds to the two aspects of reproductive risk associated with HD. He suggested that individuals could become either more or less accepting of risk when they came into contact with others in their social context, in accordance with a wide range of findings concerning the "risky shift" phenomenon which states that individuals modify their acceptance of risk in social contexts (Lamm and Myers, 1978). This is consistent with the finding from this study that responsibility involved both individual assessment and social evaluation.

Responsibility was shown to be both a private and a public concern. People had to convince themselves that their decision-making involved acting responsibly to the issues that HD raised for reproduction, and this provided the incentive for cognitive restructuring, or redefining, which could operate in various ways and lead to different outcomes. For example, a few participants redefined themselves as infertile, enabling them to utilize gamete donation in order to have a child and assure themselves that they had acted responsibly because they had bypassed the genetic risk. Redefining was consistent with NDM accounts depicting personal decision-making as making creative and adaptive responses to difficult dilemmas (Payne, Bettman and Johnson, 1993). Creativity involved reconciling discourses of genetic determinism and control within a framework of genetic connection in such a way as to present themselves as acting responsibly

This chapter has presented an account of how decision-making is experienced at a time when it has become possible to clarify uncertainty about risk for HD and utilize strategies to avoid passing this on to the next generation but when no effective treatments or cure yet exists for HD.

The model can be used to study uptake options such as preimplantation genetic diagnosis (PDG) that have subsequently become available. PGD enables people to have children who are known to be gene-negative and avoids ethical issues associated with terminating a pregnancy. Eggs are harvested, fertilized *in vitro and tests* performed on single cells biopsied at the eight-cell embryo (day 3 of development. Only those testing gene-negative are implanted. The main impediments to PGD are its expense, the low efficiency of *in vitro* fertilization (IVF), with only 20% to 30% of couples achieving pregnancy per IVF cycle (Pickering et al.) in the first 100 PGD cycles performed at the Guy's and St. Thomas' Centre in London. Some programs offer non disclosing PDG for couples who do not wish to undergo HD presymptomatic testing. The parental gene status is not revealed during the protocol and only gene negative embryos are implanted.

It is suggested that when effective treatments for HD become a reality concepts of responsibility will change again. Discourses of control could, as Rothman (1998) speculates, generate a more proactive form of decision-making:

> 'It's genetic' might very quickly not be a throwing-up-of-your-sleeves kind of problem. 'It's genetic' might be coming to mean, so let's fix it, let's engineer it, let's construct it to order. Let us make the determination, and let us predetermine (Rothman, 1998: 210-211).

The factors identified by Slovic (1987) can be used to predict, as shown in Table 4 below, how perceptions of genetic risk might change.

Questions arise about the extent to which it is possible to generalize from the findings from the study. Previous research shows the need for caution when attempting to generalise about responses to risks categorised as belonging to a particular domain. Variations that risks demonstrate on attributes other than risk category contribute to how they are perceived. Hypothetical scenarios constructed within the general category of rail accidents enabled researchers to conclude that responses varied according to specific attributes of trains, cargoes, accidents, locations, and causes of accidents (Kraus and Slovic, 1988). Similar findings were shown in connection with risk perception within other risk domains such as car faults (Slovic, MacGregor and Kraus, 1987), medicines (Slovic, Kraus et al, 1989), and unwanted environmental features (Slovic, 1992).

Attributes of genetic risk which seem likely to affect how they are perceived in relation to reproduction include mode of inheritance, perceived severity, possibilities for surveillance, treatment, cure and genetic testing, and age of onset. For example, varied modes of inheritance produce quite different scenarios of reproductive risk decision-making - some of which are briefly sketched out below.

Carriers of *recessive* disorders, such as cystic fibrosis, often remain unaware of the risk until *after* becoming parents. Frequently it is not until after the birth of an affected baby, or the development of symptoms in a previously healthy child, that they perceive their carrier status.[7] Parental anxiety arises from having to cope with the different and extra demands that an affected child creates. Both parents of children with recessive disorders are, and will remain, symptomless carriers. Carriers may also experience increased anxiety on learning of their status (Mennie et al, 1993; Marteau and Anionwu, 1996), which could influence subsequent reproductive decision-making. Reproductive dilemmas arise around whether or

---

[7] This may or may not be their first child. It is only when a child inherits an *affected* copy of the gene from *each* parent that this occurs. Otherwise they are either unaffected or symptomless carriers.

not to have *further* children who have a one in four chance of inheriting the disorder (Snowdon and Green, 1994, 1997). As was suggested at the beginning of this chapter, these scenarios differ from those associated with *late onset dominant* disorders, such as HD. A crucial difference is that those who inherit one mutated copy of a dominant gene are likely to *develop* the associated disorder. Thus, unlike carriers of recessive disorders, they face risks in connection with their own health or premature mortality. This can impact on their ability to sustain a mothering or fathering role.

Gender considerations arise in relation to another group of disorders - *X-linked* ones - which are passed on by females but mostly affect males.[8] The risk scenario that emerges for these conditions is that having males is perceived as more risky than having females (Parsons and Clarke, 1993; Kay, E. & Kingston H. (2002).

How people frame reproductive decision-making in relation to one late-onset disorder such as HD will not necessarily correspond to other late onset disorders.

Table 4. Drawing on Slovic's (1987) factors to predict the impact of genetic testing & treatments for HD on perceptions of genetic risk

Factor	Genetic risk pre-testing	Genetic risk post-testing
Dread risk	Uncontrollable	Controllable
	Dread	Less dread
	Involuntary	Voluntary
Unknown risk	Not observable	Observable
	Impact delayed	Impact predictable
Number of people	High risk for future generations	Avoidable risk for future generations

The likelihood of the parent becoming affected depends on the degree of penetrance shown by the particular mutation.[9] Some, such as that associated with HD show almost 100% penetrance whilst others, such as the BRAC1 gene mutation, associated with some inherited breast and ovarian cancers, and that associated with early onset Alzheimer's disease, show a much lower rate. Clarifying risks for these diseases is further complicated in that they also arise sporadically, creating a general population risk for those with no family history. This means that those shown not to carry genes for familial breast, ovarian and bowel cancers can still develop these disorders but those found to carry the genes may not. Other differences exist between these cancers and HD. As mentioned above, those who will develop HD cannot yet benefit from the surveillance measures or treatments which exist for cancer sufferers, and they are likely to develop psychiatric as well as physical symptoms. The first comparative study of the psychological effects of predictive testing of neurodegenerative and cancer syndrome late-onset autosomal dominantly inherited genetic disorders lend support to this in that the groups at-risk for neurodegenerative disorders showed more distress than those at-risk

---

[8] Fragile X primarily affects males but one third of carrier females show some degree of intellectual impairment (Davies, 1989).
[9] Penetrance refers to the likelihood of the disorder manifesting.

for the cancer syndromes (Dudok de Wit, 1997).[10] Families with neurodegenerative disorders were also more focused upon support for the individual at-risk and partner whilst in families with cancer the focus was upon the post-test treatment options and their outcome (Dudok de Wit et al, 2000).

It may well be that findings in relation to decision-making and HD could have as much, if not more, to contribute to debates about reproductive decision-making in non-genetic situations that have aroused public concern about ability to sustain a parenting role. These include mothers and fathers with learning difficulties (Booth and Booth, 1994), parents who are HIV positive (Bor and Elford, 1998), teenage single mothers (Rains, Davies and McKinnon, 1998) parents who are lesbian and gay (Dunne, 2000), who face deafness (Gregory, 1991), and the experiences of women wishing to become impregnated under unconventional circumstances such as being post-menopausal or wanting to utilize sperm from deceased partners.

Acknowledgments

Thanks are due to the participating families for entrusting me with their thoughts and experiences some of which are touched on here; to the Medical Research Council for funding the study under their GAHH initiative and to my supervisors, Jo Green and Martin Richards, for their encouragement and guidance. The Huntington's Disease Association and Addenbrookes Regional Clinical Genetics Centre shared their expertise with me and helped recruit participants to the study. I am also grateful to Martin Richards, Nina Hallowell, Helen Statham, Robert Osborne and Helena Willen for their helpful comments on subsequent drafts of this chapter. I am currently funded by The Wellcome Trust (Award Reference: 065207, 2002-2005) under their biomedical ethics programme.

References

Adam, S., Wiggins, S., Whyte, P., Bloch, M., Shokeir, M. H. K., Soltan, H., et al. (1993). Five year study of prenatal testing for Huntington's disease: demand, attitudes, and psychological assessment. *Journal of Medical Genetics, 30*, 549-56.

Arkes, H. R. & Blumer, C. (1985). The psychology of sunk costs. *Organizational Behaviour and Human Decision Processes, 35*, 124-40.

Barette, J. & Marsden, C. D. (1979). Attitudes of families to some aspects of Huntington's chorea. *Psychological Medicine, 9*, 327-36.

Bartlett, J. (1994). *Will you be mother? Women who choose to say no.* London: Virago.

Beach, L. R. (1990). *Image theory: Decision making in personal and organisational contexts.* Chichester: Wiley.

Beach, L. R. (1993a). *Making the right decision: organizational culture, vision, and planning.* Englewood Cliffs, NJ: Prentice Hall.

---

[10] Risk perception was also a function of experience of the particular disorder and the perceived impact it had had on the testee which meant that individual differences in psychological distress between testees at-risk for a particular disorder were larger than the similarities.

Beach, L. R. (1993b). Image theory: Personal and organizational decisions. In G. A. Klein, J. Orasanu, R. Calderwood, C. E. Zsambok (eds.), *Decision making in action: models and methods*, 148-57. Norwood, NJ: Ablex.

Beach, L. R. (1996). *Decision making in the workplace: a unified perspective*. Mahwah, NJ: Lawrence Erlbaum Associates.

Beach, L. R., Townes, B. D., Campbell, F. L. & Keating, G. W. (1976). Developing and testing a decision aid for birth planning decisions. *Organizational Behaviour and Human Performance, 15,* 99-116.

Beach, L. R. & Lipshitz, R. (1993). Why classical decision theory is an inappropriate standard for evaluating and aiding most human decision making. In G. A. Klein, J. Orasanu, R. Calderwood & C.E. Zsambok (eds.), *Decision making in action: models and methods*, 21-35. Norwood, NJ: Ablex Publishing Corporation.

Beattie, J., Baron, J., Hershey, J. C. & Spranca, M. D. (1994). Psychological determinants of decision attitude. *Journal of Behavioural Decision Making, 7,* 129-44.

Beck, U. & Beck-Gernsheim, E. (1995). *The normal chaos of love*. Cambridge: Polity Press.

Becker, M. H. (1974). The Health Belief Model and sick role behaviour. *Health Education Monographs, 2,* 409-19.

Bekker, H., Thornton, J., Airey, M., Connelly, J., Hewison, J., Lilleyman, et al (1999). Informed decision making: an annotated bibliography and systematic review. *UK: Health Technology Assessment*, No. *3,* National Health Service, Research & Development.

Bloch, M. & Hayden, M. R. (1990). Opinion: predictive testing for Huntington disease in childhood: challenges and implications. *American Journal of Human Genetics, 46,* 1-4.

Booth, T. & Booth, W. (1994). *Parenting under pressure: mothers and fathers with learning difficulties*. Buckingham: Open University Press.

Bor, R. & Elford, J. (1998). *The family and HIV today: recent research and practice*. London: Cassell.

Bostrom, A., Fischhoff, B. & Granger Morgan, M. (1992). Characterizing mental models of hazardous processes: a methodology and application to radon. *Journal of Social Issues, 48,* 85-100.

Boulton, M. & Fitzpatrick, R. (1997). Evaluating qualitative research. *Evidence Based Health Policy and Management, 1,* 83-5.

Broadstock, M. & Michie, S. (2000). Processes of decision making: theoretical and methodological issues. *Psychology and Health, 15,* 191-204.

Bruner, J. (1986). *Actual minds, possible worlds*. Cambridge, MA: Harvard University Press.

Bruner, J. (1990). *Acts of meaning*. Cambridge, MA: Harvard University Press.

Busfield, J. & Paddon, M. (1977). *Thinking about children: sociology and fertility in post-war England*. Cambridge: Cambridge University Press.

Callahan, S. (1979). An ethical analysis of responsible parenthood. In A. M. Capron, M. Lappe, & R. F. Murray (eds.), *Genetic counseling: facts, values and norms. Birth Defects: Original Articles Series*, XV, 217-38. New York: Alan R. Liss Inc.

Campbell, E. (1985). *The childless marriage: an exploratory study of couples who do not want children*. London: Tavistock Publications.

Cannon-Bowers, J. A., Salas, E. & Pruitt, J. S. (1996). Establishing the boundaries of a paradigm for decision-making research. *Human Factors, 38,* 193-205.

Carter, C. O. & Evans, K. (1979). Counselling and Huntington's chorea. *Lancet*, ii, 470-2.

Conrad, P. & Gabe, J. (1999). (eds) *Sociological Perspectives on the New Genetics*. Oxford & Malden, Ma.: Blackwell Publishers Ltd.

Cox, S.M. (Ed.) (2002). *Personal Perspectives on Genetic Testing for Huntington Disease*. Cambridge, Ontario: Huntington Society of Canada.

Cox, S. M. (2003). Stories in Decisions: How At Risk Individuals Decide to Request Predictive Testing for Huntington Disease. *Qualitative Sociology, 26,* 257-280.

Creswell, J. W. (1998). *Qualitative inquiry and research design: choosing among five traditions.* Thousand Oaks: Sage Publications Inc.

Currie, D. (1988). Re-thinking what we do and how we do it: a study of reproductive decision making. *Review of Canadian Sociology and Anthropology, 25,* 231-53.

Dallos, R. (1997). *Interacting stories: narratives, family beliefs and therapy.* London: Karnac.

Davenport, C. B. & Muncey, E. B. (1916). *Huntington's Chorea in relation to heredity and eugenics.* Eugenics Record Office, Bulletin No.7, 195-220.

Decruyenaere, M., Evers-Kiebooms, G., Boogaerts, A., Cassiman, J. J., Cloostermans, T., Demyttenaere, K., et al (1996). Prediction of psychological functioning one year after the predictive test for Huntington's disease and impact of the test result on reproductive decision making. *Journal of Medical Genetics, 33,* 737-43.

Douglas, M. (1985). *Risk acceptability according to the social sciences.* New York: Russell Sage Foundation.

Downing, C. (2002). *Reproductive decision-making in families at risk for Huntington's disease: perceptions of responsibility.* An unpublished dissertation submitted to the University of Cambridge for the Degree of Doctor of Philosophy

Downing, C. (2005). Negotiating responsibility: case studies of reproductive decision-making and prenatal genetic testing in families facing Huntington disease. *Journal of Genetic Counseling, Vol. 14, 3,* 219-234.

Dowrick, S. & Grundberg, S. (1980). *Why children?* London: The Women's Press.

Drake, H., Reid, M. & Marteau, T. (1996). Attitudes towards termination for fetal abnormality: comparisons in three European countries. *Clinical Genetics, 49,* 134-40.

Dudok de Wit, A. C. (1997). To know or not to know: the psychological implications of presymptomatic DNA testing for autosomal dominant inheritable late onset genetic disorders. PhD thesis, Rotterdam, Erasmus University of Rotterdam.

Dunne, G. A. (2000). Opting into motherhood: lesbians blurring the boundaries and transforming the meaning of parenthood and kinship. *The Journal of Gender and Society, 14,* 11-35.

Erikson, E. H. (1963). *Childhood and society.* (2nd ed.). New York: Norton.

Evers-Kiebooms, G., Swerts, A. & van den Berghe, H. (1990). Partners of Huntington patients: implications of the disease and opinions about predictive testing and prenatal diagnosis. *Genetic Counseling, 39,* 151-9.

Evers-Kiebooms, G., Denayer, L., Decruyenaere, M. &Van den Berghe, H. (1993). Community attitudes towards prenatal testing for congenital handicap. *Journal of Reproductive and Infant Psychology, 11,* 21-30.

Festinger, L. (1957). *A theory of cognitive dissonance,* Evanston, IL: Row Peterson.

Finch, J. and Mason, J. (1993). *Negotiating family responsibilities.* London: Tavistock Routledge.

Fletcher, J. F. (1980). Knowledge, risk, and the right to reproduce: a limiting principle. In A. Milunsky & G.J. Annas (eds.), *Genetics and the law II,* 131-5. New York: Plenum Press.

Fox, G.L. (1982). *The childbearing decision: fertility attitudes and behavior.* Beverley Hills: Sage Publications Inc.

Gifford, S. M. (1986). The meaning of lumps: a case study of the ambiguities of risk. In C. R. Janes, R. Stall & S. M. Gifford (eds.), *Anthropology and Epidemiology: interdisciplinary approaches to the study of health and disease,* 213-46. Boston: Reidel.

Gladwin, C. H. (1989). *Ethnographic decision tree modeling.* Newbury Park: Sage Publications.

Glaser, B. G. & Strauss, A. L. (1966). *Awareness of dying.* London: Weidenfeld and Nicolson.

Green, J. (1990). *Calming or harming? A critical review of psychological effects of fetal diagnosis on pregnant women.* London: The Galton Institute.

Green, J. M., Statham, H. & Snowdon, C. (1992). Screening for fetal abnormalities: attitudes and experiences. In T. Chard & M. P. M. Richards (eds.), *Obstetrics in the 1990's: current controversies,* 65-89. London: McKeith Press.

Green, J. M., Statham, H. & Snowdon, C. (1994). *Pregnancy: A testing time.* Report of the Cambridge Prenatal Screening Study. Centre for Family Research, University of Cambridge.

Gregory, S (1991). Challenging motherhood: mothers and their deaf children. In A. Phoenix, A. Woollett, & E. Lloyd (eds.), *Motherhood: meanings, practices and ideologies,* 123-42. London: Sage Publications Ltd.

Hallowell, N. (1999). Doing the right thing: genetic risk and responsibility. *Sociology of Health and Illness, 21,* 597-621.

Hamburg, K., Johansson, E., Lindgren, G. & Westman, G. (1994). Scientific rigour in qualitative research - examples from a study of women's health in family practice. *Family Practice, 11,* 176-81.

Hans, M. B & Koeppen, A. H. (1980). Huntington's chorea: its impact on the spouse. *The Journal of Mental and Nervous Disease, 168,* 209-14.

Harper, P. (1996a). The epidemiology of Huntington's disease. In P. S. Harper (ed.), *Huntington's disease.* (2nd edn.). 201-39. London: W.B. Saunders Company Ltd.

Harper, P. (1996b). The natural history of Huntington's disease. In P. S. Harper (ed.), *Huntington's disease.* 2nd edn. 123-136. London: W.B. Saunders Company Ltd.

Harper, P. S., Walker, D. A., Tyler, A., Newcombe, R. O. & Davies, K. (1979). Huntington's chorea: the basis for long term prevention. *Lancet, 8138,* 346-9.

Harper. P.S., Lim, C. & Craufurd, D. (2000). Ten years of presymptomatic testing for Huntington's disease: the experience of the UK Huntington's Disease Prediction Consortium. *Journal of Medical Genetics, 37,* 567-71.

Hayden, M. (2000). Predictive testing for Huntington's disease: the calm after the storm. *Lancet, 356,* 1944-5.

Hayes, C. V. (1992). Genetic testing for Huntington's disease - a family issue. *The New England Journal of Medicine, 327,* 1449-51.

Heider, F. (1946). Attitudes and cognitive organization. *Journal of Psychology, 21,* 107-12.

Heider, F. (1958). *The psychology of interpersonal relations.* New York: John Wiley.

Hill, S. A. (1994). Motherhood and the obfuscation of medical knowledge: the case of sickle cell disease. *Gender and Society, 8,* 29-47.

Huys, J., Evers-Kiebooms, G. & d'Ydewalle, G. (1992). Decision making in the context of genetic risk: the use of scenarios. *Birth Defects:Original Article Series, 28,* 17-20.

Johnson, B. T. & Eagly, A. H. (1989). Effects of involvement on persuasion: a meta-analysis. *Psychological Bulletin, 106,* 290-314.

Jungermann, H. (1997). Advice giving and taking. Paper presented at the 16th bi-annual Conference on Subjective Probability Utility and Decision Making (SPUDM), Leeds.

Kahneman, D. (1991). Judgment and decision making: a personal view. *Psychological Science, 2,* 142-45.

Kahneman, D. & Tversky, A. (1979). Prospect theory: an analysis of decision under risk. *Econometrica, 49,* 263-91.

Kahneman, D., Slovic, P. & Tversky, A. (1982). *Judgement under uncertainty: Heuristics and biases.* Cambridge: Cambridge University Press.

Kahneman, D. & Tversky, A. (1984). Choices, values and frames. *American Psychologist, 39,* 341-50.

Karlsson, G. (1987). A phenomenological psychological method: theoretical foundations and empirical applications in the field of decision making and choice. PhD thesis, Stockholm, University of Stockholm.

Kates, R. W. & Kasperson, J. X. (1983). Comparative risk analysis of technological hazards. *Proceedings of the National Academy of Science, U.S.A., 80,* 7027-38.

Kay, E. & Kingston H. (2002). Feelings associated with being a carrier and characteristics of reproductive decision making in women known to be carriers of X-linked conditions. *Journal of Health Psychology, 7* 169-81.

Kelly, G. A. (1955). *The psychology of personal constructs.* New York: Norton.

Kemdal, A. B. & Montgomery, H. (1997). Perspectives and emotions in personal decision making. In R. Ranyard, W. R. Crozier and O. Svenson (eds.),. *Decision making: cognitive models and explanations,* 72-89. London: Routledge.

Kenen, R. (1994). The Human Genome Project: creator of the potentially sick, potentially vulnerable and potentially stigmatized? In L. Robinson (ed.), *Life and death under high technology medicine,* 49-64. Manchester: Manchester University Press.

Kenen, R. H. (1996). The at-risk health status and technology: a diagnostic invitation and the 'gift' of knowing. *Social Science and Medicine, 42,* 1545-53.

Kessler, S. (1988). Invited essay on the psychological aspects of genetic counseling: V. Preselection: a family coping strategy in Huntington disease. *American Journal of Medical Genetics, 31,* 617-21.

Kessler, S. (1989). Psychological aspects of genetic counseling. VI. A critical review of the literature dealing with education and reproduction. *American Journal of Medical Genetics, 34,* 340-53.

Kiernan, K., E. (1989). Who remains childless? *Journal of Biosocial Science, 21,* 387-98.

Kitcher, P. (1996). *The lives to come: the genetic revolution and human possibilities.* London: Penguin Books.

Klein, G. (1997). Naturalistic decision making: where are we going? In C. E. Zsambok & G. Klein (eds.)., *Naturalistic decision making,* 383-97. Mahwah, NJ: Lawrence Erlbaum Associates.

Klein, G. (1999). *Sources of power: how people make decisions.* Cambridge, MA: The M.I.T. Press.

Klein, G. A., Orasanu, J., Calderwood, R. & Zsambok, C. E. (eds.) (1993). *Decision making in action: models and methods.* Norwood, NJ: Ablex Publishing Corporation.

Korer, J. and Fitzsimmons, J. S. (1985). The effect of Huntington's chorea on family life. *British Journal of Social Work, 15,* 581-97.

Kraus, N. N. and Slovic, P. (1988). Taxonomic analysis of perceived risk: modelling individual and group perceptions within homogenous hazard domains. *Risk Analysis, 8,* 435-55.

Krimsky, S. and Golding, D. (eds.) (1992). *Social theories of risk.* Westport, CT: Praeger.

Lamm, H. and Myers, D. G. (1978). Group induced polarization of attitudes and behavior. In L. Berkovitz (ed.), *Advances in experimental social psychology, 11,* 145-95. New York: Academic Press.

Lee, R. M. (1993). *Doing research on sensitive topics.* London: Sage Publications Ltd.

Leventhal, H., Meyer, D., and Nerenz, D. (1980). The common-sense representation of illness danger. *Medical Psychology, 2,* 7-30.

Leventhal, H., Nerenz, D. R. & Steele, D. J. (1994). Illness representations and coping with health threats. In A. Baum, S. E. Taylor & J. E. Singer (eds.), *Handbook of psychology and health,* Vol. IV: Social aspects of health. 210-52. Hillsdale, NJ: Lawrence Erlbaum Associates.

Leventhal, H., Benyamini, Y., Brownlee, S., et al (1997). Illness representations. In K.J. Petrie & J.A. Weinman, *Perceptions of health and illness,* 19-45. Amsterdam: Harwood Academic Publishers.

Lippman, A. (1992). Mother matters: a fresh look at prenatal genetic testing. *Issues in Reproductive and Genetic Engineering, 5,* 141-54.

Lipshitz, R. (1993). Converging themes in the study of decision making in realistic settings. In G. A. Klein, J. Orasanu, R. Calderwood & C. Zsambok (eds.), *Decision making in action: models and methods.* 103-37. Norwood, NJ: Ablex Publishing Corporation..

Lopes, L. L. (1987). Between hope and fear: the psychology of risk. *Advances in Experimental Social Psychology, 20,* 255-95.

Maat-Kievit, A., Vegter-van der Vlis, M., Zoeteweij, M., Losekoot, M., van Haeringen, A., Kanhai, H. et al. (1999). Experience in prenatal testing for Huntington's disease in the The Netherlands: Procedures, results and guidelines (1987-1997). *Prenatal Diagnosis, 19,* 450-57.

Mansfield, P. & Collard, J. (1988). *The beginning of the rest of your life: a portrait of newly-wed marriage.* London: The Macmillan Press Ltd.

Marshall, S. E. (1995). Choosing the family. In C. Ulanowsky (ed.), *The family in the age of biotechnology,* 105-117. Aldershot: Avebury.

Marteau, T. M., Drake, H. & Bobrow, M. (1994). Counselling after diagnosis of fetal abnormality: the differing approaches of obstetricians, clinical geneticists and genetic nurses, *Journal of Medical Genetics, 31,* 865-67.

Marteau, T. & Anionwu, E. (1996). Evaluating carrier testing: objectives and outcomes. In T.Marteau and M. Richards, (eds.), *The troubled helix: social and psychological implications of the new human genetics,* 123-29. Cambridge: Cambridge University Press.

Mason, M. C. (1993). *Male infertility: men talking.* London: Routledge.

Mastromauro, C. A., Myers, R. H. & Berkman, B. (1987). Attitudes towards presymptomatic testing in Huntington's disease. *American Journal of Medical Genetics, 26,* 271-82.

Mays, N. & Pope, C. (1995). Rigour and qualitative research. *British Medical Journal, 311,* 109-12.

McAllister, F. and Clarke, L. (1996). *Choosing childlessness.* London: Family Policy Studies Centre.

McAllister, M. (1999). Predictive testing for Hereditary Non-Polyposis Colorectal Cancer (HNPCC): a theory of engagement. Unpublished PhD thesis. Centre for Family Research, University of Cambridge, Cambridge.

Mennie, M. E., Compton, M. E., Gilfillan, A., Liston, W. A., Pullen, I., Whyte, et al (1993). Prenatal screening for cystic fibrosis: psychological effects on carriers and their partners. *Journal of Medical Genetics, 30,* 543-48.

Miller, S. M. (1987). Monitoring and blunting: validation of a questionnaire to assess styles of information seeking under threat. *Journal of Personality and Social Psychology, 52,* 345-53.

Montgomery, H. (1993). The search for a dominance structure in decision making: examining the evidence. In G. A. Klein, J. Orasanu, R. Calderwood, C. E. Zsambok (eds.), *Decision making in action: models and methods,* 182-87. Norwood, NJ: Ablex Publishing Corporation.

Montgomery, H. (1999). NDM: the challenge to JDM. Book review of *Naturalistic Decision Making* (1997), C. E. Zsambok & G. Klein (eds.). Mahwah, NJ: Lawrence Erlbaum Associates. In *Journal of Behavioural Decision Making, 12,* 340-42.

Montgomery, H. & Svenson, O. (eds.) (1989). *Process and structure in human decision making.* Chichester, NY: John Wiley & Sons Ltd.

Morell, C. M. (1994). *Unwomanly conduct: the challenges of intentional childlessness.* New York: Routledge.

Nelson, H. L. and Nelson, J. L. (1995). *The patient in the family: an ethics of medicine and family.* London: Routledge.

Orasanu, J. & Connolly, T. (1993). The reinvention of decision making. In G. A. Klein, J. Orasanu, R. Calderwood, C. E. Zsambok (eds.)., *Decision making in action: models and methods,* 3-20. Norwood, NJ: Ablex Publishing Corporation.

Palmlund, I. (1992) Social drama and risk evaluation. In S. Krimsky and D. Golding (eds.), *Social theories of risk,* 197-212. Westport, CT: Praeger.

Parsons, E. & Clarke, A. J. (1993) Genetic risk: women's understanding of carrier risks in Duchenne muscular dystrophy. *Journal of Medical Genetics, 30,* 562-66.

Payne, J. (1978). Talking about children: an examination of accounts about reproduction and family life. *Journal of Biosocial Science, 10,* 367-74.

Payne, J. W., Bettman, J. R. & Johnson, E. J. (1993). *The adaptive decision maker.* Cambridge: Cambridge University Press.

Pfeffer, N. & Woollett, A. (1983). *The experience of infertility.* London: Virago.

Proctor, H. (1981). Family construct psychology. In S. Walrond-Skinner (ed.), *Family therapy and approaches.* London: Routledge & Kegan Paul.

Quaid, K. & Wesson, M. K, (1995). Exploration of the effects of predictive testing for Huntington disease on intimate relationships. *American Journal of Medical Genetics, 45,* 46-51

Rains, P., Davies, L. & McKinnon, M. (1998). Taking responsibility: an insider view of teen motherhood. *Families in Society: The Journal of Contemporary Human Services, 79,* 308-19.

Read, J. (1995). *Counselling for fertility problems.* London: Sage Publications Ltd.

Reid, M. & Armstrong, D. (1998). Guidelines on evaluating qualitative research proposals in health services research. Unpublished guidelines prepared for the Cancer Research Council, Educational and Psychosocial Research Committee, London.

Richards, L. (1978). *Having families: marriage, parenthood and social pressure in Australia.* Harmondsworth: Penguin Books Ltd.

Richards, M.P.M. (1993). The new genetics: some issues for social scientists. *Sociology of Health and Illness, 15,* 567-86.

Richards, M. P. M. & Green, J. M. (1993). Attitudes toward prenatal screening for fetal abnormality and detection of carriers of genetic disease: a discussion paper. *Journal of Reproductive and Infant Psychology, 11,* 49-56.

Richards, T. J. & Richards, L. (1994). Using computers in qualitative research. In N.K. Denzin and Y.S. Lincoln (eds) *Handbook of qualitative research,* 445-62. Thousand Oaks: Sage Publications.

Robertson, A. (2000). Embodying risk, embodying political responsibility: women's accounts of risks for breast cancer. *Health, Risk and Society, 2,* 219-35.

Robinson, G. E., Carr, M. L., Olmsted, M. P. & Wright, C. (1991). Psychological reactions to pregnancy loss after prenatal diagnostic testing: preliminary results. *Journal of Psychosomatic Obstetrics and Gynecology 12,* 181-92.

Rogers, W. (1994). Changing health-related attitudes and behaviour: the role of preventative health psychology. In J. J. Harvey, E. Maddox, R. F. McGlynn, R. F. & C. D. Stoltenberg (eds.), *Social perception in clinical and counseling psychology,* Vol. 2, pp, 91-112. Lubbock, TX: Texas Technical University Press.

Roos, R. A., Vegter van der Vlis, M., Tibben, A., Skraastad, M. I. & Pearson, P.L. (1990). [Procedure and initial results of presymptomatic DNA studies in Huntington's Chorea]. *Ned. Tijdschr. Geneeskd., 134,* 704-7.

Rothman, B. K. (1986). *The tentative pregnancy: prenatal diagnosis and the future of motherhood.* New York: Viking.

Rothman, B. K. (1998). *Genetic maps and human imaginations. The limits of science in understanding who we are.* New York: W. W. Norton & Company.

Rotter, J. B. (1966). Generalized expectancies for internal versus external control of reinforcement. *Psychological Monographs, 609,* (whole).

Sandelowski, M. & Jones, L. C. (1996). 'Healing fictions': stories of choosing in the aftermath of the detection of fetal abnormalities. *Social Science and Medicine, 42,* 353-61.

Seale, C. (1999). *The Quality of Qualitative Research.* London: Sage Publications Ltd.

Shakespeare, J. (1992). Communication in Huntington's disease families. Unpublished paper presented at the Third European Meeting on Psychosocial Aspects of Genetics, Nottingham, University of Nottingham.

Shiloh, S. (1996). Decision-making in the context of genetic risk. In T. Marteau & M. Richards (eds.), *The troubled helix: social and psychological implications of the new human genetics,* 82-103. Cambridge: Cambridge University Press.

Simon, H. A. (1955). Behavioural Model of Rational Choice. *Quarterly Journal of Economics, 69,* 99-108.

Simpson, S. A. & Harding, A. E. (1993). Predictive testing for Huntington's disease: after the gene. *Journal of Medical Genetics, 30,* 1036-38.

Simpson, S.A. and Harper, P.S. (2001). Prenatal testing for Huntington's disease: experience within the UK 1994-1998. Letter to the Editor. *Journal of Medical Genetics, 38,* 333-35.

Sloan, T. S. (1986). *Deciding: Self-deception in life choices.* New York: Methuen.

Slovic, P. (1987). Perception of risk. *Science, 236,* 280-85.

Slovic, P. (1992). Perception of risk: reflections on the psychometric paradigm. In S. Krimsky & D. Golding (eds.), *Social theories of risk,* 117-52. Westport, CT: Praeger.

Slovic, P., Lichstenstein, S. & Fischhoff, B. (1984). Modeling the societal impact of fatal accidents. *Management Science, 30,* 464-74.

Slovic, P., MacGregor, D. & Kraus, N. N. (1987). Perception of risk from automobile safety defects. *Accident Analysis and Prevention, 19,* 359-73.

Slovic, P., Kraus, N. N., Lappe, H., Letzel, H. & Malmfors, T. (1989). Risk perception of prescription drugs: report on a survey in Sweden. *Pharmaceutical Medicine, 4,* 43-65.

Smith, J.A., Stephenson, M. & Quarrel, O. (1999). Factors influencing the decision whether to take the genetic test for Huntington's disease: an interpretative phenomenological analysis. Unpublished paper presented at the Qualitative Methods and Genetics Group, Centre for Family Research, University of Cambridge.

Sobel, S. K. & Cowan, D. B. (2000). Impact of genetic testing for Huntington's disease on the family system. *American Journal of Medical Genetics, 90,* 49-59.

Song, M. (1998). Hearing competing voices: sibling research. In J. Ribbens and R. Edwards (eds.), *Feminist dilemmas in qualitative research,* 103-18. London: Sage Publications Ltd.

Strathern, M. (1995). New families for old? In C. E. Ulanowsky (ed.), *The family in the age of biotechnology,* 27-45. Aldershot: Avebury.

Strauss, A. and Corbin, J. (1998). *Basics of qualitative research: techniques and procedures for developing grounded theory.* (2nd edn.) Thousand Oaks: Sage Publications Inc.

Svenson, O. (1991). Differentiation and consolidation theory of human decision making. Paper presented at the 13th SPUDM Conference, Fribourg, Switzerland.

Svenson, O. (1996). Decision making and the search for fundamental psychological regularities: what can be learnt from a process perspective? *Organizational Behaviour and Human Decision Processes, 65,* 252-67.

Svenson, O. and Maule, A. J. (eds.) (1993). *Time pressure and stress in human judgement and decision making.* New York: Plenum Press.

Taylor, S.D. (2004). Predictive genetic test decisions for Huntington's Disease: context, appraisal and new moral imperatives. *Social Science and Medicine, 58,* 137- 149.

Teigen and Brun (1997). Anticipating the future. In R. Ranyard, W. R. Crozier & O. Svenson (eds.), *Decision making: cognitive models and explanations,* 112-127. London: Routledge.

Thom, D. and Jennings, M. (1996). Human pedigree and the 'best stock': from eugenics to genetics. In T. Marteau & M. Richards (eds.), *The troubled helix: social and psychological implications of the new human genetics,* 211-234. Cambridge: Cambridge University Press.

Tibben, A., Frets, P. G., van de Kamp, J., J. P., Niermeijer, M. F., Vegter-van de Vlis, M., Roos, R. A. C., et al. (1993b). On attitudes and appreciation, 6 months after predictive DNA testing for Huntington disease in the Dutch program. *American Journal of Medical Genetics, 48,* 103-11.

Tolmie, J. L., Davidson, H. R., May, H. M., McIntosh, K., Paterson, J. S. & Smith, B. (1995). The prenatal exclusion test for Huntington's disease: experience in the west of Scotland. *Journal of Medical Genetics, 32,* 97-101.

Tversky, A. & Kahneman, D. (1973). Availability: a heuristic for judging frequency and probability. *Cognitive Psychology, 5,* 207-32.

Tversky, A. and Shafir, E. (1992). Choice under conflict: the dynamics of deferred decision. *Psychological Science, 3,* 358-61.

Tymstra, T. J. (1989). The imperative character of medical technology and the meaning of "anticipated decision regret". *International Journal of Technology Assessment in Health Care, 5,* 207-13.

Van der Steenstraten, I.M., Tibben, A., Roos, R.A.C., van de Kamp, J.J.P. and Niermeijer, M.R. (1994). Predictive testing for Huntington disease: nonparticipants compared with participants in the Dutch program. *American Journal of Human Genetics, 55,* 618-25.

Verplanken, B. & Svenson, O. (1997). Personal involvement in human decision making: conceptualisations and effects on decision processes. In R. Ranyard, W. R. Crozier & O. Svenson (eds.), *Decision making:Cognitive models and explanations,* 40-57. London: Routledge.

Wallston, K. A. (1992). Hocus-pocus, the focus isn't strictly on locus: Rotter's social learning theory modified for health. *Cognitive Therapy and Research, 16,* 183-99.

Wallston, K. A. (1997). Perceived control and health behaviour. In A. Baum, S. Newman, J. Weinman, R. West & C. McManus (eds.), *Cambridge Handbook of psychology, health and medicine,* 151-54. Cambridge: Cambridge University Press.

Wallston, K. A., Wallston, B. S. & DeVellis, R. (1978). Development of the Multidimensional Health Locus of Control (MHLC) scale. *Health Education Monographs, 6,* 161-70.

Wartofsky, M. W. (1982). Medical knowledge as a social product: rights, risks, and responsibilities. In W. B. Bondeson, H. T. Englehardt, S. F. Spicker, & J. M.White (eds.), *New knowledge in the biomedical sciences: some moral implications of its acquisition, possession and use,* 113-30. Dordrecht: D. Reidel Publishing Company.

Wertz, D. C., Sorenson, J. R., and Heeren, T. C. (1984). Genetic counseling and reproductive uncertainty. *American Journal of Medical Genetics, 18,* 79-88.

Wexler, N. S. (1979). Russian roulette: the experience of being at risk for Huntington's disease: harbinger of the new genetics. In S. Kessler (ed.), *Genetic counseling: psychological dimensions,* 199-220. New York: Academic Press.

Wexler, N. (1980). "Will the circle be unbroken?" Sterilizing the genetically impaired. In A. Milunsky & G. J. Annes (eds.), *Genetics and the law II,* 313-329. New York: Plenum Press.

Wexler, N. S. (1992). Clairvoyance and caution: repercussions from the Human Genome Project. In D. J. Kevles & L. Hood (eds.), *The code of codes: scientific and social issues in the Human Genome Project,* 211-43. Cambridge, MA: Harvard University Press.

Willen, H. (1994). The childbearing decision: motivational aspects, process characteristics and consequences of the decision. Ph.D.thesis, Department of Psychology, Goteborg University. *Goteborg Psychological Reports, 24,* No.1.

Willen, H. & Montgomery, H. (2000). Creating a realisable option on the ruins of a wrecked marriage. In C. M. Allwood & M. Selart (eds.), *Decision-making: social and creative dimensions.* Dordrecht: Kluwer Academic Publishers.

Yow, V. R. (1994). *Recording oral history: a practical guide for social scientists.* Thousand Oaks: Sage Publications Inc.

Zsambok, C. E. and Klein, G. (eds.) (1997). *Naturalistic decision making.* Mahwah, NJ: Lawrence Erlbaum Associates.

In: Encyclopedia of Neuroscience Research
Editors: Eileen J. Sampson and Donald R. Glevins

ISBN 978-1-61324-861-4
© 2012 Nova Science Publishers, Inc.

Chapter XXXIX

The Control of Adult Neurogenesis by the Microenvironment and How This May be Altered in Huntington's Disease

Wendy Phillips[*,1,2] *and Roger A. Barker*[1,2]

[1] Cambridge Centre for Brain Repair, E.D. Adrian Building, Forvie Site, Cambridge, England, CB2 2PY

[2] Department of Neurology, Addenbrooke's Hospital, Hills Road, Cambridge, England, CB2 2QQ

Abstract

Neurogenesis is the processes whereby newborn neurons are formed and occurs in mammals in adulthood in specialised areas, known as 'neurogenic niches'. The subventricular zone, and subgranular zone of the dentate gyrus of the hippocampus are such specialised neurogenic niches and newborn neurons formed here contribute to learning and memory, but neurogenesis may occur elsewhere in the brain to a more limited extent. The neurogenic niche is composed of specialised glial cells, basal lamina, ependymal cells, neurotransmitter complement and vasculature. Neurogenesis is a complex process, involving many different cell types, and several stages of neuronal maturation; and each component may be affected by many different microenvironmental perturbations. External perturbations, like seizures and lesions alter the microenvironment and in turn, alter neurogenesis. Chronic disease can also affect neurogenesis and the inherited neurodegenerative condition Huntington's disease may do so through an alteration of the microenvironment. We will use the example of Huntington's disease to explore how changes in the microenvironment might impact on neurogenesis and, thus identify potential therapeutic targets.

---

[*] Corresponding author: Dr. Wendy Phillips, Cambridge Centre for Brain Repair, E.D. Adrian Building, Forvie Site, Cambridge, CB2 2PY. E-mail: wp212@cam.ac.uk; Tel: +44 1223 331160; Fax: +44 1223 331174.

1. Neurogenesis

1.1. Introduction

Neurogenesis is the formation of new neurons, from stem cells (which are self-renewing, multipotent and have longevity) or progenitor cells (which have a more limited potential) (Weiss *et al.* 1996). Neurogenesis in the adult mammalian brain (Altman and Das 1965;Kaplan 1981) occurs in specialised 'neurogenic niches': the subventricular zone (SVZ) and subgranular zone (SGZ) of the dentate gyrus (DG) of the hippocampus (Gage 2000). These areas are said to undergo 'constitutive neurogenesis', i.e. neurogenesis occurs at all times under basal conditions, whereas other brain areas may undergo 'non-constitutive neurogenesis' when the microenvironment is altered by lesions and seizures, for example (Parent 2003). This is contentious, with some authors claiming the existence of constitutive neurogenesis in the non-classical areas. Furthermore, neuronal precursor cells (NPCs) (a mixed or unknown population containing stem and progenitor cells) can be isolated from areas throughout the brain *in vitro* (Palmer *et al.* 1999;Weiss *et al.* 1996;Tropepe *et al.* 2000;Lie *et al.* 2002).

The distinction between constitutive and non-constitutive neurogenesis is an important one. An area of constitutive neurogenesis implies that it is functionally important for the animal under normal circumstances, and the existence of a specialised microenvironment, which allows neurogenesis to continue into adulthood. The function of neuronal precursors that exist in areas of non-constitutive neurogenesis is unknown. The precursors may be quiescent, produce neurons constitutively at a difficult-to-detect level, or produce glia constitutively.

Neurogenesis clearly occurs developmentally but it seems more difficult to envisage this process occurring in adulthood, and it took some time for this idea to become accepted. Neurogenesis declines with age in adult mammals, with the microenvironment of the aged brain becoming more hostile (Kuhn *et al.* 1996;Shetty *et al.* 2004;Limke and Rao 2002;Tropepe *et al.* 1997). The neurogenic niches retain many aspects of the embryonic brain within a small, defined area. The niche contains specialised glial cells, basal lamina, ependymal cells, a specific neurotransmitter milieu and vasculature, which are permissive for continued neurogenesis in a 'hostile' environment.

1.2. The Functional Importance of Adult Neurogenesis

Neuroblasts from the anterior SVZ migrate up the rostral migratory stream (RMS), towards the olfactory bulb (OB), and then differentiate into granule or periglomelular interneurons (Luskin 1993;Emsley *et al.* 2005). These neurons integrate, form synapses and are functional, being required for olfactory discrimination (Carleton *et al.* 2003;Gheusi *et al.* 2000).

Neurons born in the SGZ migrate to the granule cell layer (GCL), integrate into existing circuitry and are functional (van Praag *et al.* 2002). Neurogenesis in the DG is necessary for hippocampal dependent learning and memory (Shors *et al.* 2002;Shors *et al.* 2001;Winocur *et* al. 2006). Basal neurogenesis in the DG may be involved in emotion and the 'cognitive-affective' interface (Sapolsky 2004;Kempermann 2002).

Neurogenesis occurs in humans *in vivo* (Eriksson *et al.* 1998;Sanai *et al.* 2004) and *in vitro* (Arsenijevic *et al.* 2001;Roy *et al.* 2000;Kukekov *et al.* 1999). Neurons from human SVZ *in vitro* display maturing electrophysiological properties over 4 weeks in culture (Moe *et al.* 2005). Bromodeoxyuridine (BrdU) is a thymidine analogue that labels dividing cells during the S-phase of mitosis and can be used as a marker for cell proliferation (Miller and Nowakowski 1988). Cell proliferation has been demonstrated *in vivo* in the DG after patients were given BrdU during treatment for head and neck cancer (Eriksson *et al.* 1998). Proliferating cells and, possibly, immature neurons have been found in human Huntington's disease (HD) SVZ (Curtis *et al.* 2003;Curtis *et al.* 2005b;Curtis *et al.* 2005a;Pearson *et al.* 2005) and in the human OB (Bedard and Parent 2004), although there is no well-defined RMS in the human brain (Sanai *et al.* 2004).

The function of 'non-constitutive' neurogenesis is more elusive. Neurogenesis may occur constitutively in the primate neocortex (Gould *et al.* 1999), rodent visual cortex (Kaplan 1981), substantia nigra (SN) (Zhao *et al.* 2003), dorsal vagal complex (Bauer *et al.* 2005) and piriform cortex and amygdala (Bernier *et al.* 2002), although this is controversial (Frielingsdorf *et al.* 2004;Ehninger and Kempermann 2003;Koketsu *et al.* 2003;Lie *et al.* 2002;Kornack and Rakic 1999). Neurogenesis increases in the striatum in response to ischaemia and excitotoxic lesions (Arvidsson *et al.* 2002;Parent *et al.* 2002b;Phillips *et al.* 2005;Tattersfield *et al.* 2004), in the cortex in response to apoptosis (Magavi *et al.* 2000), in the hypothalamus in response to growth factors (Kokoeva *et al.* 2005) and in the caudal SVZ and hippocampus in response to ischaemia and seizures (Nakatomi *et al.* 2002;Parent *et al.* 1997;Phillips *et al.* 2005). Neurogenesis occurring after lesions may contribute to self-repair after injury (Nakatomi *et al.* 2002;Arvidsson *et al.* 2002).

1.3. The Components of the Neurogenic Niche

The stem cell of the SVZ (type-B cell) is a specialised astrocyte (Doetsch *et al.* 1999;Morshead and van der Kooy 2004), which gives rise to the rapidly proliferating transit amplifying cell, the type-C cell, that in turn, form neuroblasts (type-A cell), which migrate along the lateral ventricle (LV) encased in tubes formed from the processes of type-B cells. Ependymal cells line the LV and extend multiple ciliary processes into the LV, and the type B cell extends a short cilia into the LV (Doetsch 2003). Ependymal cells also secrete Noggin, a bone morphogenic protein (BMP) inhibitor, which promotes a neurogenic lineage in the SVZ (Lim *et al.* 2000).

A tentative nomenclature has been developed for neurogenesis in the DG, pending further experimental clarification (Kempermann *et al.* 2004). Type 1 cells account for around 5% of cell divisions among the nestin-expressing cells of the DG, are the putative stem cell but are unlikely to be the same type of cell as the type B cell of the SVZ (Seaberg and van der Kooy 2002). The nestin and glial fibrillary acidic protein (GFAP)-expressing radial glia-like stem cell (type-1 or B cell) gives rise to the bipolar transit amplifying type 2a (doublecortin (DCX) negative) and b (DCX-positive) cells, which in turn give rise to type 3 (or G) cells, after around 3 days (Seri *et al.* 2001;Kempermann *et al.* 2004), but full maturation takes several weeks. SGZ astrocytes secrete neurogenesin-1, which acts as a BMP inhibitor, paralleling the function of Noggin in the SVZ (Ueki *et al.* 2003).

Endothelial cells release soluble factors, which are crucial for neurogenesis (Shen *et al.* 2004). NPCs in the SGZ are closely associated with capillaries and a proportion of NPCs express endothelial cell markers (Palmer *et al.* 2000). Consistent with a 'vascular niche' for neurogenesis, angiogenic factors such as vascular endothelial derived growth factor (VEGF), insulin-like growth factor (IGF), fibroblast growth factor (FGF) and erythropoetin promote neurogenesis; and neurogenic factors such as brain derived neurotrophic factor (BDNF), glial cell line derived neurotrophic factor (GDNF) and nerve growth factor (NGF) promote angiogenesis (reviewed by Park *et al.* 2003; Greenberg and Jin 2006).

Astrocytes are also important to neurogenesis, as co-culture of NPCs with astrocytes from a neurogenic area (hippocampus) promotes a neuronal fate, but not when astrocytes from a non-neurogenic area are used (Song *et al.* 2002). Astrocytes in neurogenic areas secrete growth factors that can enhance neurogenesis (Hagg 2005), as well as neurogenic factors (such as neurogenesin-1) (Ueki *et al.* 2003). Certain cytokines promote neurogenesis and migration of NPCs (Belmadani *et al.* 2006; Battista *et al.* 2006). Furthermore, astrocytes instruct newborn neurons to form functional synapses (Ullian *et al.* 2001), and support the migration of neuroblasts (Mason *et al.* 2001).

Neurotransmitters affect neurogenesis and the SVZ has a distinct complement of neurotransmitter input, such as an overlap of dopaminergic and serotonergic input (Hoglinger *et al.* 2004; Hagg 2005) (table 1.1). Type 2 cells in the DG receive gamma amino butyric acid (GABA)-ergic excitatory input, which leads to upregulation of NeuroD, a transcription factor important to neurogenesis (Miyata *et al.* 1999; Tozuka *et al.* 2005).

The basal lamina is rich in laminin and collagen-1, which may concentrate cytokines or growth factors from surrounding cells in the neurogenic niche (Mercier *et al.* 2002; Alvarez-Buylla and Lim 2004; Hagg 2005).

Neurogenic niches contain molecules normally associated with embryonic development: Notch and cognate receptor Jagged, sonic hedgehog (SHH), BMPs, Wnt, ephrins (Alvarez-Buylla and Lim 2004). Notch and BMP signalling may keep neural stem cells (NSCs) quiescent (Alvarez-Buylla and Lim 2004). SHH overexpression increases SGZ proliferation and neurogenesis (Lai *et al.* 2003), and in animals lacking the SHH co-receptor, Smoothened, SGZ and SVZ neurogenesis is reduced (Machold *et al.* 2003). In the SGZ, NPCs express receptors and signalling components for Wnt proteins, overexpression of Wnt3 increases neurogenesis, and blockade of Wnt signalling dramatically reduces neurogenesis *in vivo* (Lie *et al.* 2005). From the SGZ to the GCL, neuroblasts are guided by persistent radial glial elements (Seri *et al.* 2001) and possibly by stromal cell derived factor-1 (Lu *et al.* 2002). Type A neuroblasts from the SVZ have an elongated morphology and a leading process containing a growth cone (Wichterle *et al.* 1997), and they may contain locomotory mechanisms used by growing axons such as collapsing-response mediated protein 4 (Nacher *et al.* 2000).

Table 1.1. Some molecular factors affecting neurogenesis

Factor	Effect on neurogenesis in DG	Effect in SVZ/ OB
GABA	↑ (Tozuka *et al.* 2005)	↓ (Nguyen *et al.* 2003)
Glutamate	via NMDA R ↓ (Nacher *et al.* 2003), via AMPA R ↑ (Bai *et al.* 2003), via mGluR ↓ (Yoshimizu and Chaki 2004)	↓ (Brazel *et al.* 2005)
Dopamine	↑ (Hoglinger *et al.* 2004;Van Kampen *et al.* 2004;Baker *et al.* 2004)	↑ (Hoglinger *et al.* 2004;Van Kampen *et al.* 2004;Baker *et al.* 2004)
Serotonin	↑ via 5HT$_{1A/2A}$ (Banasr *et al.* 2004)	↑ via 5HT$_{1A/2C}$ (Banasr *et al.* 2004)
Acetylcholine	↑ (Cooper-Kuhn *et al.* 2004)	↑ (Cooper-Kuhn *et al.* 2004)
Noradrenaline	↑ (Kulkarni *et al.* 2002)	nNOS ↓ (Packer *et al.* 2003), eNOS ↑ (Chen *et al.* 2005)
Nitric oxide	nNOS ↓ (Packer *et al.* 2003) but e- and iNOS ↑ (Cardenas *et al.* 2005)	
Neuropeptide Y (NPY)	↑ (Howell *et al.* 2005;Howell *et al.* 2003)	
Adenosine	↓ A2A receptor mRNA (Tarditi *et al.* 2006)	
Estrogen	↑ (Tanapat *et al.* 1999)	↑ (Smith *et al.* 2001)
Prolactin		↑ (Shingo *et al.* 2003)
Gluco-corticoids	↓ (Cameron and Gould 1994;Wong and Herbert 2004)	
Thyroid hormone	↑ (Desouza *et al.* 2005)	↑ (Lemkine *et al.* 2005)
Erythropoetin		↑ (Shingo *et al.* 2001)
BDNF	↑ (Lee *et al.* 2002;Bull and Bartlett 2005)	↑ (Pencea *et al.* 2001;Benraiss *et al.* 2001)
IGF	↑ (Aberg *et al.* 2000)	
VEGF	↑ (Fabel *et al.* 2003)	
FGF	↑ (Kuhn *et al.* 1997)	↑ (Kuhn *et al.* 1997)
CNTF	↑ (Emsley and Hagg 2003)	↑ (Emsley and Hagg 2003)
PDGF		↑ (Mohapel *et al.* 2005)

1.4. Alteration of the Microenvironment by External Perturbation and How this Impacts on Neurogenesis

Ischaemia, seizures, excitotoxic lesions, trauma, environmental enrichment, dietary restriction, growth factor infusion and neurological disease all affect neurogenesis (reviewed by Phillips *et al.* 2006). These perturbations are clearly diverse, alter the microenvironment in different ways, and alter neurogenesis in different ways. How neurogenesis is altered depends on the species of animal and even the strain, husbandry and experimental methods (Kempermann *et al.* 1997;Abrous *et al.* 2005;Kempermann *et al.* 1998a;Prickaerts *et al.* 2004). The concept that different stimuli can alter neurogenesis in different ways can be illustrated at the cellular and whole animal level. For example, stimulation of microglia using interferon (INF)γ and interleukin (IL)-4 render them protective, whereas stimulation with lipopolysaccharide (LPS) renders them neurotoxic (Butovsky *et al.* 2005). Furthermore, activation of microglia using IL-4 promotes neurogenesis and activation of microglia using INFγ promotes oligodendrogenesis from adult NPCs (Butovsky *et al.* 2006). In the R6/2 transgenic mouse model of HD, neurogenesis does not increase after small excitotoxic lesions or seizures (Phillips *et al.* 2005), but some aspects of neurogenesis 'improve' following environmental enrichment (Lazic *et al.* 2006), neurogenesis in the DG increases after treatment with a selective serotonin reuptake inhibitor in the R6/1 mouse (Grote *et al.* 2005) and neurogenesis increases dramatically in the R6/2 mouse after FGF infusion (Jin *et al.* 2005).

Cell death, growth factors, hormonal alterations, glial activation, neurotransmitter changes, and angiogenesis can all occur after different perturbations. Each of these components may alter neurogenesis in different ways. Furthermore, cells die by different mechanisms: isolated apoptosis may be permissive for neurogenesis (Magavi *et al.* 2000), whereas necrosis might be inhibitory, possibly reducing the survival of newborn neurons (Arvidsson *et al.* 2002). It is possible that the core of a lesion is non-permissive but the penumbra is permissive (Arvidsson *et al.* 2002;Yoshimi *et al.* 2005). Cell death may be acute, as in ischaemia, or chronic, as in HD; and this may produce different microenvironmental perturbations, with different effects on neurogenesis.

Ischaemia of the striatum increases neurogenesis in the SVZ and SGZ despite a lack of cell death in these areas (Jin *et al.* 2001), as well as in the striatum itself (Arvidsson *et al.* 2002;Parent *et al.* 2002b). New neurons migrate from the SVZ to the site of striatal ischaemia and may also proliferate *in situ*. Most newborn neurons die and very few of the mature cells killed by the lesion are replaced by newborn neurons (Arvidsson *et al.* 2002). Neurogenesis increases in the caudal SVZ ('posterior periventricle') in response to transient global ischaemia, which injures the hippocampus (Nakatomi *et al.* 2002). Epidermal growth factor (EGF) and FGF administration further increases neurogenesis, producing a functional improvement of the animal in the Morris water maze (Nakatomi *et al.* 2002).

Seizures increase neurogenesis in the DG (Parent *et al.* 1997;Bengzon *et al.* 1997), and in the SVZ (Parent *et al.* 2002a). It may be that neurogenesis after seizures is stimulated by cell death, but electrical stimulation can increase neurogenesis without cell death (Deisseroth *et al.* 2004;Derrick *et al.* 2000), as can small doses of systemic excitotoxin (Morton and Leavens 2000) and electroconvulsive therapy (ECT)- induced seizures (Abrous *et al.* 2005).

Dietary restriction increases neurogenesis, possibly via increases in BDNF (Lee *et al.* 2000). Interestingly, dietary restriction can improve the phenotype of HD mice (Mattson *et al.* 2004), raising the possibility that increased neurogenesis might be responsible.

Olfactory learning increases neurogenesis in the OB (Rochefort *et al.* 2002) and wheel running and environmental enrichment increases neurogenesis only in the DG (Brown *et al.* 2003). In normal animals, environmental enrichment increases neurogenesis and 'enhances' existing neurones, mediated in part by growth factors such as VEGF, BDNF, IGF, and angiogenesis (van Praag *et al.* 2000).

Neurogenesis declines with age (Kuhn *et al.* 1996;Kempermann *et al.* 1998b). When SVZ NPCs from aged animals are grown in culture, they proliferate at the same rate and differentiate into the same proportion of neurons as NPCs from young animals (Tropepe *et al.* 1997). This suggests that the *in vivo* microenvironment may be responsible for the reduction in neurogenesis with age. Glucocorticoids increase with age and the decline in neurogenesis with age can be restored to that of younger animals by adrenalectomy (Nichols *et al.* 2001;Montaron *et al.* 1999;Cameron and Gould 1994). Similarly, levels of FGF, EGF and IGF decline with age and restoration of these growth factors restores neurogenesis in the aged brain (Lichtenwalner *et al.* 2001;Jin *et al.* 2003). Dopamine and noradrenaline levels in the hippocampus decline with age, in parallel with decline in performance in the Morris water maze (Stemmelin *et al.* 2000).

Depression is associated with low levels of neurogenesis, although it does not necessarily follow that there is a causal relationship (Sapolsky 2004;Vollmayr *et al.* 2003). Depressed patients have smaller hippocampi than controls (Sheline *et al.* 2003;MacQueen *et al.* 2003), but this could be due to cell death and neurite retraction and, again does not imply that decreased neurogenesis and depression are causally related. It is interesting, however, that depressed women who have suffered abuse have smaller hippocampi than depressed women without abuse (Vythilingam *et al.* 2002), raising the possibility that the early abuse triggered the small hippocampus that predisposed to depression, making it potentially amenable to early treatment (MacQueen *et al.* 2003;Sheline *et al.* 2003). Depression is associated with perturbed serotonergic transmission and selective serotonin reuptake inhibitors such as fluoxetine increase neurogenesis via the $5HT_{1A}$ receptor, although hippocampal neurogenesis in $5HT_{1A}$ receptor knock-out mice is unimpaired implying the receptor is not an absolute requirement for neurogenesis (Malberg and Duman 2003;Malberg *et al.* 2000;Santarelli *et al.* 2003). Stress and depression are associated with high cortisol levels, which in turn reduce neurogenesis (Sapolsky 2000;Sonino and Fava 2001;Gould *et al.* 1997;Cameron and Gould 1994), and the effect of cortisol on neurogenesis may be via reductions in BDNF (Cosi *et al.* 1993).

Alzheimer's disease (AD) is the most common cause of dementia in both younger patients and the elderly (Lobo *et al.* 2000;Harvey *et al.* 2003). Interestingly, the hippocampus degenerates, particularly the CA1 and subiculum but the DG is relatively spared (Bobinski *et al.* 1997). Neurogenesis is impaired in the DG of some AD mouse models (Haughey *et al.* 2002;Wen *et al.* 2004;Wang *et al.* 2004), possibly due to lower expression of BDNF in AD (Phillips *et al.* 1991), loss of cholinergic neurons and a loss of function effect of normal soluble amyloid precursor protein (APP) and presenilin-1 (PS1) (Caille *et al.* 2004;Hitoshi *et al.* 2002). Neurogenesis is however, increased in one mouse model of AD that has lower glutamate transmission (Jin *et al.* 2004a), and in AD patients (Jin *et al.* 2004b).

Parkinson's disease (PD) is a common neurodegenerative disease and patients exhibit slowness of movement (bradykinesia), rigidity and tremor as well as cognitive and psychiatric disturbance (Foltynie *et al.* 2002). Mouse models of PD, with less dopaminergic transmission, and PD patients, have low levels of neurogenesis in the SVZ, DG and OB (Baker *et al.* 2004;Hoglinger *et al.* 2004), but may have higher levels of neurogenesis in the SN (Yoshimi *et al.* 2005).

2. Huntington's Disease

2.1. Huntington's Disease: History

HD is an inherited neurodegenerative disorder characterised by movement disorder, dementia and psychiatric disturbance. It occurs usually in middle age and is relentlessly progressive with no treatments proven to halt or reverse the disease (Bates *et al.* 2002).

The discovery of the gene and intense research in recent years has provided some hope for curative therapies, including cellular repair, to overcome this distressing condition. The disease is crucially important also because it has served as a paradigmatic disorder, for many themes including molecular genetics; mechanisms of neurodegeneration; protein abnormalities and aggregates in neurodegenerative disorders; selective vulnerability and late-onset; and, perhaps, neurogenesis in chronic disease.

The eponymous George Huntington, a general practitioner in Long Island, New York, described HD in 1872, from his own observations and those of his father and grandfather (reviewed by Harper in Bates *et al.* 2002), although the disease had been definitively described earlier by Waters (1841). He recognised the autosomal dominant pattern of heredity; "once the thread is broken... the grandchildren... of the original shakers may rest assured that they are free from the disease" and "it never skips a generation to manifest in another; once having yielded its claims, it never regains them". The cognitive impairments were described; "in many amounting to insanity", including the fronto-striatal-based loss of judgement "They are men of about 50 years of age, but never let an opportunity to flirt with a girl go past unimproved. The effect is ridiculous in the extreme". Psychiatric disturbances, a "nervous temperament" and particularly suicide were also recognised by Huntington. The adult onset is described as "its coming on... only in adult life... while those who pass the fortieth year without symptoms of the disease are seldom attacked". The progressive and destructive nature of the disorder was chillingly recounted, "It is spoken of by those in whose veins the seeds of the disease are known to exist, with a kind of horror, and not at all alluded to except through dire necessity... (The disease develops) gradually but surely, increasing by degrees, and often occupying years in its development, until the hapless sufferer is but a quivering wreck of his former self".

Meynert (1877) originally proposed the striatum as the principal pathological site in HD, confirmed later in 1908 by Jelgersma, who also described global brain atrophy.

The autosomal dominant nature of HD, as well as many other aspects of HD genetics was recognised definitively in 1908 by Punnett, and by Jelliffe.

The field of HD research has benefited from collaborative research groups such as, the Hereditary Disease Foundation (founded by Dr Milton Wexler), The Huntington's Study

Group, EuroHD (a collaboration of European HD research groups) and the Network of European CNS Transplantation and Restoration (NECTAR; founded in 1990, focusing on two main programs, one pertaining to transplantation in PD and the other in HD, named NEST-HD). The HD Collaborative Research Group conducted linkage analysis of large HD kindreds in Venezuela over 10 years, leading to the cloning of the gene, as well as detailed clinical studies. The cloning of the gene that leads to HD paved the way for development of animal models of the disease based on the genetic abnormality (Mangiarini *et al.* 1996;Rubinsztein 2002).

2.2. Huntington's Disease: Genetics

The genetic abnormality in HD is an expanded triplet repeat of the bases cysteine, adenine, and guanine (CAG) (which encode glutamine, Q) on exon 1 of the 'interesting transcript' (IT15) gene on chromosome 4. There are many disorders caused by triplet repeats, with similar genetic characteristics, including fragile X syndrome, Friedreich's ataxia, myotonic dystrophy, spinocerebellar ataxias (SCA) 8 and 12, and the 'polyglutamine' diseases, spinal and bulbar muscular atrophy (SBMA), dentatorubro-pallidoluysian atrophy, SCA1-3, 6, 7 and 17 (Cummings and Zoghbi 2000;Koeppen 2005;Rubinsztein 2002).

The gene was localised to the short arm of chromosome 4 in 1983 using linkage to the marker 'G8' (Gusella *et al.* 1983), and the CAG repeat in the IT15 gene was identified in 1993 using positional cloning techniques (The Huntington's Disease Collaborative Research Group, 1993). The gene codes for huntingtin (htt), a ubiquitous protein, which may act as a scaffold in many different intracellular reactions (Harjes and Wanker 2003). Normally, the htt gene has less than 36 CAG repeats, with most people having in the order of 16 repeats (Bates *et al.* 2002;Rubinsztein *et al.* 1996;Snell *et al.* 1993). The mutant gene, which has greater than 39 CAG repeats, encodes for htt with an elongated glutamine chain, near the N-terminal. Some people have 'intermediate alleles' i.e. repeats longer than normal but less than 40, which display incomplete penetrance. The intermediate allele is more likely to result in clinical disease if it occurs in a person from an HD family than from the general population (Goldberg *et al.* 1995), and this may be due to defective DNA repair mechanisms in these family members (Bates *et al.* 2002). The disease is fully penetrant if the patient inherits greater than 39 CAG repeats (41 Q) (Rubinsztein *et al.* 1996). It is worth pointing out that there are two extra glutamines translated relative to the CAG codons due to the CAA-CAG sequence following the poly-Q stretch (Bates *et al.* 2002).

Because the elongated CAG repeat is unstable, it tends to expand as it is inherited, particularly down the paternal line (possibly due to CAG instability during spermatogenesis), a process known as 'genetic anticipation' (MacDonald *et al.* 1993;Ridley *et al.* 1988). A similar phenomenon accounts for the observation that juvenile-onset cases (that have long repeat lengths) are more often paternally inherited (Siesling *et al.* 1997;Telenius *et al.* 1993).

The number of CAG repeats correlates inversely with age at onset (Andrew *et al.* 1993;Snell *et al.* 1993). The age at onset can be more accurately predicted from CAG repeat length in juvenile-onset cases, which typically have greater than 50 CAG repeats and a severe and non-classical phenotype (Telenius *et al.* 1993). Elderly-onset cases have lower numbers of CAG repeats, and the repeat length does not accurately predict age at onset (Kremer *et al.* 1993). Overall, 50-69% of the variance of age at onset can be accounted for by CAG repeat length, with other genetic factors and environmental influences also being important (Rubinsztein *et al.* 1997;Andrew *et al.* 1993;Wexler *et al.* 2004;van Dellen *et al.* 2005).

2.3. Clinical Phenotype

Disease normally occurs at around 40-50 years of age, but the range is wide, and disease duration is commonly 15-20 years. Drug treatment is symptomatic only.

Patients exhibit a variety of movement disorders: as Huntington himself said, the disease is "unstable and whimsical". In general, patients tend to display chorea (a writhing dance-like movement) but as the disease progresses, they may lose their chorea and become dystonic, rigid and hyper-reflexic. Those with more severe disease may display the latter phenotype at the initial or early stages especially in juvenile onset disease.

During the early stages of the disease, usually 3 years in duration, patients may exhibit subtle cognitive and psychiatric changes, and minor motor abnormalities such as restlessness, impaired finger tapping, and slowed saccadic eye movements (Ali *et al.* 2006). Patients may present with cognitive and psychiatric symptoms, and without motor symptoms. Conjugate eye movements are often abnormal early on and can provide a clue to the diagnosis. Mid stages are dominated by extrapyramidal dysfunction, commonly chorea. During late stage disease the patient is often rigid, dystonic and bradykinetic. Dysarthria (problems enunciating words) and dysphagia (problems with eating and swallowing) are common, and dementia universal in late stages.

Juvenile onset patients exhibit more bradykinesia, dystonia, dysarthria, clumsiness, coarse tremor, and may suffer from seizures. Cognitive decline is early and severe. 10% of adult onset cases also present with this phenotype, known as the 'Westphal variant' but these adult patients tend to be younger and possess longer CAG repeats.

Cognitive dysfunction is typically a subcortical frontal executive syndrome, with slowness of thinking (bradyphrenia), poor planning and organisation, a lack of judgement and poor mental flexibility (Ho *et al.* 2003;Lawrence *et al.* 1998;Lawrence *et al.* 2000). 'Set shifting' is affected early i.e. patients are unable to shift their attention from one aspect of a cognitive task to another (Lawrence *et al.* 1998). Spatial and working memory is reduced, as well as lexical retrieval, as demonstrated by reduced letter fluency (Lawrence *et al.* 2000). Perception of disgust is affected specifically and even presymptomatically (Sprengelmeyer *et al.* 2006).

> Psychiatric symptoms include depression, anxiety, disinhibition, aggressive behaviour and a tendency to suicide (Sorensen and Fenger 1992). Psychiatric symptoms are common and disabling but not necessarily progressive (Bates *et al.* 2002).

The disease is accompanied by a number of other physiological manifestations including sleep-wake cycle disruption (Silvestri *et al.* 1995; Morton *et al.* 2005b), and weight changes (Hamilton *et al.* 2004; Djousse *et al.* 2002). As Huntington alluded to in an address to the New York Medical Society in 1909 "..two women, mother and daughter, both tall, thin, almost cadaverous, both bowing, twisting, grimacing."

2.4. Huntingtin (Htt)

Htt is a large 348kDa protein, with 3144 amino acids. It is ubiquitously expressed, particularly in neurons (Sharp *et al.* 1995; Sapp *et al.* 1997; Wilkinson *et al.* 1999). In the brain, high levels of endogenous htt are found in cortical pyramidal cells, cerebellar Purkinje cells and calbindin-positive cells of the striatum (Gutekunst *et al.* 1995; Ferrante *et al.* 1997). Subcellularly, htt is found predominantly in the cytoplasm in somatodendritic regions and less so in axons, and associates with microtubules and vesicles (Gutekunst *et al.* 1995; Gutekunst *et al.* 1998).

Htt is essential for development because lack of the gene is embryonically lethal (Duyao *et al.* 1995). HD-/- embryonic stem (ES) cells, however, develop normally into neurons (Metzler *et al.* 1999), and because normal extra-embryonic tissue can rescue homozygote HD-/- mutants from lethality, htt may deliver nutrition to developing cells, particularly given the function of htt in cellular transport (Harjes and Wanker 2003; Dragatsis *et al.* 1998). Htt also has a role in later brain development because gene targeted mice that express <50% normal htt display severe mid- and hind-brain abnormalities and die perinatally (White *et al.* 1997).

2.5. How Does Mutant Htt Causes Disease?

Htt is likely to cause disease mainly through a gain of toxic function. There are several lines of evidence for this, including:

1. Chromosomal deletions and translocations in the region of the HD gene do not lead to HD (Bates *et al.* 2002; Ambrose *et al.* 1994). Thus haploinsufficiency (whereby a single copy of the normal gene leads to insufficient protein product) is unlikely to account for disease.

2. Homozygous HD patients survive to adulthood (Squitieri *et al.* 2003).

3. Inactivation of the HD gene is embryologically lethal (Duyao *et al.* 1995).

4. Neurodegeneration is reversible in a conditional transgenic mouse when the mutant gene is switched off (Yamamoto *et al.* 2000).

5. Heterozygote knock-out mice are normal (Duyao *et al.* 1995).

6. Transgenic mouse models retain the endogenous gene, implying that the insertion of the mutant gene causes the disease. This does not of course preclude the possibility that the mutant protein interferes with the normal protein, thus a gain and loss of function may cause the disease.

7. The poly-Q stretch is pathogenic because poly-Q stretches alone, or fragments of exon-1 when inserted into a mouse genome are sufficient to cause disease, and indeed cause a more severe phenotype than full-length htt mouse models (table 2.2).

2.6. Do Aggregates Cause Disease?

Normal htt is soluble but mutant htt forms insoluble fibrillar aggregates both *in vivo* and *in vitro* (Davies *et al.* 1997;Scherzinger *et al.* 1997). Htt fibrillar aggregates are found in humans (DiFiglia *et al.* 1997), and transgenic mouse models (Davies *et al.* 1997), as well as other models of HD (Bates *et al.* 2002). Inclusions are present in the nucleus, as well as the cytoplasm, and in axons and axon terminals ('neuropil aggregates') (Li *et al.* 2000;DiFiglia *et al.* 1997).

Perutz *et al.* (1994) developed the 'polar zipper' model to explain the structure of htt aggregates. He proposed that, once the number of glutamines in the htt protein exceeds 41, stable hairpins are formed, which associate with each other and, in turn, precipitate into amyloid-like fibrils (Perutz *et al.* 1994).

There are several lines of evidence for aggregates being the toxic species, although some of which are circumstantial:

1. The threshold of aggregation equals the pathological threshold of CAG repeats, and the rate of aggregation correlates with CAG repeat length (Chen *et al.* 2002). Similarly, aggregate load correlates with phenotype progression (Davies *et al.* 1997;Morton *et al.* 2000).

2. Preformed aggregates are highly toxic when directed to the nucleus (Yang *et al.* 2002).

3. Aggregates might be resistant to unfolding and thus proteasome degradation, and aggregates themselves might inhibit proteasome activity (Jana *et al.* 2001), with accumulation of misfolded proteins toxic to the cell.

4. Aggregates may sequester vital proteins, ablating or altering their function. For example, cAMP response element binding protein (CREB)- binding protein (CBP), is found in aggregates and impairment of CBP function may lead to many downstream effects on transcription (Steffan *et al.* 2000).

5. A reduction in aggregate formation improves the phenotype of transgenic mice (Sanchez *et al.* 2003).

6. Yamamoto *et al.* (2000) showed that switching off the mutant gene in a conditional transgenic mouse model resulted in aggregate clearance, with reversal of the abnormal phenotype.

The situation may be more complex, however. Reducing the proportion of cells with aggregates actually increases cell death (although this may have been due to reduction in ubiquinisation during experimentation) (Saudou *et al.* 1998); increasing inclusion formation prevents proteasome dysfunction in transfected PC12 cells (Bodner *et al.* 2006); and inclusion body formation increases cellular survival (Arrasate *et al.* 2004). Aggregates are rarely found in those neurons most susceptible to disease (Gutekunst *et al.* 1999;Kuemmerle *et al.* 1999), suggesting that there is not a simple linear relationship between aggregate number and neurodegeneration. Furthermore, inclusions may be initially protective, clearing the abnormal protein and its toxic products, but their presence becomes deleterious once the inclusions recruit essential cellular proteins (Morton *et al.* 2000). Aggregates can also reduce the soluble fraction of mammalian target of rapamycin (mTOR), which inhibits the disposal mechanism, autophagy; and this may have beneficial effects for the cell (Ravikumar *et al.* 2004). The expression of an N-terminal fragment with 120Q (under the human htt promoter) has many neuronal nuclear inclusions (NII) but no neurodegenerative phenotype (Slow *et al.* 2005). This may negate the aggregate theory, or it may be the precise structure of fragment/ aggregate that confers toxicity.

2.7. Do Htt Fragments Cause Disease?

Htt is cleaved by aspartic endopeptidases, caspases and calpains (Wellington *et al.* 2002;Lunkes *et al.* 2002;Gafni *et al.* 2004). Fragments of mutant htt appear before large aggregates, coincident with the disruption of cellular processes such as gene expression and protein transcription (Saudou *et al.* 1998;Arrasate *et al.* 2004). Inhibition of caspase cleavage eliminates toxicity of mutant htt (Wellington *et al.* 2000;Chen *et al.* 2000;Ona *et al.* 1999). N-terminal cleavage products, which include the elongated poly-Q tracts, are more toxic than the full-length protein (Jana *et al.* 2001; Wellington *et al.* 2002;Lunkes *et al.* 2002). Consistent with this, transgenic mice expressing full-length exon-1, or the full-length gene have a less severe phenotype than those mice expressing N-terminal fragments of the gene (table 2.2). N-terminal htt fragments initiate aggregation dependent on the length of poly-Q (Martindale *et al.* 1998;Chen *et al.* 2002; van Dellen *et al.* 2005;Dyer and McMurray 2001). Full-length mutant htt is cleaved in the cytoplasm, and N-terminal fragments are translocated to the nucleus, where they become more toxic (Wellington *et al.* 2000;Yang *et al.* 2002). htt fragments can interact with mitochondria (Panov *et al.* 2002), furthering the risk of excitotoxic damage. Fragments accumulate with age, as a consequence of reduced proteosomal function with age (Zhou *et al.* 2003), and this may account for the late onset of the disease.

2.8. Does Loss of Function of Htt Play a Role in Pathogenesis?

Inactivation of the htt gene in adulthood results in a neurodegenerative phenotype (Dragatsis *et al.* 2000). htt interacts directly and through co-factors, with many different proteins involved with endocytosis, vesicle transport, cell signalling and transcription. htt therefore, has many wide ranging effects e.g. cytoskeletal organisation, synaptic plasticity, an increase in translation and transport of BDNF, a reduction in caspase activation, and is even involved in haemopoesis and spermatogenesis (Harjes and Wanker 2003;Zuccato *et al.* 2005;Zuccato *et al.* 2005;Zuccato *et al.* 2001;Rigamonti *et al.* 2001;Metzler *et al.* 2000). Since the poly-Q domain is important for protein-protein interactions, mutations in this region produce alterations in binding (Harjes and Wanker 2003). A pure loss of function effect is unlikely to account for the HD phenotype, although slowly developing late-life loss of function is a possibility and has not been ruled out thus far.

2.9. Excitotoxicity

Animals treated with excitotoxins and mitochondrial toxins recapitulate many phenotypical and pathological features of HD, and were indeed used to produce the first animal models of HD (Ferrante *et al.* 1993;Beal *et al.* 1993). There is some evidence for impaired energy metabolism in HD, and degeneration of mitochondria, although this may be a secondary phenomenon (Brennan, Jr. *et al.* 1985;Goebel *et al.* 1978;Jenkins *et al.* 1993;Tabrizi *et al.* 2000;Guidetti *et al.* 2001).

Glutamate is an essential mediator of excitotoxicity (Beal 1992). Glial glutamate uptake is reduced in HD, which may increase synaptic glutamate levels (Behrens *et al.* 2002;Lievens *et al.* 2001). Wild type (WT) htt may modulate the post-synaptic density protein 95 (PSD-95) that is involved in the post-synaptic clustering of N-methyl-D-aspartate (NMDA) receptors and mutant htt alters this binding to PSD-95, resulting in sensitisation of these receptors to glutamate, and thus possible toxicity (Sun *et al.* 2001).

2.10. Pathology

Pathological grade in patients is assigned according to the Vonsattel scale (Vonsattel *et al.* 1985). Focal degeneration occurs particularly in the caudate nucleus, which may atrophy years before the onset of symptoms (Aylward *et al.* 2000), but is widespread especially in the late stages of the disease (Vonsattel *et al.* 1985). Moderate astrogliosis occurs and tends also to predominate in the striatum (Myers *et al.* 1991). There is a dorsal-to-ventral, anterior-to-posterior and medial-to-lateral progression of neuronal death in the caudate nucleus (Vonsattel *et al.* 1985).

Other brain areas also dysfunction and degenerate early in the disease, such as the insula, hippocampus, cortex and amygdala (Rosas *et al.* 2003;Thieben *et al.* 2002), and the significance of this is becoming increasingly recognised, and may underlie some of the cognitive and psychiatric problems of HD (section 2.14). Hypothalamic atrophy (Kremer 1992) may underlie some of the metabolic symptoms and signs seen in HD (Pratley *et al.* 2000).

Caspases are upregulated and activated in HD (Friedlander 2003), and terminal deoxynucleotidyl transferase biotin-dUTP nick end labelling (TUNEL) staining has been reported in HD brains (Dragunow *et al.* 1995;Portera-Cailliau *et al.* 1995). Necrosis occurs at least in late stage HD (Vonsattel *et al.* 1985).

2.11. Why Is the Striatum Vulnerable to Degeneration?

Volume loss in the caudate nucleus is striking for its severity and relative selectivity, although other brain areas do dysfunction and lose cells early in the disease (Rosas *et al.* 2003;Thieben *et al.* 2002). The cause of this selectivity is unclear but there are several possibilities (list 2.1). For example, the rich input of glutamate to the striatum from the cortex, coupled with the increased sensitivity of the NMDA receptor subtype, NR2B, in the striatum in HD (Zeron *et al.* 2004d; Chen *et al.* 1999) and the increased vulnerability to excitotoxic stress of the susceptible neurons (Mitchell *et al.* 1999) may all contribute to make the striatum more vulnerable to degeneration. In addition, striatal glial cells may fail to take up glutamate (Lievens *et al.* 2001;Behrens *et al.* 2002), thus increasing the concentration of glutamate in the striatum.

N-terminal htt fragments preferentially accumulate in the striatum (Li *et al.* 2000) and this, along with tissue specific proteolysis and ubiquinisation (Mende-Mueller *et al.* 2001) may provide a basis for selective neurodegeneration.

CAG repeats are unstable and tend to expand particularly in the striatum (Kennedy and Shelbourne 2000;Telenius *et al.* 1994). The mechanism for this somatic instability may be due to excessive repair mechanisms because in R6/1 mice lacking the mismatch repair gene (*msh2*) there is no such somatic instability (Manley *et al.* 1999). It is not clear, however, that somatic instability and longer CAG repeats in the striatum produce deleterious effects.

It is possible that cellular dysfunction occurs initially in the cortex rather than the striatum, as has been the traditional view. NII are found initially in the cortex (Kuemmerle *et al.* 1999) and there may be dysfunction of striatal glutamate afferents (Cepeda *et al.* 2003) or failure of cortical release of neurotrophic factors, such as BDNF (Zuccato *et al.* 2001).

As well as selective vulnerability of the striatum, compared to the rest of the brain, distinct subpopulations within the striatum degenerate. In the striatum, D_2-expressing medium, spiny, GABA-ergic projection (MSP) neurons selectively degenerate. Large cholinergic and medium aspiny inter-neurons containing somatostatin (SS), neuropeptide Y (NPY), and reduced nicotinamide adenine dinucleotide phosphate (NADPH) diaphorase are relatively resistant to the disease process (Ferrante *et al.* 1985;Ferrante *et al.* 1987a;Ferrante *et al.* 1987b). It may be the case that stimulation of NADPH diaphorase neurons by glutamate from corticostriatal afferents generate harmful free radicals and kill surrounding MSP neurons (Bates *et al.* 2002). MSP neurons contain relatively high levels of the NR1A/2B receptor subunits that show enhanced sensitivity to glutamate receptor activation (Zeron *et al.* 2004).

List 2.1. Putative Reasons for Selective Vulnerability of the Striatum
Vulnerability to glutamate-mediated excitotoxicity:

- Heavy glutamatergic cortical afferent projections
- Striatal glia fail to remove synaptic glutamate
- Sensitivity of NR2B to glutamate
- High [glutamate] activates adjacent resistant neurons, which release free radicals and kill MSP neurons
- Paucity of calcium binding proteins in vulnerable subpopulation of MSP neurons
- High-energy requirements of vulnerable neurons
- Accumulation of particularly toxic species of mutant htt
- htt cleaved differently in the striatum
- More N-terminal fragments in the striatum
- Differential ubiquinisation
- Somatic mosaicism
- Murder' by the cortex (failure to release neurotrophic factors or release of excessive glutamate)

It must be remembered that other brain areas also dysfunction and degenerate early in the disease, which may have been somewhat overlooked due to the 'striatum-centric' view of HD (section 2.14).

2.12. Why Does HD Have a Late-Onset?

The symptoms of HD might be the result of many years of cumulative neuronal damage: due to accumulation of NII (Chen *et al.* 2002), accumulation of toxic proteins secondary to proteasome dysfunction (Zhou *et al.* 2003), a decline in glial glutamate transporter function (Behrens *et al.* 2002), or failure of compensatory mechanisms (Morton *et al.* 2000). Finally, as neurogenesis declines with age (Kuhn *et al.* 1996), it could be speculated that impaired neurogenesis contributes to the disease process.

2.13. The Circuitry of the Basal Ganglia Is Disrupted by the Loss of GABA-Ergic Projection Neurons

The basal ganglia (BG) are comprised of the striatum (caudate and putamen); SN; subthalamic nucleus (STN); and globus pallidus (GP). The striatum receives inputs from the cortex, SN pars compacta (SNc), dorsal raphe nucleus, amygdala, and thalamus. The internal segment of the GP (GPi) and the SN pars reticulata (SNr) are the main output nuclei of the BG, and send inhibitory GABA-ergic efferents to the thalamus, superior colliculus and pedunculopontine nucleus. There are many different neuronal populations in the striatum: MSP neurons (enkephalin-rich or substance-P dynorphin-rich), medium-sized aspiny NADPH diaphorase/SS/NPY-rich interneurons, large cholinergic interneurons, large and medium sized calretinin and substance P-rich interneurons, and parvalbumin-rich interneurons (Mitchell *et al.* 1999).

The thalamus is controlled via the 'direct' and 'indirect' pathways from the striatum (figure 2.1). The D_1/substance P-containing projection neurons are excited by the dopaminergic input from the SNc, as well as the excitatory input from the cortex. They directly inhibit the GPi/SNr, which releases the thalamus from tonic inhibition, resulting in cortical activation. This is the 'direct' pathway. The D_2/enkephalin-containing projection neurons are inhibited by dopaminergic input from the SNc, but activated by cortical input. When activated, the GP externa (e) is inhibited, releasing the excitatory glutamatergic STN, as well as the GPi/SNr. Thus, there is more excitatory input to the GPi/SNr, providing more inhibition to the thalamus, and less cortical activation. This is the 'indirect' pathway, and when the D_2-expressing striatal projection neurons are activated, serves as a 'brake' on the 'direct' pathway.

In HD, the D_2/enkephalin-containing projection neurons are most vulnerable to degeneration (Mitchell *et al.* 1996) and so it is hypothesised that the direct pathway becomes relatively overactive (figure 2.1B), activating the thalamus and thence the cortex, leading to chorea (Albin *et al.* 1989). Loss of D_1/substance P-containing neurons later in the disease may lead to underactivity of the cortex and to the more bradykinetic features seen in late stages of the disease (Glass *et al.* 2000).

The cortex and BG nuclei are linked by parallel loops: motor, occulomotor, dorsolateral prefrontal, orbito-frontal, and anterior cingulate (Alexander *et al.* 1990;Tekin and Cummings 2002) (figure 2.2). Thus the BG and cortex are closely linked and are involved in emotional and cognitive control (including reward, judgement and executive function) as well as motor control. Psychiatric and cognitive dysfunction may also be due to the dysfunction and atrophy in the limbic system (Rosas *et al.* 2003;Thieben *et al.* 2002) (section 2.14). The emotional and cognitive dysfunction, as opposed to the motor dysfunction, is arguably the most important symptoms of HD, and is therefore vitally important that we understand how it arises.

2.14. Dysfunction and Degeneration of Extra-Striatal Areas in HD

Latterly, evidence is accumulating that abnormalities occur outside of the basal ganglia and its frontal connections. Thus, volume loss in HD brains can be found in asymptomatic gene-positive individuals in the nucleus accumbens; orbito-frontal cortex, particularly the left olfactory sulcus; bilateral grey matter in the posterior insula, on the left extending ventromedially into the amygdala; and right posterior parietal sulcus (Thieben *et al.* 2002). The hippocampus, brainstem, and cortex have also been found to atrophy early in the disease in another study (Rosas *et al.* 2003). Atrophy of the primary sensorimotor and higher-order association cortex is found in early stage disease, with a posterior-anterior gradient as time progresses (Rosas *et al.* 2002).

Functional neuroimaging, which allows one to study neuronal dysfunction, as opposed to neuronal loss, has also revealed early extrastriatal abnormalities. For example, [^{11}C]-raclopide positron emission tomography (PET), which is used to measure postsynaptic dopamine D_2 receptor binding, has been shown in early-stage HD to be reduced in extrastriatal areas involved in mood and cognition: the amygdala, and frontal and temporal cortex (Pavese *et al.* 2003). Functional imaging studies have also suggested that the HD brain may recruit other brain areas, including the hippocampus, to compensate for striatal dysfunction, but this compensation breaks down with time (Rosas *et al.* 2004).

It is therefore difficult to attribute cognitive and psychiatric changes in HD exclusively to the striatum for the following reasons:

1. The striatum has connections with other brain areas, in particular the cortex and limbic system. It is therefore difficult to attribute dysfunction to one structure in isolation because these areas form circuits and disruption to any part of it can disrupt the whole circuit irrespective of the site of damage/ dysfunction.

2. Neuronal dysfunction may occur before neuronal loss and this is more difficult to measure accurately, and so subtle extrastriatal abnormalities may be missed in standard investigations.

3. There are relatively few anatomical, physiological and neuropsychological studies concentrating on other brain areas, although there is some more recent evidence that extrastriatal areas are involved in cognitive decline in HD.

The 'limbic system' is loosely defined but here we shall take it to be the medial temporal lobe, including the hippocampus and amygdala; structures that are involved in emotion and affective processing. We have included the hippocampus because of its putative role in cognition and mood, particularly pertaining to neurogenesis (Sapolsky 2004;Kempermann 2002).

Neurogenesis has been described in the piriform cortex (Pekcec et al. 2006) and amygdala (Bernier et al. 2002;Bernier and Parent 1998) but the existence of newborn neurons in some of these regions may be transient (Takemura 2005;Pekcec et al. 2006). It has been shown that the numbers of DCX-positive cells are reduced in the piriform cortex and absent in the insular cortex in R6/2 mice, which may represent a reduction in neurogenesis or plasticity ion these regions (Phillips et al. 2006).

Table 2.1. Neurotransmitter changes in R6/2 mice and HD patients. A summary of neurotransmitter changes in the R6 transgenic mouse model and in humans. The neurotransmitters refer to those in the striatum unless otherwise indicated

Neurotransmitter	R6 transgenic mouse model	Human
GABA	→ (Reynolds et al. 1999)	↓ in striatum, cortex and hippocampus (Reynolds and Pearson 1987)
Serotonin	↓ in striatum and hippocampus (Reynolds et al. 1999)	↑ in cortex/ striatum not hippocampus (Reynolds and Pearson 1987;Kish et al. 1987) ↓ (Waeber and Palacios 1989)
Glutamate	→ (Reynolds et al. 1999) ↓ (Nicniocaill et al. 2001)	↓ (Reynolds and Pearson 1987)
Dopamine	↓ (Reynolds et al. 1999)	→ (Reynolds and Garrett 1986), ↓ (Kish et al. 1987)
Acetylcholine Choline acetyltransferase	↓ striatum (Vetter et al. 2003)	↓ striatum and hippocampus (Spokes 1980)
Noradrenaline	→ in striatum, ↓↓ in hippocampus (Reynolds et al. 1999)	↑ (Spokes 1980)

Cognitive and psychiatric dysfunction in HD is likely due to dysfunction of striatal, cortical and limbic structures; and these areas are inter-connected and dysfunction in one area is likely to affect the others. However, it is possible that there is independent dysfunction of the limbic system (through reduced neurogenesis or plasticity, or intrinsic pathology), which contributes to the cognitive and psychiatric dysfunction in HD.

2.15. Neurochemistry of HD

The level of many neurotransmitters is perturbed in HD and in mouse models, and this is detailed in table 2.1.

2.16. Summary of Pathophysiological Mechanisms That May Cause Disease: The Common Final Pathways?

The following list illustrates a selection of pathophysiological mechanisms, which may contribute in different degrees to pathology; as primary or secondary phenomena. They may arise via loss of function of normal htt (by insufficient protein product, or sequestration within aggregates), dysfunction of other proteins by mutant htt, sequestration of other proteins into aggregates, or compensatory mechanisms.

1. *Disruption of synaptic function.* Pre- and postsynaptic receptor distribution is altered in R6/2 mice (Cha *et al.* 1998), and receptors show altered function in HD, often before overt disease (Zeron *et al.* 2004). Low frequency synaptic transmission at hippocampal synapses in R6/2 mice is normal, but high frequency synaptic transmission is abnormal (Murphy *et al.* 2000). Protein and mRNA trafficking to synapses is impaired possibly due to the presence of neuropil aggregates and disruption of cytoskeletal structures (Li *et al.* 2001).

2. *Disruption of vesicle transport.* Huntingtin activating protein-1 (HAP1) interacts with both htt and with proteins involved in vesicle transport (Rubinsztein 2002). For example, both HAP-1 and htt enhance transport of BDNF along microtubules. BDNF transport is impaired by reducing the levels of normal htt (a loss of function), or by the mechanical effects of aggregates (a gain of function) (Gauthier *et al.* 2004;Ross 2004).

3. *Disruption of transcription.* Mutant htt is found in the nucleus as well as cytoplasm (Wheeler *et al.* 2000;Hodgson *et al.* 1999). mRNA associated with inflammatory or cell cycle functions are increased; and mRNA associated with signal transduction, ion channels, transcription, G-protein coupled receptors, metabolism, calcium signalling, retinoid signalling and cell structure are decreased in HD mice (Luthi-Carter *et al.* 2000). Many of the genes downregulated in HD are CRE-responsive, a feature common to other poly-Q diseases (Rubinsztein 2002). An important change of transcription in HD is reduction of transcription of BDNF, and this may produce disease through a 'loss-of-function' effect (Zuccato *et al.* 2005;Zuccato *et al.* 2001). Expression of receptors, particularly the G-protein coupled metabotrobic glutamate receptors, D_1 (Cha *et al.* 1998) and cannabinoid $(CB)_1$ (Denovan-Wright and Robertson 2000) is decreased in HD mice, which may also impact on synaptic function. Htt also binds to histone deacetylase and histone deacetylase inhibitors improve the HD phenotype (Rubinsztein 2002;Ferrante *et al.* 2003).

4. *Altered plasticity.* Long-term potentiation (LTP) is reduced early in mouse models of HD, which could contribute to learning impairments in these mice (see Bates *et al.* 2002;Murphy *et al.* 2000;Mangiarinini *et al.* 1996).

5. Disruption of protein-removing machinery e.g. proteasomes (Jana et al. 2001). Disruption of proteasomal activity is reversible however, because abnormal protein and aggregates can be cleared from neurons in the reversible HD mouse under the control of an inducible promoter (Yamamoto et al. 2000).

2.17. Mouse Models of HD

Excitotoxic models (table 2.2), whereby neurotoxins are administered to produce a striatal lesion, were first developed in the 1970s (Borlongan *et al.* 1995;Mason and Fibiger 1978;Coyle and Schwarcz 1976;Sanberg *et al.* 1989). Kainic acid (KA) is a non-selective glutamate receptor agonist, and intrastriatal injections induce striatal pathology and a behavioural phenotype resembling HD (Mason and Fibiger 1978). Quinolinic acid (QA), a specific NMDA-receptor agonist, spares NADPH diaphorase neurons, and therefore more closely resembles the pathology of HD (Sanberg *et al.* 1989). 3-nitroproprionic acid (3NP) and malonate are mitochondrial toxins, and systemic administration produces a mouse model of HD (Beal *et al.* 1993;Borlongan *et al.* 1995).

Genetic models are vitally important because they allow study of early stage disease, mimic the disease better than excitotoxic models, can be genetically manipulated and are useful in the testing of therapeutic interventions.

The first transgenic mouse model of HD was developed in 1996, with the insertion of exon-1 of the human htt gene (encoding for the first 69 amino acids), with expanded CAG repeats, plus the HD gene promoter into the mouse genome (Mangiarini *et al.* 1996). The mice, therefore, retained their endogenous htt expression. Four transgenic models with expanded CAG repeats were created and the R6/2 model has one functioning repeat length, with 144 CAG repeats (a range of 141 to 157 repeats). The R6/1 model has fewer CAG repeats (116) and a less severe phenotype (Mangiarini *et al.* 1996).

R6/2 mice display subtle cognitive and then motor deficits from 3-4 weeks of age (Carter *et al.* 1999;Lione *et al.* 1999). At 8 weeks, symptoms and signs are more overt, with the mice displaying abnormal clasping behaviour when suspended by their tail, as well as tremor, ataxia, myoclonus (a jerking shock-like movement) and stereotyped movements (Mangiarini *et al.* 1996). Weight loss occurs from 8 weeks, and the mice show urinary incontinence and have reduced fertility. At 12 weeks, a proportion of mice become diabetic (Hurlbert *et al.* 1999), and mice exhibit severe weight loss. Pathologically, neuronal loss is limited and focal striatal loss is not exhibited until peri-death (Mangiarini *et al.* 1996). R6 models exhibit non-apoptotic, non-necrotic neurodegeneration at 14 weeks, with condensation of neurons into small dark bodies, and plasma membrane ruffling (Turmaine *et al.* 2000). Loss of brain weight occurs at 5 weeks; and at 12 weeks, the brain weighs around 20% less than that of WT mice. NII can be found in the cortex as early as 3.5 weeks of age and appear in the striatum at 4.5 weeks of age (Davies *et al.* 1997).

R6/1 mice display a similar neurological phenotype, but the motor onset is delayed to around 14 weeks of age, and lifespan is longer (32-40 weeks of age) than R6/2 mice.

The benefits of using the R6/2 model include the progressive neurological and cognitive phenotype, which is well characterised; and the short life span of the mouse, which allows rapid completion of experiments. The disadvantages are that the mice express exon-1, as opposed to the full-length protein, the mice continue to express endogenous htt, and the pathological changes are not as extensive as human pathology.

Table 2.2. Current mouse models of HD

Type of model	Promoter	Phenotype	NII	Pathology	References
Excitotoxic	Not applicable	Phenotype resembles HD	No	Striatal degeneration, sparing of NAPDH diaphorase neurons (for QA model)	(Coyle and Schwarcz 1976;Sanberg et al. 1989;Borlongan et al. 1995;Beal et al. 1991)
Transgenic: exon-1 R6 lines	Human HD promoter	Severe	Early, widespread	Little cell loss	(Mangiarini et al. 1996)
Transgenic: longer exon-1 e.g. N171-82Q	Human prion promoter	Variable, less severe than R6, onset ~3 months	Yes	More cell loss than R6 mice	(Schilling et al. 2004)
Transgenic: full length mutant protein e.g. YAC Q72, cDNA (48,89 CAG repeats)	Human HD promoter (YAC), CMV (cDNA)	Hyperactive at 6 months	Late	Striatal atrophy at 12 months	(Hodgson et al. 1999;Reddy et al. 1998)
Knock-in	Mouse HD promoter	Mild, 'Detloff mice' have more severe phenotype (Lin et al. 2001)	Some have NII (Lin et al. 2001; Rubinsztein 2002)	Little cell loss, more cell loss with astrogliosis in Detloff mice (Lin et al. 2001)	(Rubinsztein 2002)

Table 2.3. Neurogenesis in the DG in HD mouse models

Study	Proliferation	Neurogenesis (DCX)	Morphology	Neural differentiation	Newborn neuron survival	Effect of intervention
Lazic et al 2004 (R6/1)	↓	↓	ns	nd	nd	
Gil et al 2005 (R6/2)‡	↓	↓	ns	↔	↓	ns
Phillips et al 2005(R6/2)*	↓†	↔	↓	↔	↔	no upregulation after seizures
Grote et al 2005 (R6/1)	↔	ns	ns	↓	↓	↑ differentiation and survival (fluoxetine)
Jin et al 2005 (R6/2)	↔	ns	ns	ns	ns	↔ (subcut FGF)
Lazic et al 2006 (R6/1)	↓, ↔ #	↓	↓	↔	↓	↑

‡ This is a follow-up study with similar findings to Gil et al 2004
* Mice from Morton colony, environmentally enriched
† Statistics for *a priori* analysis not given in original publication: $F_{(1,6)} = 11.163$; $p < 0.02$
# There were fewer BrdU-positive cells in 10 week old R6/1 mice compared with WT mice but the same number in 25 week old R6/1 compared with WT mice
ns = not studied
nd = not detected.

Also, because of the long repeats, the transgenic models may recapitulate juvenile onset HD, as opposed to the adult onset disease and the short life span of the mouse reduces their value in exploring specifically the gradual disease process.

Yeast artificial chromosome (YAC) mice express the full-length htt protein, display a progressive phenotype over a longer time course and do show similar pathological changes to the human disease (table 2.2). Knock-in mice, such that CAG repeats are inserted directly into the endogenous gene, display limited motor abnormalities, and the pathological changes do not parallel human pathology (table 2.3).

3. Neurogenesis in Huntington's Disease

3.1. Endogenous Neurogenesis in HD

An attractive therapeutic option and alternative to transplantation of stem-like cells is to stimulate the brain's own stem cell population (Phillips and Barker 2006;Michell *et al.* 2004). There are, however, still problems with this approach including drug delivery and control of neurogenesis (Phillips *et al.* 2006). Furthermore, is it possible to stimulate the stem cells in an HD brain? The microenvironment in the R6/2 mouse brain is abnormal and this may be responsible for a low level of neurogenesis in these mice (Phillips *et al.* 2005). This abnormal, 'hostile' microenvironment may preclude successful neural transplantation as well as endogenous stimulation, which may need to be overcome using growth factors or by changing the hormonal and neurotransmitter milieu in HD (Jin *et al.* 2005; Morton *et al.* 2005a).

3.2. Neurogenesis in the DG of HD mice

Several studies have shown that neurogenesis is reduced in the DG of transgenic HD mice (Gil *et al.* 2004;Gil *et al.* 2005;Lazic *et al.* 2004;Grote *et al.* 2005; Phillips *et al.* 2005;Lazic *et al.* 2006) (table 2.3). Whether or not reduced DG neurogenesis in R6/2 mice is a primary or secondary phenomenon, it is likely that changes in DG neurogenesis will impact on cognition and mood (Kempermann *et al.* 2002) in HD. Indeed, the reduction in DG neurogenesis correlates with cognitive and affective symptoms of R6/1 mice (Grote *et al.* 2005). The failure of upregulation of neurogenesis after seizures (Phillips *et al.* 2005) also has implications for handling induced seizures in R6/2 mice (Mangiarini *et al.* 1996); the mechanism of impaired neurogenesis; and the ability of the diseased brain to self-repair and to adapt to novel stimuli.

> Phillips *et al.* (2005) found a smaller difference in neurogenesis between R6/2 and WT mice compared to Gil *et al.* (2005). Neurogenesis in the DG was lower in R6/2 mice compared to WT in a group of mice including those that had suffered experimentally-induced seizures but not using post-hoc comparisons to compare neurogenesis basally between genotypes (Phillips *et al.* 2005), although cell proliferation was lower in R6/2 mice under basal conditions using an *a priori* analysis (Phillips *et al.*, unpublished). Thus, it seems likely that there is a true difference in neurogenesis between R6/2 and WT mice, but the difference is smaller in the Phillips *et al.* (2005) study. This could be due to the use of mice from the Morton colony (Lione *et al.* 1999;Carter *et al.* 2000) versus Jackson lab colony. The former

have longer CAG repeat lengths, greater longevity, and fewer seizures; and the method of maintaining the strain is different for the two colonies. Additionally, the Morton colony mice are housed with plastic toys and running equipment, which may be deemed environmental enrichment; and even minimal environmental enrichment ameliorates disease progression in R6/2 mice (Hockly *et al.* 2002). Additionally, Lazic *et al.* (2006) found similar basal rates of cell proliferation between R6/1 and WT mice at 25 weeks of age, although reduced proliferation in R6/1 mice at 10 weeks of age. Indeed, we have preliminary data using Jackson lab mice, which shows a larger difference in neurogenesis between genotypes.

The differentiation fate and survival of NPCs were intact in R6/2 mice in one study (Phillips *et al.* 2005) but not in another (Gil *et al.* 2005). Lazic *et al.* (2004) could not demonstrate BrdU/NeuN co-localisation in R6/1 mice at 2 weeks but Gil *et al.* (2005) found that the 2-week survival of BrdU-positive cells was lower in R6/2 compared to WT mice. Similarly, Lazic *et al* (2006) found reduced newborn neuron survival in the DG in R6/1 mice after 4 weeks. All these studies showed that percentage co-localisation of BrdU with NeuN was no different between genotypes.

The reasons for the difference in survival rates between studies are unclear but may involve subtle differences in mouse colonies such as the frequency of handling-induced seizures and CAG repeat length, different methods of quantification, and differences in BrdU dosing regimens and time of sacrifice (Prickaerts *et al.* 2004) (see above).

3.3. Morphology of Newborn Neurons in R6 Mice

Newborn neurons are abnormal in R6 mice (Phillips *et al.* 2005;Lazic *et al.* 2006) and this is consistent with the reduced sprouting seen in the sciatic nerve of R6/2 mice (Ribchester *et al.* 2004); reduced dendritic area and soma size of medium spiny striatal neurons of R6/2 mice (Klapstein *et al.* 2001); and abnormal spine density of R6/1 mice (Spires *et al.* 2004a) and HD patients (Ferrante *et al.* 1991). Cortical and striatal neurons in cDNA mice have thickened proximal dendrites and fewer dendritic spines (Guidetti *et al* 2001). Laforet *et al.* showed that medium sized striatal and cortical neurons from HD100 mice displayed retraction and 'disorientation' of apical dendrites, as well as 'shrunken neurons' in the hippocampus (Laforet *et al.* 2001). Iannicola *et al.* reported that hippocampal neurites in a transgenic HD mouse are shortened (Iannicola *et al.* 2000).

Environmental enrichment 'enhances' the morphological characteristics of newborn granule neurons, in both WT and R6/1 mice, but this effect is not of sufficient magnitude to overcome differences in genotype (Lazic *et al.* 2006). Environmental enrichment rescues the hippocampal deficit of BDNF and cortical deficit of dopamine (Spires *et al.* 2004b) in R6/1 mice, which may contribute to some improvement of morphological features but this is not sufficient to normalise morphology of newborn neurons. Consistent with this, environmental enrichment does not normalise spine density in R6/1 mice (Spires *et al.* 2004b) and BDNF is not essential for morphological changes in neurons because mossy fibre sprouting can occur in BDNF knock-out mice (Bender *et al.* 1998). Interestingly, environmental enrichment increases neurogenesis in R6/1 mice (Lazic *et al.* 2006), and ameliorates a hippocampal-dependent cognitive decline in HD mice (Grote *et al.* 2005), although the two observations are not necessarily causally linked.

FGF affects neuronal morphology and migration in the DG (Lowenstein and Arsenault 1996) and is one of the most potent proliferative agents for neuronal progenitors (Gensburger *et al.* 1987). It is possible, then, that reduced levels of FGF in R6/2 mice may be responsible for reduced proliferation of NPCs and abnormal morphology of newborn neurons in R6/2 mice. FGF levels have not been studied in R6/2 mice; FGF is upregulated in the striatum and midbrain in HD patients (Tooyama *et al.* 1993), but the level of FGF in the hippocampus in humans is unknown.

The altered morphology of newborn granule neurons in the R6/2 mouse may also be due to abnormal cytoskeletal reorganisation and dendritic growth (reviewed by Harjes and Wanker, 2003; Jarabek *et al.* 2004). This is consistent with the observation that DCX-positive cells are reduced in the piriform cortex of R6/2 mice, which may reflect a general reduction in structural plasticity (Phillips *et al.* 2006).

Following 3 weeks of repeated emotional stress, there are fewer long thin spines and more short stubby spines in the prefrontal cortex of normal mice (Radley and Morrsion 2005). Spines sequester calcium and it may be that these morphological changes provide a neuroprotective mechanism from excitotoxic insults (discussed in section 3.8). This may however, be at the expense of a functionally impaired neuron. It is possible that the observed morphological changes are in reaction to a 'hostile' microenvironment in R6/2 mice (Phillips *et al.* 2005) with consequent functional impairment.

These morphological changes of newborn neurons in R6 mice may contribute to the cognitive impairment of R6/2 mice. Newborn neurons are more amenable to LTP (Wang *et al.* 2000;Schmidt-Hieber *et al.* 2004) but LTP is reduced in R6/2 mice (Murphy *et al.* 2000), and it is tempting to speculate that fewer or abnormal newborn neurones may contribute to this deficit in LTP.

Ageing mice also have morphological abnormalities in newborn neurons. Rao *et al.* (2005) showed that rats from 12 months of age had a higher ratio of more immature DCX-positive cells in the DG, less branching and shorter dendritic lengths. Migration of immature granule neurons from the SGZ to the GCL was also reduced in older mice but the neurons reached a similar position in the GCL as in their younger counterparts after several months. van Praag *et al* (2005), however, did not find morphological differences in newborn neurons between young and aged rats. As may be the case for R6/2 mice, such morphological changes with age may be postulated to cause functional abnormalities in newborn neurons.

3.4. Putative Mechanisms for Reduced Neurogenesis in the R6/2 DG

There are many possible mechanisms for the reduction in DG neurogenesis in transgenic HD mice, listed in 3.1 below.

List 3.1. Putative Reason for Impaired Neurogenesis in the DG Of R6/2 Mice

Relevant references and further discussion can be found within the body of the text.

1. A decrease of factors permissive for neurogenesis

    - Growth factors: BDNF, FGF, IGF, VEGF
    - Neurotransmitters: serotonin, dopamine, acetylcholine, noradrenaline

2. Abnormal components of the neurogenic niche

- Glia
- Vasculature
- Basement membrane
- Neurones (autocrine and paracrine factors)
- NPCs (abnormalities undetectable by neurosphere assay)

3. An increase in factors non-permissive for neurogenesis

- Glucocorticoids
- Inflammatory cytokines

BDNF is important for basal neurogenesis in the DG (Lee *et al.* 2000;Lowenstein and Arsenault 1996), and levels are reduced in the hippocampus of R6 mice (Spires *et al.* 2004b). FGF is also important for neurogenesis (Lowenstein and Arsenault 1996), and although FGF levels are increased in the striatum, midbrain and SVZ of HD patients (Tooyama *et al.* 1993), the levels of FGF in the DG in HD patients or R6/2 mice are unknown. Other growth factors that are required for neurogenesis may be deficient in HD, contributing to the observed impairment of neurogenesis in R6/2 mice.

Serotonin plays an important role in neurogenesis (see Hagg 2005), and serotonin levels are reduced in the hippocampus from 8 weeks of age (in the striatum from 12 weeks) in the R6/2 mouse (Reynolds *et al.* 1999). The SVZ and SGZ receive dopaminergic fibres, dopamine increases neurogenesis and the overlap of dopaminergic and serotonergic signalling in neurogenic areas may contribute to a neurogenic niche (Hoglinger *et al.* 2004;Hagg 2005). Levels of dopamine in the R6/2 hippocampus (and striatum) are, however low, although not to the same extent as other transmitters such as serotonin or noradrenaline (Reynolds *et al.* 1999). Depletion of noradrenaline reduces DG NPC proliferation but does not affect differentiation and survival (Kulkarni *et al.* 2002). R6/2 mice have a particularly marked decline of noradrenaline in the hippocampus, but noradrenaline levels are intact in the striatum (Reynolds *et al.* 1999). Thus, a paucity of noradrenaline could explain all the observed abnormalities seen in one study: reduced proliferation in the DG, intact differentiation and survival, intact neurogenesis in the striatum and a perturbed microenvironment (Phillips *et al.* 2005).

Soluble and membrane bound factors released from glia are critically important for neurogenesis (Song *et al.* 2002), and microglial activity may also influence neurogenesis (Ekdahl *et al.* 2003;Monje *et al.* 2002). Microglial activation inhibits neurogenesis (Ekdahl *et al.* 2003;Monje *et al.* 2002) and the numbers and morphology of microglia are altered in HD (Sapp *et al.* 2001;Ma *et al.* 2003). Interleukin-6 may be involved in this inflammation-induced downregulation of neurogenesis (Monje *et al.* 2003). Blockade of inflammation induced by seizures and cell death or LPS infusion (Ekdahl *et al.* 2003;Monje *et al.* 2003) restores neurogenesis. However, microglia do release trophic factors and may have a beneficial role during inflammation (Liu and Hong 2003). It may therefore be the degree, timing or nature of any inflammatory response that is important for neurogenesis and neuronal survival following brain insults. In the Phillips *et al.* (2005) study, microglia did not negatively correlate with neurogenesis and there was no difference in the amount of microglia between condition or genotypes, suggesting that the increase in neurogenesis following seizures is not mediated nor accompanied by increases in density of microglia, although the population of microglia may fluctuate at different time points. Neuronal death produces microglial activation (reviewed by Lui and Hong, 2003) and it is possible that there was no significant microglial activation following seizures in that study because there was no cell loss seen. This possibility is supported by Ekdahl *et al* (2003), who found that rats with generalised seizures and more cell loss displayed the highest numbers of activated microglia. Alternatively, differences in microglial markers used (ED1 in the Ekdahl study), difference in species (rats in the Ekdahl study), timing of seizure induction to sacrifice, or study design may account for different findings. There was no effect of age on microglia in the Phillips *et al.* (2005) study, in contrast to Ma *et al* (2004), who showed that microglial numbers reduced with age particularly in the R6/2 mouse, although the age difference used in the Phillips *et al.* study was smaller, the hippocampus was not studied by Ma *et al.* and the microglial antigen and method of quantification used was different. Ma *et al.* (2004) described a reduction in numbers of microglia in the cortex of R6/2 mouse, but not in other brain regions. In contrast to Ma *et al.* (2004), Sapp *et al.* (2001) found increased levels of microglia in HD patients. This may be due to species differences or different levels of microglia at different stages of the disease. Phillips *et al.* (2005) did not find differences in the amount of microglia between genotypes, but this does not rule out the possibility of qualitative differences in microglia affecting neurogenesis, and morphological differences of microglia are indeed present in R6/2 mice (Ma *et al.* 2004). Mature neurons, as well as newborn granule neurons in the hippocampus are qualitatively abnormal in HD (Ribchester *et al.* 2004;Spires *et al.* 2004a;Klapstein *et al.* 2001) and the same may also be true of glia. Thus, a functional disturbance of glia in HD may perpetuate the disease by impairing neurogenesis, as well as having wider pathological implications.

The vasculature is an important component of the neurogenic niche (Palmer *et al.* 2000). It is interesting that angiogenesis does not increase in aged rats in response to running (van Praag *et al.* 2005), and angiogenesis may also be impaired in R6/2 mice, due to reduced BDNF for example.

3.5. Neurogenesis in the DG Cannot Be Upregulated in R6/2 Mice in Response to Seizures

Neurogenesis increases in the DG after KA-induced seizures but not in R6/2 mice (Phillips *et al.* 2005). This suggests R6/2 mice have a lower capacity to mobilise the neurogenic process in response to injury, potentially reducing their capacity for endogenous brain repair. Interestingly, C57BL/6 mice (part of the hybrid strain that the R6/2 is bred from) are resistant to cell loss from intrahippocampal injections of standard doses of KA (Schauwecker 2002), and this may be due to strain differences in distal dendritic calcium regulation (Shuttleworth and Connor 2001). It may be speculated that R6/2 mice are resistant to cell loss in response to KA- induced seizures, because they have morphological differences in their dendrites (section 3.3), or differences in calcium handling (Hansson *et al.* 2001b).

Putative factors that might cause impairment of neurogenesis basally in R6/2 mice may also be responsible for a failure of up-regulation of neurogenesis in response to seizures. However, it is worth considering the microenvironmental perturbations post-seizure and whether this may be different in R6/2 mice. The mRNA and protein levels of many growth factors and cytokines are increased after seizures (Jankowsky and Patterson 2001). Transcription is abnormal in HD (van Dellen *et al.* 2005) and it is possible that up-regulation of growth factors is impaired post-seizure, accounting for the failure of up-regulation of neurogenesis.

Transmitter release is impaired during periods of intense synaptic activity in R6/2 mice (Murphy *et al.* 2000), thus during seizures, pre-synaptic terminals from R6/2 mice might release less neurotransmitters, possibly curtailing the neurogenic response.

Activity-dependent neuronal excitation is coupled to NPC proliferation (Deisseroth *et al.* 2004), and there may be less neuronal excitation in the R6/2 hippocampus because there is less LTP (Murphy *et al.* 2000), less newborn neurons, morphologically abnormal newborn neurons and disruption of synaptic activity. If there is indeed less neuronal activity in the R6/2 DG, this excitation coupling may provide an explanation for why neurogenesis does not increase in response to activity-dependent stimuli.

3.6. Neurogenesis in the SVZ and Striatum of HD Mice

Cell proliferation in the SVZ is the same between R6 mice and WT littermates (Phillips *et al.* 2005;Gil *et al.* 2005;Lazic *et al.* 2004), or increased in R6/2 mice (Jin *et al.* 2005;Batista *et al.* 2006). Similarly, numbers of DCX-positive cells (Phillips *et al.* 2005) and BrdU-positive cells (Phillips *et al.*, unpublished) in the striatum are similar between genotypes.

DCX-positive cells have been noted previously in the normally non-neurogenic striatum by Nacher *et al.* 2001, who proposed these cells were undergoing microtubule reorganisation linked to axonal outgrowth or synaptogenesis. It is possible, however, that the observed DCX-positive cells in the striatum and corpus callosum represent newborn neurons. It may be that neurogenesis is not detected in these regions under normal circumstances because neurogenesis is occurring at a very low level, easily missed by intermittent BrdU injections.

Migration of NPCs to a site of injury is unimpaired in R6/2 mice, and indeed DCX-positive cells from R6/2 mice may migrate towards a site of injury from the contralateral side, across the corpus callosum (Phillips *et al.* 2005). It may be speculated that neuroblasts travel across this structure, in stimulated and even under basal conditions (Aboody *et al.* 2000;Magavi *et al.* 2000;Gould *et al.* 1999). NPCs exist in human subcortical white matter (Nunes *et al.* 2003) and Takemura (2005) showed that NPCs and immature neurons are found in the adult rat subcortical white matter, beneath the temporal cortex. After hippocampal lesions, newborn cells can be found in the infracallosal region, migrating towards the site of injury (Parent *et al.* 2006). Also, neuroblasts may be recruited from the RMS into the degenerating striatum (Batista *et al.* 2006).

Migration of precursors to sites of pathology and stimulation has been described previously, and there may be many factors involved in directing this migration. For example, transplanted NPCs migrate towards gliomas (Aboody *et al.* 2000) and ischaemic lesions (Veizovic *et al.* 2001); and endogenous NPCs migrate towards sites of injury (Magavi *et al.* 2000;Arvidsson *et al.* 2002;Tattersfield *et al.* 2004). The exact mechanism underlying such migration is unknown but may include glioma-cell-produced extracellular matrix (Ziu *et al.* 2006); chemokines (Belmadani *et al.* 2006); stem cell factor (Sun *et al.* 2004;Jin *et al.* 2002);

growth factors (Jankowsky and Patterson 2001); and astrocyte-derived factors (Mason *et al.* 2001).

3.7. Neurogenesis in the SVZ and Striatum Cannot Be Upregulated in Response to an Excitotoxic Striatal Lesion

Neurogenesis in WT mice increases after a QA-induced striatal lesion but not in R6/2 mice, although interpretation of the comparison between genotypes is complex, given that R6/2 mice have smaller lesions in response to the same dose of excitotoxin (Phillips *et al.* 2005). This again suggests that R6/2 mice have a lower capacity to mobilise the neurogenic process in response to injury, potentially reducing their capacity for endogenous brain repair. Parent *et al.* (2002b) however, noted that while SVZ NPC proliferation consistently increased after extensive infarcts involving both striatum and cortex, smaller infarcts limited to the striatum produced a smaller, more localised and more variable increase in proliferation. Interestingly, both SVZ proliferation and numbers of BrdU-NeuN-positive cells in the striatum correlate with lesion size, but only in WT mice (Phillips *et al.* unpublished). Thus, there may be different signals released after striatal injury, from WT and R6/2 mice; or a different response to the same signals.

Following lesions to the striatum, growth factors increase, blood vessels and astrocytes release permissive factors; and apoptosis occurs, all of which serve to increase neurogenesis (Sun *et al.* 2004;Jin *et al.* 2002;Gotts and Chesselet 2005;Magavi *et al.* 2000). R6/2 mice may fail to increase neurogenesis due to a delay in mobilisation of the neurogenic response, for similar reasons to those discussed above, or due to the same mechanisms that render the mice resistant to excitotoxic lesions (section 3.8).

3.8. Resistance of R6/2 Mice to Striatal Lesions

HD transgenic mice are resistant to excitotoxic lesions (table 3.1). The protective mechanism against excitotoxins in HD mice is influenced by the background strain, mouse model (i.e. CAG repeat length and htt length) and age. 129/SvEMS and FVB/N mice are sensitive to cell death from intrahippocamal KA but BALBB/C and C57BL/6 mice are resistant to cell death at normal doses (Schauwecker and Steward 1997). Mouse models expressing mutant htt fragments are resistant to cell loss from excitotoxins; mice expressing partial length mutant htt show no change in susceptibility to excitotoxins (Petersen *et al.* 2002); and mice expressing full-length mutant htt are more sensitive to excitotoxins (Zeron *et al.* 2004c). Interestingly, the YAC72 mice are bred on a FVB/N background, whereas R6/2 mice are bred on a C57BL/6 background, and it is possible that this may contribute to the sensitivity to excitotoxins, as well as the presence of mutant htt. In htt fragment mouse models, resistance to excitotoxins develops over time e.g. R6/1 mice develop a resistance from 6 to 8 weeks, coincident with the formation of NII (Hansson *et al.* 2001b;Hansson *et al.* 2001a). Beyond a certain age, resistance to excitotoxins seems to break down: R6/2 mice are resistant to cell loss from 3NP at 10 weeks (Hickey and Morton 2000) but more sensitive by 12 weeks of age (Bogdanov *et al.* 1998). This pattern was seen to a certain extent by Phillips *et al.* (2005), with resistance to seizures tending to occur in 9 week old but not 13-week-old R6/2 mice.

Resistance to cell death from excitotoxins in certain mouse models of HD is likely to be related to the expression of fragments of mutant htt, as opposed to full-length mutant htt. Resistance to excitotoxins develops coincident with the appearance of NII (Hansson *et al.* 2001b), and which are few in number in YAC mice (Hodgson *et al.* 1999). Furthermore, in 'shortstop' mice (which have many NII but have no behavioral abnormalities), there is also resistance to excitotoxins (Slow *et al.* 2005). NII may therefore be protective (Saudou *et al.* 1998; Hickey and Morton 2000), may induce compensatory changes in the cell, or may be an epiphenomenon.

Table 3.1. Resistance to lesions in mouse models of HD

Toxin	HD/ model and age (weeks)	Resistance	Reference
QA	R6/1 (18), R6/2 (6)	Resistant	(Hansson *et al.* 1999)
3NP	R6/2 (12)	Sensitive	(Bogdanov *et al.* 1998)
NMDA agonist	R6/2 (6 to 15 weeks), 71 CAG, 94 CAG knock-in (7 to 15 weeks)	↑ cell swelling in older mice	(Levine *et al.* 1999)
Systemic KA	R6/2 (3, 9)	Resistant	(Morton and Leavens 2000)
3NP	R6/2 (7, 10)	Resistant	(Hickey and Morton 2000)
Malonate	R6/2 (6,12), R6/1 (6, 18)	Partial resistance (not in 6 week R6/1)	(Hansson *et al.* 2001a)
QA, NMDA agonist, AMPA agonist	R6/1 (resistance from 8 weeks, tested until 36 weeks), R6/2 (3 weeks until 12 weeks)	Resistant to QA and NMDA agonist	(Hansson *et al.* 2001b)
DA, 6OHDA	R6/1 (3, 16)	Partial resistance at 16 weeks	(Petersen *et al.* 2001)
QA, NMDA, AMPA	YAC72 (6, 10 months)	Sensitive to QA, NMDA	(Zeron *et al.* 2004b)
Ischaemia	R6/1 (18)	Resistant	(Schiefer *et al.* 2002)
QA	cDNA (18, 46, 100 CAG repeats)	No change	(Petersen *et al.* 2002)

R6/2 mice have a normal complement of NMDA receptors, anti-apoptotic proteins, immediate early genes, glutamate release from synaptosomes and evoked calcium influx to neurons (Hansson *et al.* 2001b; MacGibbon *et al.* 2002). However, MSP neurons in R6 mice have an increased capacity to handle calcium loads, likely related to an improved storage capacity and they exhibit higher basal calcium levels than in WT mice (Hansson *et al.* 2001b). The NR1 subunit of the NMDA receptor in N171-82Q mice is hypophosphorylated, which may reduce NMDA-evoked currents; neuronal nitric oxide synthase (NOS) and PSD-

95 (which are required for NMDA R function), and citron (a protein involved in dendritic spine function), are all reduced in the transgenic mouse, providing more mechanisms for resistance (Jarabek *et al.* 2004). Hansson *et al.* (1999) postulated that low-grade sustained excitotoxicity produced during the disease process causes resistance to excitotoxins by a mechanism akin to 'ischaemic tolerance'. This is an attractive suggestion because it may explain the time window of resistance, the putative threshold by which R6/2 mice succumb more quickly to insults (Morton and Leavens 2000) and the late onset of the disease. Such a mechanism may even underlie the observed reduction of neurogenesis in the DG and failure to upregulate DG, SVZ and striatal neurogenesis (Lazic *et al.* 2004;Gil *et al.* 2005;Phillips *et al.* 2005). Repeated stress causes calcium-sequestering dendritic spines to become short and stubby, as a neuroprotective mechanism (Radley and Morrsion 2005); this may occur in R6 mice in response to high basal calcium levels and might cause neuronal dysfunction and morphological abnormalities. Interestingly, Levine *et al.* (1999) showed that striatal neurons from R6/2 mice have increased sensitivity of NMDA receptors, with resulting cell swelling. An excess of cell death was not, however, seen. It is possible that striatal neurons from R6/2 mice have increased sensitivity to neuronal dysfunction after excitotoxins, but these cells are resistant to cell death. Increased cell swelling, however, occurred in mice of an older age (greater than 9 weeks, up to 15 weeks of age) and this may be when the neuroprotection breaks down (Bogdanov *et al.* 1998).

Levels of FGF are high in striatal astrocytes in HD brains, providing another mechanism for neuroprotection in the striatum in response to insults (Tooyama *et al.* 1993). NMDA receptor mediated excitotoxicity is dependent on dopamine and adenosine 2A receptors (Reynolds *et al.* 1998;Popoli *et al.* 2002). Dopamine and purine metabolism is disturbed in R6 mice, which may also account for neuroprotection to excitotoxic insult (table 2.1).

Increased sensitivity to cell death from excitotoxins in YAC mice might be explained by NR2B subunit-mediated increased evoked current amplitude and caspase-3 activation (Zeron *et al.* 2004a).

3.9. SVZ Proliferation in Humans Compared to HD Mice

Neurogenesis in the DG of HD patients has not yet been studied, but SVZ proliferation is increased in HD patients, increasing with disease severity (Curtis *et al.* 2005a;Curtis *et al.* 2005b;Curtis *et al.* 2003;Pearson *et al.* 2005). In HD patients, the SVZ is thicker and contains more proliferating cell nuclear antigen (PCNA)-positive cells than control brains, particularly in the central and caudal regions, adjacent to the caudate nucleus (Curtis *et al.* 2005a). Five percent of PCNA-positive cells co-localise with βIII-tubulin (a marker for immature neurons) and neuronal nuclear antigen (NeuN) (a marker for mature neurons), nearer the striatum, but most co-localise with GFAP, nearer the LV. This could be interpreted as high levels of gliosis, but could also indicate that these dividing cells were GFAP-expressing stem-like cells, particularly because type B cells are specifically increased in HD (Curtis *et al.* 2005b). These studies used PCNA to measure neurogenesis but PCNA can be expressed by non-proliferating cells and is not valid as a marker for neurogenesis (Rakic 2002). It is possible that the increase in PCNA immunoreactivity observed in these studies reflects gliogenesis, particularly as the proportion of βIII-tubulin cells was not increased in the HD SVZ (Curtis *et al.* 2005a).

It would be interesting to study the DG in HD patients given the data from animal studies and because abnormal DG neurogenesis may contribute to some of the cognitive and psychiatric disorders seen in HD. It would be interesting to study the striatum in humans, because this is the main site of pathology in HD; and the OB, as it depends on SVZ neurogenesis and neuroblast migration, and olfaction is reduced in HD patients (Hamilton et al. 1999;Moberg et al. 1987).

The reason for the discrepancy between humans and mouse models is unknown. One possibility is that the stimulus for SVZ proliferation in humans may be neurodegeneration in the adjacent caudate nucleus, which is minimal in the R6/2 mouse (Mangiarini et al. 1996). However, SVZ proliferation did not increase in R6/2 mice after striatal injury (Phillips et al. 2005) (although the mechanism and time-course of cell death is different). It may also be that intrinsic species differences, such as anatomy (Sanai et al. 2004) and neurotransmitter and growth factor milieu (Ferrer et al. 2000;Spires et al. 2004b;Reynolds and Pearson 1987;Kish et al. 1987), account for differences in neurogenesis. A recent study, however, has shown that while NPCs in the SVZ may be unchanged in HD mice, neural stem cells may be increased compared to WT mice (Batista et al. 2006).

The human SVZ 'stem cell niche' is composed of an 'astrocyte ribbon', and there is no RMS with chains of migrating neuroblasts to the OB although scattered neuroblasts can be found between the SVZ and olfactory cortex in humans (Sanai et al. 2004). The SVZ, RMS and OB are more developed in rodents, which is not surprising given the important role of olfaction in the rodent. The lack of an RMS does not preclude a role for SVZ neurogenesis in humans: new neurons destined for the OB may still play a role in humans, the SVZ may deliver neurons to the amygdala and piriform cortex (Bernier et al. 2002) and hippocampus (Bull and Bartlett 2005;Nakatomi et al. 2002), and may contribute to self-repair and 'upkeep' in adjacent structures. Furthermore, neurogenesis is ongoing in the human olfactory system (Bedard and Parent 2004). However, the anatomical differences between humans and rodents may provide an as yet unidentified explanation for the differences in SVZ neurogenesis between humans and rodents with HD.

Neurotransmitters affect neurogenesis (table 1.1) and there are species differences in neurotransmitters (table 2.1). Serotonin increases neurogenesis (Hagg 2005), but serotonin levels are reduced in the R6/2 striatum from 12 weeks of age (Reynolds et al. 1999) and increased in human HD cortex and striatum (Reynolds and Pearson 1987). Increased striatal serotonin might account for increased SVZ proliferation in HD patients. Noradrenaline contributes to NPC proliferation (Kulkarni et al. 2002); R6/2 mice have very low levels of noradrenaline in the hippocampus, but normal levels in the striatum (Reynolds et al. 1999); and noradrenaline is increased in the HD striatum (Spokes 1980). These alterations of noradrenaline levels could therefore explain reduced NPC proliferation in the R6/2 DG, normal R6/2 SVZ proliferation, and high SVZ proliferation in human HD patients.

The level of striatal FGF is unknown in mice, but is very high in the SVZ adjacent to the caudate nucleus in HD patients, which may provide a mechanism for increased SVZ proliferation in HD patients (Tooyama et al. 1993).

Finally, it is possible that SVZ proliferation may actually increase in mouse models, but their lifespan is so short that they die before this proliferation can be detected. Indeed, Jin et al. (2006) reported a modest increase in SVZ proliferation in 14-and-a-half and 15-and-a-half week old R6/2 mice. Again, neural stem, as opposed to precursor, cells may increase with age in R6/2 mice in the SVZ (Batista et al. 2006).

Table 3.2. possible reasons for discrepancy between mouse and human SVZ proliferation

Mouse	Human
No cell loss in caudate nucleus (until late stages)	Atrophy of caudate nucleus
Well defined SVZ and RMS	Astrocyte ribbon
Serotonin normal or reduced	Serotonin increased
Noradrenaline normal in striatum	Noradrenaline increased in striatum
Level of FGF in SVZ unknown	FGF increased in the SVZ
Mice might die before SVZ proliferation is detected	Longer lifespan in human

3.10. Neurogenesis in the DG versus SVZ

R6 and WT mice show similar levels of neurogenesis in the SVZ, but neurogenesis is reduced in the DG of R6/2 mice compared with WT mice (Phillips *et al.* 2005;Gil *et al.* 2005;Lazic *et al.* 2004). The SVZ and DG are both 'neurogenic niches' but there are differences between the two regions, suggesting that different factors mediate neurogenesis in each region. For example, environmental enrichment increases neurogenesis in the DG but not SVZ (Brown *et al.* 2003). This is partly mediated by BDNF (Spires *et al.* 2004b) and BDNF levels are higher in the hippocampus compared to striatum (Katoh-Semba *et al.* 1997). Thus, the DG may be more vulnerable to deficits of BDNF and so neurogenesis is impaired in this region rather than the SVZ. Other growth factors mediate the effect of environmental enrichment and physical exercise on DG neurogenesis including IGF and VEGF (van Praag *et al.* 2000), and it is possible that R6/2 mice have reduced levels of these growth factors, which might impact preferentially in the DG.

Table 3.3. Possible reasons for discrepancy between mouse SVZ and mouse DG proliferation

Mouse SVZ	Mouse DG
Possibly less sensitive to deprivation of certain growth factors e.g. BDNF, IGF, FGF	Possibly more sensitive to the deprivation of certain growth factors e.g. BDNF, IGF, FGF
Apposition to CSF	Not apposed to CSF
Noradrenaline levels intact	Noradrenaline levels reduced
True stem cells *in situ*	NPCs may have to migrate to the DG from the SVZ

There are clear anatomical differences between the DG and SVZ, and there may be functional differences in the microenvironment in these regions, which make only DG neurogenesis vulnerable in the R6/2 mouse. One patent difference between the two regions is that the SVZ is exposed to CSF. The neural stem cell of the SVZ has a single cilium that protrudes into the ventricle (Doetsch 2003). It is possible that these cilia sample the CSF and perhaps take up a factor from the CSF. Thus, there may be a factor in the R6/2 mouse, which preserves neurogenesis in the SVZ, while neurogenesis in the DG has no such protection and therefore declines in the face of a 'hostile' microenvironment.

There may be intrinsic differences in the potency of stem-like cells (Bull and Bartlett 2005;Seaberg and van der Kooy 2002). It is even possible that NPCs in the DG have migrated from the caudal SVZ (Bull and Bartlett 2005). The lack of proliferation, neurogenesis and upregulation of neurogenesis seen in the R6/2 DG may be a consequence of retarded migration of precursors from the caudal SVZ, particularly given that retarded migration was seen in the DG itself (Phillips *et al.* 2005).

R6/2 mice have a particularly marked decline of noradrenaline in the hippocampus, but noradrenaline levels are intact in the striatum (Reynolds *et al.* 1999). Furthermore, there is a noradrenergic input into the DG but not the SVZ, making it unlikely that noradrenaline mediates SVZ neurogenesis (Hagg 2005). Thus a paucity of noradrenaline (a promoter of neurogenesis, particularly NPC proliferation) may account for reduced neurogenesis in the DG. Indeed, enhancing levels of noradrenaline improves the phenotype of R6/2 mice (Morton *et al.* 2005a).

There are many components that contribute to a neurogenic niche, which may be different in the two areas that may, in turn, account for the different sensitivity of neurogenesis in the SVZ and DG.

3.11. Consequences of Impaired Neurogenesis

Whether or not reduced DG neurogenesis in R6/2 mice is a primary or secondary phenomenon, it is likely that changes in DG and piriform/ insular neurogenesis/ plasticity (Phillips *et al.* 2006) will impact on cognition and mood (Kempermann *et al.* 2002) in HD. Psychiatric and cognitive symptoms in HD are devastating and distressing both for patient and carer and it is vital, therefore, that these symptoms are tackled, both pathophysiologically and therapeutically.

If neurogenesis cannot be upregulated in the R6/2 DG, this may render the R6/2 mouse less able to cope with novel stimuli, and to adapt, with implications for the cognitive and psychiatric disorder of HD but also potentially with a lack of regenerative capacity in response to neuronal dysfunction and loss. A note of caution must be sounded, however, with concluding that neurogenesis cannot be upregulated in the SVZ in R6/2 mice. First, there was some response of neurogenesis to QA-treatment in R6/2 mice, but this did not reach statistical significance and may be increased further with larger lesions (Phillips *et al.* 2005). Second, there was some response of neurogenesis to environmental enrichment in the DG of R6/2 mice (Lazic et al. 2006); and a large response to FGF (Jin *et al.* 2005). Thirdly, interpretation is complex, given that R6/2 mice are resistant to excitotoxic injury. Finally, proliferation of SVZ stem, as opposed to precursor, cells may increase with age in R6/2 mice (Batista *et al.* 2006).

3.12. Relevance of Studies of Neurogenesis in Mouse Models to HD Patients

Neurogenesis and key pathological features such as resistance to excitotoxicity differ between transgenic models of HD and even strains of normal mice. It is therefore not surprising that there are differences in neurogenesis between rodent and humans. The lack of

RMS in humans has been discussed, and does not imply that neurogenesis is a phenomenon that does not apply to humans. Furthermore, the main effect on neurogenesis in the R6/2 mouse model is in the DG, where neurogenesis has been shown *in vivo* in humans.

That cognitive and psychiatric disorders in HD may be due to impaired neurogenesis is possible because DG neurogenesis may be involved in depression (section 1.4); piriform and insular cortex neurogenesis/ plasticity is reduced in R6/2 mice (Phillips *et al.* 2006); the DG, insula and piriform cortex have interconnections with other parts of the limbic system; and these structures are at the interface between emotion and learning.

It would of course be pertinent to study neurogenesis and morphology of newborn neurons in HD patients, and DCX-immunoreactivity would be one method of doing this. The disadvantage of using DCX on post-mortem specimens would be that only late stage brains tend to be available for examination.

3.13. Consequences of Abnormal Microenvironment in HD

Because of the possibility that newborn neurons in R6/2 mice are functionally abnormal, increasing endogenous NPC proliferation may not be sufficient to provide therapeutic benefit, and it may be necessary to also modulate the microenvironment into which newborn cells (be it transplanted, or endogenous) reside.

3.14. Therapeutic Potential

That neurodegeneration is reversible, at least in a transgenic mouse model (Yamamoto *et al.* 2000), gives hope that therapies given after the onset of disease may induce regression of the disease. Current treatments for HD are symptomatic only, and the disease continues relentlessly. Transplantation of fetal tissue has provided hope that the disease can be arrested or the damaged brain can even be repaired (by recreating circuitry, or providing trophic support). There are ethical, financial and logistic problems with this approach and stimulating the brain's own stem cells may be an attractive option to brain repair. Since degeneration proceeds over a long time period, neuron replacement can be done on a similar time scale. Pathology in these diseases is widespread, but NPCs can migrate to sites of pathology (Aboody *et al.* 2000;Lee *et al.* 2005). There are, however, many pitfalls with this approach. One, the functional consequences of impaired neurogenesis in HD mice and humans is unknown. Two, increasing the numbers of newborn neurons (by infusion of growth factors, for example) might result in functionally abnormal neurons (in a continued non-permissive microenvironment), excessive numbers, and cells that proliferate but do not survive, for example. Infusion of growth factors may have to be tailored, delivered as a 'cocktail' (one to mediate proliferation, one maturation, one survival, and so on), delivered with other substances such as Noggin, and the microenvironmental milieu may have to be altered to support the newly formed neurons. Finally, increasing neurogenesis may be detrimental, promoting tumours, kindling, impaired clearance of memories, and so on. More research is warranted into this area, certainly before trials on humans are considered.

3.15. Conclusion

The discovery that neurogenesis persists in the adult mammalian brain has lead to much excitement in the scientific community: both in terms of potential therapies and in the implication that this has for brain plasticity. That disturbed neurogenesis and plasticity is involved in psychiatric disorders is of immense importance and implies that by manipulating neurogenesis, these distressing disorders may be alleviated.

Regarding HD, most studies have shown that neurogenesis is impaired in the R6 mouse DG but intact in the SVZ; but precursor cell proliferation in the human SVZ is increased. Furthermore, neurogenesis or plasticity may be impaired in the piriform and insular cortex (Phillips *et al.* 2006), and speculatively of course, this may be responsible in part for atrophy of the insular cortex seen in HD patients (Thieben *et al.* 2002). Thus, it seems that impaired neurogenesis may be implicated in the cognitive and psychiatric disorder of HD and not the striatal degeneration. Indeed, striatal repopulation post-injury with newborn neurons is very limited (Arvidsson *et al.* 2002). SVZ/ striatal neurogenesis cannot be upregulated after injury in R6/2 mice (albeit with the reservations previously discussed) (Phillips *et al.* 2005), and this might have some implications for slow neurodegeneration. Thus, impairments in neurogenesis could contribute to the neurodegenerative process, and cognitive and psychiatric problems seen in HD, as well as an inability to cope with and adapt to stressors and novel stimuli.

Much of the work published on neurogenesis in HD is observational and it will be important in time, to assess the functional consequences of altered neurogenesis and plasticity in HD mice and humans, and whether enhancement of neurogenesis would arrest the disease. Thus, studies of neurogenesis in neurodegenerative disease could allow us to discover important pathological mechanisms and open up avenues for therapeutic intervention by either altering the microenvironmental milieu, enhancing conditions for cellular transplantation or enhancing endogenous neurogenesis, and with this the potential for curative therapies.

Abbreviations

3NP	3-nitropropionic acid
5HT	5-hydroxytryptamine (serotonin)
AD	Alzheimer's disease
APP	amyloid precursor protein
BDNF	brain derived neurotrophic factor
BG	basal ganglia
BMP	bone morphogenic protein
BrdU	bromodeoxyuridine
CA	*cornu ammonis*
CAG	cysteine adenine guanine
CBP	CREB binding protein
CMV	cytomegalovirus
CNTF	ciliary neurotrophic factor
CREB	cyclic adenosine monophosphate response element binding protein

CSF	cerebrospinal fluid
DCX	doublecortin
DG	dentate gyrus
ECT	electroconvulsive therapy
EGF	epidermal growth factor
ES	embryonic stem cell
FGF	fibroblast growth factor
GABA	gamma amino butyric acid
GCL	granule cell layer
GDNF	glial cell derived neurotrophic factor
GFAP	glial fibrillary acidic protein
GP	globus pallidus
GPe	globus pallidus externa
GPi	globus pallidus interna
HAP1	huntingtin associated protein1
HD	Huntington's disease
htt	huntingtin
IGF	insulin-like growth factor
IL	interleukin
INF	interferon
IT15	interesting transcript 15
KA	kainic acid
LPS	lipopolysaccharide
LTP	long-term potentiation
LV	lateral ventricle
mRNA	messenger RNA
MSP	medium spiny projection
mTOR	mammalian target of rapamycin
NADPH	nicotinamide adenine dinucleotide phosphate (reduced)
NCAM	neural cell adhesion molecule
NECTAR	Network of European CNS Transplantation and Restoration
NeuN	neuronal nuclear antigen
NGF	nerve growth factor
NII	neuronal intranuclear inclusions

NMDA	N-methyl-D-asparate
NOS	nitric oxide synthase
NPC	neural precursor cell
NPY	neuropeptide Y
NSC	neural stem cell
OB	olfactory bulb
PCNA	proliferating cell nuclear antigen
PD	Parkinson's disease
PDGF	platelet-derived growth factor
PET	positron emission tomography
PFC	prefrontal cortex
PS1	presenilin-1
PSA-NCAM	polysialyted neural cell adhesion molecule
PSD-95	postsynaptic density protein 95
Q	glutamine
QA	quinolinic acid
RMS	rostral migratory stream
RNA	ribonucleic acid
SBMA	spinobulbar muscular atrophy
SCA	spinocerebellar ataxia
SGZ	subgranular zone
SHH	sonic hedgehog
SN	substantia nigra
SNc	substantia nigra pars compacta
SNr	substantia nigra pars reticulara
SS	somatostatin
STN	subthalamic nucleus
SVZ	subventricular zone
TUNEL	Terminal deoxynucleotidyl transferase biotin-dUTP nick end labelling
VEGF	vascular endothelial derived growth factor
WT	wild type
YAC	yeast artificial chromosome

References

Aberg, MA; Aberg, ND; Hedbacker, H; Oscarsson, J; Eriksson, PS. Peripheral infusion of IGF-I selectively induces neurogenesis in the adult rat hippocampus. *J. Neurosci.,* 2000, 20, 2896-2903.

Aboody, KS; Brown, A; Rainov, NG; *et al.* Neural stem cells display extensive tropism for pathology in adult brain: evidence from intracranial gliomas. *Proc. Natl. Acad. Sci. U. S. A.* 2000, 97, 12846-12851.

Abrous, DN; Koehl, M; Le, MM. Adult neurogenesis: from precursors to network and physiology. *Physiol. Rev.,* 2005, 85, 523-569.

Albin, RL; Young, AB; Penney, JB. The functional anatomy of basal ganglia disorders. *Trends Neurosci.,* 1989, 12, 366-375.

Alexander, GE; Crutcher, MD; DeLong, MR. Basal ganglia-thalamocortical circuits: parallel substrates for motor, oculomotor, "prefrontal" and "limbic" functions. *Prog. Brain Res.,* 1990, 85, 119-46., 119-146.

Ali, FR; Michell, AW; Barker, RA; Carpenter, RH. The use of quantitative oculometry in the assessment of Huntington's disease. *Exp. Brain Res.,* 2006, 169, 237-245.

Altman, J; Das, GD. Post-natal origin of microneurones in the rat brain. *Nature.,* 1965, 207, 953-6., 953-956.

Alvarez-Buylla, A; Lim, DA. For the long run: maintaining germinal niches in the adult brain. *Neuron.,* 2004, 41, 683-686.

Ambrose, CM; Duyao, MP; Barnes, G; *et al.* Structure and expression of the Huntington's disease gene: evidence against simple inactivation due to an expanded CAG repeat. *Somat Cell Mol. Genet.,* 1994, 20, 27-38.

Andrew, SE; Goldberg, YP; Kremer, B; *et al.* The relationship between trinucleotide (CAG) repeat length and clinical features of Huntington's disease. *Nat. Genet.,* 1993, 4, 398-403.

Arrasate, M; Mitra, S; Schweitzer, ES; Segal, MR; Finkbeiner, S. Inclusion body formation reduces levels of mutant huntingtin and the risk of neuronal death. *Nature.,* 2004, 431, 805-810.

Arsenijevic, Y; Villemure, JG; Brunet, JF; *et al.* Isolation of multipotent neural precursors residing in the cortex of the adult human brain. *Exp. Neurol.,* 2001, 170, 48-62.

Arvidsson, A; Collin, T; Kirik, D; Kokaia, Z; Lindvall, O. Neuronal replacement from endogenous precursors in the adult brain after stroke. *Nat. Med.,* 2002, 8, 963-970.

Aylward, EH; Codori, AM; Rosenblatt, A; *et al.* Rate of caudate atrophy in presymptomatic and symptomatic stages of Huntington's disease. *Mov. Disord.,* 2000, 15, 552-560.

Bai, F; Bergeron, M; Nelson, DL. Chronic AMPA receptor potentiator (LY451646) treatment increases cell proliferation in adult rat hippocampus. *Neuropharmacology.,* 2003, 44, 1013-1021.

Baker, SA; Baker, KA; Hagg, T. Dopaminergic nigrostriatal projections regulate neural precursor proliferation in the adult mouse subventricular zone. *Eur. J. Neurosci.,* 2004, 20, 575-579.

Banasr, M; Hery, M; Printemps, R; Daszuta, A. Serotonin-induced increases in adult cell proliferation and neurogenesis are mediated through different and common 5-HT receptor subtypes in the dentate gyrus and the subventricular zone. *Neuropsychopharmacology.,* 2004, 29, 450-460.

Bates, G; Harper, P; Jones, L. *Huntington's Disease.* Oxford, 2002.

Batista, CM; Kippin, TE; Willaime-Morawek, S; Shimabukuro, MK; Akamatsu, W; van der, KD. A progressive and cell non-autonomous increase in striatal neural stem cells in the Huntington's disease R6/2 mouse. *J. Neurosci.,* 2006, 26, 10452-10460.

Battista, D; Ferrari, CC; Gage, FH; Pitossi, FJ. Neurogenic niche modulation by activated microglia: transforming growth factor beta increases neurogenesis in the adult dentate gyrus. *Eur. J. Neurosci.,* 2006, 23, 83-93.

Bauer, S; Hay, M; Amilhon, B; Jean, A; Moyse, E. In vivo neurogenesis in the dorsal vagal complex of the adult rat brainstem. *Neuroscience.,* 2005, 130, 75-90.

Beal, MF. Mechanisms of excitotoxicity in neurologic diseases. *FASEB J.,* 1992, 6, 3338-3344.

Beal, MF; Brouillet, E; Jenkins, B; Henshaw, R; Rosen, B; Hyman, BT. Age-dependent striatal excitotoxic lesions produced by the endogenous mitochondrial inhibitor malonate. *J. Neurochem.,* 1993, 61, 1147-1150.

Beal, MF; Ferrante, RJ; Swartz, KJ; Kowall, NW. Chronic quinolinic acid lesions in rats closely resemble Huntington's disease. *J. Neurosci.,* 1991, 11, 1649-1659.

Bedard, A; Parent, A. Evidence of newly generated neurons in the human olfactory bulb. *Brain Res. Dev. Brain Res.,* 2004, 19, 151, 159-168.

Behrens, PF; Franz, P; Woodman, B; Lindenberg, KS; Landwehrmeyer, GB. Impaired glutamate transport and glutamate-glutamine cycling: downstream effects of the Huntington mutation. *Brain.,* 2002, 125, 1908-1922.

Belmadani, A; Tran, PB; Ren, D; Miller, RJ. Chemokines regulate the migration of neural progenitors to sites of neuroinflammation. *J. Neurosci.,* 2006, 26, 3182-3191.

Bender, R; Heimrich, B; Meyer, M; Frotscher, M. Hippocampal mossy fiber sprouting is not impaired in brain-derived neurotrophic factor-deficient mice. *Exp. Brain Res.,* 1998, 120, 399-402.

Bengzon, J; Kokaia, Z; Elmer, E; Nanobashvili, A; Kokaia, M; Lindvall, O. Apoptosis and proliferation of dentate gyrus neurons after single and intermittent limbic seizures. *Proc. Natl. Acad. Sci. U. S. A.,* 1997, 94, 10432-10437.

Benraiss, A; Chmielnicki, E; Lerner, K; Roh, D; Goldman, SA. Adenoviral brain-derived neurotrophic factor induces both neostriatal and olfactory neuronal recruitment from endogenous progenitor cells in the adult forebrain. *J. Neurosci.,* 2001, 21, 6718-6731.

Bernier, PJ; Bedard, A; Vinet, J; Levesque, M; Parent, A. Newly generated neurons in the amygdala and adjoining cortex of adult primates. *Proc. Natl. Acad. Sci., U. S. A.,* 2002, 20, 99, 11464-11469.

Bernier, PJ; Parent, A. Bcl-2 protein as a marker of neuronal immaturity in postnatal primate brain. *J. Neurosci.,* 1998, 18, 2486-2497.

Bobinski, M; Wegiel, J; Tarnawski, M; *et al.* Relationships between regional neuronal loss and neurofibrillary changes in the hippocampal formation and duration and severity of Alzheimer disease. *J. Neuropathol. Exp. Neurol.,* 1997, 56, 414-420.

Bodner, RA; Outeiro, TF; Altmann, S; *et al.* From the Cover: Pharmacological promotion of inclusion formation: A therapeutic approach for Huntington's and Parkinson's diseases. *Proc. Natl. Acad. Sci., U. S. A.,* 2006, 103, 4246-4251.

Bogdanov, MB; Ferrante, RJ; Kuemmerle, S; Klivenyi, P; Beal, MF. Increased vulnerability to 3-nitropropionic acid in an animal model of Huntington's disease. *J. Neurochem.,* 1998, 71, 2642-2644.

Borlongan, CV; Koutouzis, TK; Randall, TS; Freeman, TB; Cahill, DW; Sanberg, PR. Systemic 3-nitropropionic acid: behavioral deficits and striatal damage in adult rats. *Brain Res. Bull.*, 1995, 36, 549-556.

Brazel, CY; Nunez, JL; Yang, Z; Levison, SW. Glutamate enhances survival and proliferation of neural progenitors derived from the subventricular zone. *Neuroscience.*, 2005, 131, 55-65.

Brennan, WA; Jr., Bird, ED; Aprille, JR. Regional mitochondrial respiratory activity in Huntington's disease brain. *J. Neurochem.*, 1985, 44, 1948-1950.

Brown, J; Cooper-Kuhn, CM; Kempermann, G; et al. Enriched environment and physical activity stimulate hippocampal but not olfactory bulb neurogenesis. *Eur. J. Neurosci.*, 2003, 17, 2042-2046.

Bull, ND; Bartlett, PF. The adult mouse hippocampal progenitor is neurogenic but not a stem cell. *J. Neurosci.*, 2005, 25, 10815-10821.

Butovsky, O; Talpalar, AE; Ben-Yaakov, K; Schwartz, M. Activation of microglia by aggregated beta-amyloid or lipopolysaccharide impairs MHC-II expression and renders them cytotoxic whereas IFN-gamma and IL-4 render them protective. *Mol. Cell Neurosci.*, 2005, 29, 381-393.

Butovsky, O; Ziv, Y; Schwartz, A; et al. Microglia activated by IL-4 or IFN-gamma differentially induce neurogenesis and oligodendrogenesis from adult stem/progenitor cells. *Mol. Cell Neurosci.*, 2006, 31, 149-160.

Caille, I; Allinquant, B; Dupont, E; et al. Soluble form of amyloid precursor protein regulates proliferation of progenitors in the adult subventricular zone. *Development.*, 2004, 131, 2173-2181.

Cameron, HA; Gould, E. Adult neurogenesis is regulated by adrenal steroids in the dentate gyrus. *Neuroscience.*, 1994, 61, 203-209.

Cardenas, A; Moro, MA; Hurtado, O; Leza, JC; Lizasoain, I. Dual role of nitric oxide in adult neurogenesis. *Brain Res. Brain Res. Rev.*, 2005, 50, 1-6.

Carleton, A; Petreanu, LT; Lansford, R; varez-Buylla, A; Lledo, PM. Becoming a new neuron in the adult olfactory bulb. *Nat. Neurosci.*, 2003, 6, 507-518.

Carter, RJ; Hunt, MJ; Morton, AJ. Environmental stimulation increases survival in mice transgenic for exon 1 of the Huntington's disease gene. *Mov. Disord.*, 2000, 15, 925-937.

Cepeda, C; Hurst, RS; Calvert, CR; et al. Transient and progressive electrophysiological alterations in the corticostriatal pathway in a mouse model of Huntington's disease. *J. Neurosci.*, 2003, 23, 961-969.

Cha, JH; Kosinski, CM; Kerner, JA; et al. Altered brain neurotransmitter receptors in transgenic mice expressing a portion of an abnormal human huntington disease gene. *Proc. Natl. Acad. Sci. U. S. A.*, 1998, 95, 6480-6485.

Chen, J; Zacharek, A; Zhang, C; et al. Endothelial nitric oxide synthase regulates brain-derived neurotrophic factor expression and neurogenesis after stroke in mice. *J. Neurosci.*, 2005, 25, 2366-2375.

Chen, M; Ona, VO; Li, M; et al. Minocycline inhibits caspase-1 and caspase-3 expression and delays mortality in a transgenic mouse model of Huntington disease. *Nat. Med.*, 2000, 6, 797-801.

Chen, N; Luo, T; Wellington, C; et al. Subtype-specific enhancement of NMDA receptor currents by mutant huntingtin. *J. Neurochem.*, 1999, 72, 1890-1898.

Chen, S; Berthelier, V; Hamilton, JB; O'Nuallain, B; Wetzel, R. Amyloid-like features of polyglutamine aggregates and their assembly kinetics. *Biochemistry.*, 2002, 41, 7391-7399.

Cosi, C; Spoerri, PE; Comelli, MC; Guidolin, D; Skaper, SD. Glucocorticoids depress activity-dependent expression of BDNF mRNA in hippocampal neurones. *Neuroreport.*, 1993, 4, 527-530.

Coyle, JT; Schwarcz, R. Lesion of striatal neurones with kainic acid provides a model for Huntington's chorea. *Nature.*, 1976, 263, 244-246.

Cummings, CJ; Zoghbi, HY. Trinucleotide repeats: mechanisms and pathophysiology. *Annu. Rev. Genomics Hum. Genet.*, 2000, 1, 281-328., 281-328.

Curtis, MA; Penney, EB; Pearson, AG; et al. Increased cell proliferation and neurogenesis in the adult human Huntington's disease brain. *Proc. Natl. Acad. Sci., U. S. A.* 2003, 100, 9023-9027.

Curtis, MA; Penney, EB; Pearson, J; Dragunow, M; Connor, B; Faull, RL. The distribution of progenitor cells in the subependymal layer of the lateral ventricle in the normal and Huntington's disease human brain. *Neuroscience.*, 2005a, 132, 777-788.

Curtis, MA; Waldvogel, HJ; Synek, B; Faull, RL. A histochemical and immunohistochemical analysis of the subependymal layer in the normal and Huntington's disease brain. *J. Chem. Neuroanat.*, 2005b, 30, 55-66.

Davies, SW; Turmaine, M; Cozens, BA; et al. Formation of neuronal intranuclear inclusions underlies the neurological dysfunction in mice transgenic for the HD mutation. *Cell.*, 1997, 90, 537-548.

Deisseroth, K; Singla, S; Toda, H; Monje, M; Palmer, TD; Malenka, RC. Excitation-neurogenesis coupling in adult neural stem/progenitor cells. *Neuron.*, 2004, 42, 535-552.

Denovan-Wright, EM; Robertson, HA. Cannabinoid receptor messenger RNA levels decrease in a subset of neurons of the lateral striatum, cortex and hippocampus of transgenic Huntington's disease mice. *Neuroscience.*, 2000, 98, 705-713.

Derrick, BE; York, AD; Martinez, JL; Jr. Increased granule cell neurogenesis in the adult dentate gyrus following mossy fiber stimulation sufficient to induce long-term potentiation. *Brain Res.*, 2000, 857, 300-307.

Desouza, LA; Ladiwala, U; Daniel, SM; Agashe, S; Vaidya, RA; Vaidya, VA. Thyroid hormone regulates hippocampal neurogenesis in the adult rat brain. *Mol. Cell Neurosci.*, 2005, 29, 414-426.

DiFiglia, M; Sapp, E; Chase, KO; et al. Aggregation of huntingtin in neuronal intranuclear inclusions and dystrophic neurites in brain. *Science.*, 1997, 277, 1990-1993.

Djousse, L; Knowlton, B; Cupples, LA; Marder, K; Shoulson, I; Myers, RH. Weight loss in early stage of Huntington's disease. *Neurology.*, 2002, 59, 1325-1330.

Doetsch, F. A niche for adult neural stem cells. *Curr. Opin. Genet. Dev.*, 2003, 13, 543-550.

Doetsch, F; Caille, I; Lim, DA; Garcia-Verdugo, JM; varez-Buylla, A. Subventricular zone astrocytes are neural stem cells in the adult mammalian brain. *Cell.*, 1999, 97, 703-716.

Dragatsis, I; Efstratiadis, A; Zeitlin, S. Mouse mutant embryos lacking huntingtin are rescued from lethality by wild-type extraembryonic tissues., 1998, 1529-39.

Dragatsis, I; Levine, MS; Zeitlin, S. Inactivation of Hdh in the brain and testis results in progressive neurodegeneration and sterility in mice. *Nat. Genet.*, 2000, 26, 300-306.

Dragunow, M; Faull, RL; Lawlor, P; *et al*. In situ evidence for DNA fragmentation in Huntington's disease striatum and Alzheimer's disease temporal lobes. *Neuroreport.*, 1995, 6, 1053-1057.

Duyao, MP; Auerbach, AB; Ryan, A; *et al*. Inactivation of the mouse Huntington's disease gene homolog Hdh. *Science.*, 1995, 269, 407-410.

Dyer, RB; McMurray, CT. Mutant protein in Huntington disease is resistant to proteolysis in affected brain. *Nat. Genet.*, 2001, 29, 270-278.

Ehninger, D; Kempermann, G. Regional effects of wheel running and environmental enrichment on cell genesis and microglia proliferation in the adult murine neocortex. *Cereb. Cortex.*, 2003, 13, 845-851.

Ekdahl, CT; Claasen, JH; Bonde, S; Kokaia, Z; Lindvall, O. Inflammation is detrimental for neurogenesis in adult brain. *Proc. Natl. Acad. Sci.*, U. S. A. 2003, 100, 13632-13637.

Emsley, JG; Hagg, T. Endogenous and exogenous ciliary neurotrophic factor enhances forebrain neurogenesis in adult mice. *Exp. Neurol.*, 2003, 183, 298-310.

Emsley, JG; Mitchell, BD; Kempermann, G; Macklis, JD. Adult neurogenesis and repair of the adult CNS with neural progenitors, precursors, and stem cells. *Prog. Neurobiol.*, 2005, 75, 321-341.

Eriksson, PS; Perfilieva, E; Bjork-Eriksson, T; *et al*. Neurogenesis in the adult human hippocampus. *Nat. Med.*, 1998, 4, 1313-1317.

Fabel, K; Fabel, K; Tam, B; *et al*. VEGF is necessary for exercise-induced adult hippocampal neurogenesis. *Eur. J. Neurosci.*, 2003, 18, 2803-2812.

Ferrante, RJ; Beal, MF; Kowall, NW; Richardson, EP; Jr.; Martin, JB. Sparing of acetylcholinesterase-containing striatal neurons in Huntington's disease. *Brain Res.*, 1987a, 411, 162-166.

Ferrante, RJ; Gutekunst, CA; Persichetti, F; *et al*. Heterogeneous topographic and cellular distribution of huntingtin expression in the normal human neostriatum. *J. Neurosci.*, 1997, 17, 3052-3063.

Ferrante, RJ; Kowall, NW; Beal, MF; Martin, JB; Bird, ED; Richardson, EP; Jr. Morphologic and histochemical characteristics of a spared subset of striatal neurons in Huntington's disease. *J. Neuropathol. Exp. Neurol.*, 1987b, 46, 12-27.

Ferrante, RJ; Kowall, NW; Beal, MF; Richardson, EP; Jr., Bird, ED; Martin, JB. Selective sparing of a class of striatal neurons in Huntington's disease. *Science.*, 1985, 230, 561-563.

Ferrante, RJ; Kowall, NW; Cipolloni, PB; Storey, E; Beal, MF. Excitotoxin lesions in primates as a model for Huntington's disease, histopathologic and neurochemical characterization. *Exp. Neurol.*, 1993, 119, 46-71.

Ferrante, RJ; Kowall, NW; Richardson, EP; Jr. Proliferative and degenerative changes in striatal spiny neurons in Huntington's disease: a combined study using the section-Golgi method and calbindin D28k immunocytochemistry. *J. Neurosci.*, 1991, 11, 3877-3887.

Ferrante, RJ; Kubilus, JK; Lee, J; *et al*. Histone deacetylase inhibition by sodium butyrate chemotherapy ameliorates the neurodegenerative phenotype in Huntington's disease mice. *J. Neurosci.*, 2003, 23, 9418-9427.

Ferrer, I; Goutan, E; Marin, C; Rey, MJ; Ribalta, T. Brain-derived neurotrophic factor in Huntington disease. *Brain Res.*, 2000, 866, 257-261.

Foltynie, T; Brayne, C; Barker, RA. The heterogeneity of idiopathic Parkinson's disease. *J. Neurol.*, 2002, 249, 138-145.

Friedlander, RM. Apoptosis and caspases in neurodegenerative diseases. *N. Engl. J. Med.*, 2003, 348, 1365-1375.

Frielingsdorf, H; Schwarz, K; Brundin, P; Mohapel, P. No evidence for new dopaminergic neurons in the adult mammalian substantia nigra. *Proc. Natl. Acad. Sci.*, U. S. A. 2004, 101, 10177-10182.

Gafni, J; Hermel, E; Young, JE; Wellington, CL; Hayden, MR; Ellerby, LM. Inhibition of calpain cleavage of huntingtin reduces toxicity: accumulation of calpain/caspase fragments in the nucleus. *J. Biol. Chem.*, 2004, 279, 20211-20220.

Gage, FH. Mammalian neural stem cells. *Science.*, 2000, 287, 1433-1438.

Gauthier, LR; Charrin, BC; Borrell-Pages, M; *et al.* Huntingtin controls neurotrophic support and survival of neurons by enhancing BDNF vesicular transport along microtubules. *Cell.*, 2004, 118, 127-138.

Gensburger, C; Labourdette, G; Sensenbrenner, M. Brain basic fibroblast growth factor stimulates the proliferation of rat neuronal precursor cells in vitro. *FEBS Lett.*, 1987, 217, 1-5.

Gheusi, G; Cremer, H; McLean, H; Chazal, G; Vincent, JD; Lledo, PM. Importance of newly generated neurons in the adult olfactory bulb for odor discrimination. *Proc. Natl. Acad. Sci.*, U. S. A. 2000, 97, 1823-1828.

Gil, JM; Leist, M; Popovic, N; Brundin, P; Petersen, A. Asialoerythropoietin is not effective in the R6/2 line of Huntington's disease mice. *BMC Neurosci.*, 2004, 5, 17., 17.

Gil, JM; Mohapel, P; Araujo, IM; *et al.* Reduced hippocampal neurogenesis in R6/2 transgenic Huntington's disease mice. *Neurobiol. Dis.*, 2005, 20, 744-751.

Glass, M; Dragunow, M; Faull, RL. The pattern of neurodegeneration in Huntington's disease: a comparative study of cannabinoid, dopamine, adenosine and GABA(A) receptor alterations in the human basal ganglia in Huntington's disease. *Neuroscience.*, 2000, 97, 505-519.

Goebel, HH; Heipertz, R; Scholz, W; Iqbal, K; Tellez-Nagel, I. Juvenile Huntington chorea: clinical, ultrastructural, and biochemical studies. *Neurology.*, 1978, 28, 23-31.

Goldberg, YP; McMurray, CT; Zeisler, J; *et al.* Increased instability of intermediate alleles in families with sporadic Huntington disease compared to similar sized intermediate alleles in the general population. *Hum. Mol. Genet.*, 1995, 4, 1911-1918.

Gotts, JE; Chesselet, MF. Vascular changes in the subventricular zone after distal cortical lesions. *Exp. Neurol.*, 2005, 194, 139-150.

Gould, E; McEwen, BS; Tanapat, P; Galea, LA; Fuchs, E. Neurogenesis in the dentate gyrus of the adult tree shrew is regulated by psychosocial stress and NMDA receptor activation. *J. Neurosci.*, 1997, 17, 2492-2498.

Gould, E; Reeves, AJ; Graziano, MS; Gross, CG. Neurogenesis in the neocortex of adult primates. *Science.*, 1999, 286, 548-552.

Grote, HE; Bull, ND; Howard, ML; *et al.* Cognitive disorders and neurogenesis deficits in Huntington's disease mice are rescued by fluoxetine. *Eur. J. Neurosci.*, 2005, 22, 2081-2088.

Guidetti, P; Charles, V; Chen, EY; *et al.* Early degenerative changes in transgenic mice expressing mutant huntingtin involve dendritic abnormalities but no impairment of mitochondrial energy production. *Exp. Neurol.*, 2001, 169, 340-350.

Gusella, JF; Wexler, NS; Conneally, PM; *et al.* A polymorphic DNA marker genetically linked to Huntington's disease. *Nature.*, 1983, 306, 234-238.

Gutekunst, CA; Levey, AI; Heilman, CJ; *et al*. Identification and localization of huntingtin in brain and human lymphoblastoid cell lines with anti-fusion protein antibodies. *Proc. Natl. Acad., Sci., U. S. A.* 1995, 92, 8710-8714.

Gutekunst, CA; Li, SH; Yi, H; Ferrante, RJ; Li, XJ; Hersch, SM. The cellular and subcellular localization of huntingtin-associated protein 1 (HAP1): comparison with huntingtin in rat and human. *J. Neurosci.*, 1998, 18, 7674-7686.

Gutekunst, CA; Li, SH; Yi, H; *et al*. Nuclear and neuropil aggregates in Huntington's disease: relationship to neuropathology. *J. Neurosci.*, 1999, 19, 2522-2534.

Hamilton, JM; Murphy, C; Paulsen, JS. Odor detection, learning, and memory in Huntington's disease. *J. Int. Neuropsychol. Soc.*, 1999, 5, 609-615.

Hamilton, JM; Wolfson, T; Peavy, GM; Jacobson, MW; Corey-Bloom, J. Rate and correlates of weight change in Huntington's disease. *J. Neurol. Neurosurg. Psychiatry.*, 2004, 75, 209-212.

Hansson, O; Castilho, RF; Korhonen, L; Lindholm, D; Bates, GP; Brundin, P. Partial resistance to malonate-induced striatal cell death in transgenic mouse models of Huntington's disease is dependent on age and CAG repeat length. *J. Neurochem.*, 2001a, 78, 694-703.

Hansson, O; Guatteo, E; Mercuri, NB; *et al*. Resistance to NMDA toxicity correlates with appearance of nuclear inclusions, behavioural deficits and changes in calcium homeostasis in mice transgenic for exon 1 of the huntington gene. *Eur. J. Neurosci.* 2001b, 14, 1492-1504.

Hansson, O; Petersen, A; Leist, M; Nicotera, P; Castilho, RF; Brundin, P. Transgenic mice expressing a Huntington's disease mutation are resistant to quinolinic acid-induced striatal excitotoxicity. *Proc. Natl. Acad. Sci.*, U. S. A. 1999, 20, 96, 8727-8732.

Harjes, P; Wanker, E. The hunt for huntingtin function: interaction partners tell many different stories. *Trends Biochem. Sci.*, 2003, 28, 425-433.

Harvey, DJ; Beckett, LA; Mungas, DM. Multivariate modeling of two associated cognitive outcomes in a longitudinal study. *J. Alzheimers Dis.*, 2003, 5, 357-365.

Haughey, NJ; Nath, A; Chan, SL; Borchard, AC; Rao, MS; Mattson, MP. Disruption of neurogenesis by amyloid beta-peptide, and perturbed neural progenitor cell homeostasis, in models of Alzheimer's disease. *J. Neurochem.*, 2002, 83, 1509-1524.

Hickey, MA; Morton, AJ. Mice transgenic for the Huntington's disease mutation are resistant to chronic 3-nitropropionic acid-induced striatal toxicity. *J. Neurochem.*, 2000, 75, 2163-2171.

Hitoshi, S; Alexson, T; Tropepe, V; *et al*. Notch pathway molecules are essential for the maintenance, but not the generation, of mammalian neural stem cells. *Genes Dev.* 2002, 16, 846-858.

Ho, AK; Sahakian, BJ; Brown, RG; *et al*. Profile of cognitive progression in early Huntington's disease. *Neurology.*, 2003, 61, 1702-1706.

Hockly, E; Cordery, PM; Woodman, B; *et al*. Environmental enrichment slows disease progression in R6/2 Huntington's disease mice. *Ann. Neurol.*, 2002, 51, 235-242.

Hodgson, JG; Agopyan, N; Gutekunst, CA; *et al*. A YAC mouse model for Huntington's disease with full-length mutant huntingtin, cytoplasmic toxicity, and selective striatal neurodegeneration. *Neuron.*, 1999, 23, 181-192.

Hoglinger, GU; Rizk, P; Muriel, MP; *et al*. Dopamine depletion impairs precursor cell proliferation in Parkinson disease. *Nat. Neurosci.*, 2004, 7, 726-735.

Howell, OW; Doyle, K; Goodman, JH; *et al*. Neuropeptide Y stimulates neuronal precursor proliferation in the post-natal and adult dentate gyrus. *J. Neurochem.*, 2005, 93, 560-570.

Howell, OW; Scharfman, HE; Herzog, H; Sundstrom, LE; Beck-Sickinger, A; Gray, WP. Neuropeptide Y is neuroproliferative for post-natal hippocampal precursor cells. *J. Neurochem.*, 2003, 86, 646-659.

Hurlbert, MS; Zhou, W; Wasmeier, C; Kaddis, FG; Hutton, JC; Freed, CR. Mice transgenic for an expanded CAG repeat in the Huntington's disease gene develop diabetes. *Diabetes.*, 1999, 48, 649-651.

Iannicola, C; Moreno, S; Oliverio, S; Nardacci, R; Ciofi-Luzzatto, A; Piacentini, M. Early alterations in gene expression and cell morphology in a mouse model of Huntington's disease. *J. Neurochem.*, 2000, 75, 830-839.

Jana, NR; Zemskov, EA; Wang, G; Nukina, N. Altered proteasomal function due to the expression of polyglutamine-expanded truncated N-terminal huntingtin induces apoptosis by caspase activation through mitochondrial cytochrome c release. *Hum. Mol. Genet.*, 2001, 10, 1049-1059.

Jankowsky, JL; Patterson, PH. The role of cytokines and growth factors in seizures and their sequelae. *Prog. Neurobiol.*, 2001, 63, 125-149.

Jarabek, BR; Yasuda, RP; Wolfe, BB. Regulation of proteins affecting NMDA receptor-induced excitotoxicity in a Huntington's mouse model. *Brain.*, 2004, 127, 505-516.

Jenkins, BG; Koroshetz, WJ; Beal, MF; Rosen, BR. Evidence for impairment of energy metabolism in vivo in Huntington's disease using localized 1H NMR spectroscopy. *Neurology.*, 1993, 43, 2689-2695.

Jin, K; Galvan, V; Xie, L; *et al*. Enhanced neurogenesis in Alzheimer's disease transgenic (PDGF-APPSw,Ind) mice. *Proc. Natl. Acad. Sci.*, U. S. A. 2004a, 101, 13363-13367.

Jin, K; LaFevre-Bernt, M; Sun, Y; *et al*. FGF-2 promotes neurogenesis and neuroprotection and prolongs survival in a transgenic mouse model of Huntington's disease. *Proc. Natl. Acad. Sci.*, U. S. A. 2005, 102, 18189-18194.

Jin, K; Mao, XO; Sun, Y; Xie, L; Greenberg, DA. Stem cell factor stimulates neurogenesis in vitro and in vivo. *J. Clin. Invest.*, 2002, 110, 311-319.

Jin, K; Minami, M; Lan, JQ; *et al*. Neurogenesis in dentate subgranular zone and rostral subventricular zone after focal cerebral ischemia in the rat. *Proc. Natl. Acad. Sci.*, U. S. A. 2001, 98, 4710-4715.

Jin, K; Peel, AL; Mao, XO; *et al*. Increased hippocampal neurogenesis in Alzheimer's disease. *Proc. Natl. Acad. Sci.*, U. S. A. 2004b, 101, 343-347.

Jin, K; Sun, Y; Xie, L; *et al*. Neurogenesis and aging: FGF-2 and HB-EGF restore neurogenesis in hippocampus and subventricular zone of aged mice. *Aging Cell.*, 2003, 2, 175-183.

Kaplan, MS. Neurogenesis in the 3-month-old rat visual cortex. *J. Comp. Neurol.*, 1981, 195, 323-338.

Katoh-Semba, R; Takeuchi, IK; Semba, R; Kato, K. Distribution of brain-derived neurotrophic factor in rats and its changes with development in the brain. *J. Neurochem.*, 1997, 69, 34-42.

Kempermann, G. Regulation of adult hippocampal neurogenesis: implications for novel theories of major depression. *Bipolar Disord.*, 2002, 4, 17-33.

Kempermann, G; Brandon, EP; Gage, FH. Environmental stimulation of 129/SvJ mice causes increased cell proliferation and neurogenesis in the adult dentate gyrus. *Curr. Biol.,* 1998a, 8, 939-942.

Kempermann, G; Jessberger, S; Steiner, B; Kronenberg, G. Milestones of neuronal development in the adult hippocampus. *Trends Neurosci.,* 2004, 27, 447-452.

Kempermann, G; Kuhn, HG; Gage, FH. Genetic influence on neurogenesis in the dentate gyrus of adult mice. *Proc. Natl. Acad. Sci.,* U. S. A. 1997, 94, 10409-10414.

Kempermann, G; Kuhn, HG; Gage, FH. Experience-induced neurogenesis in the senescent dentate gyrus. *J. Neurosci.,* 1998b, 18, 3206-3212.

Kennedy, L; Shelbourne, PF. Dramatic mutation instability in HD mouse striatum: does polyglutamine load contribute to cell-specific vulnerability in Huntington's disease? *Hum. Mol. Genet.,* 2000, 9, 2539-2544.

Kish, SJ; Shannak, K; Hornykiewicz, O. Elevated serotonin and reduced dopamine in subregionally divided Huntington's disease striatum. *Ann. Neurol.,* 1987, 22, 386-389.

Klapstein, GJ; Fisher, RS; Zanjani, H; *et al.* Electrophysiological and morphological changes in striatal spiny neurons in R6/2 Huntington's disease transgenic mice. *J. Neurophysiol.,* 2001, 86, 2667-2677.

Koeppen, AH. The pathogenesis of spinocerebellar ataxia. *Cerebellum.,* 2005, 4, 62-73.

Koketsu, D; Mikami, A; Miyamoto, Y; Hisatsune, T. Nonrenewal of neurons in the cerebral neocortex of adult macaque monkeys. *J. Neurosci.,* 2003, 23, 937-942.

Kokoeva, MV; Yin, H; Flier, JS. Neurogenesis in the hypothalamus of adult mice: potential role in energy balance. *Science.,* 2005, 310, 679-683.

Kornack, DR; Rakic, P. Continuation of neurogenesis in the hippocampus of the adult macaque monkey. *Proc. Natl. Acad. Sci.,* U. S. A. 1999, 96, 5768-5773.

Kremer, B; Squitieri, F; Telenius, H; *et al.* Molecular analysis of late onset Huntington's disease. *J. Med. Genet.,* 1993, 30, 991-995.

Kremer, HP. The hypothalamic lateral tuberal nucleus: normal anatomy and changes in neurological diseases. *Prog. Brain Res.,* 1992, 93, 249-61., 249-261.

Kuemmerle, S; Gutekunst, CA; Klein, AM; *et al.* Huntington aggregates may not predict neuronal death in Huntington's disease. *Ann. Neurol.,* 1999, 46, 842-849.

Kuhn, HG; Dickinson-Anson, H; Gage, FH. Neurogenesis in the dentate gyrus of the adult rat: age-related decrease of neuronal progenitor proliferation. *J. Neurosci.,* 1996, 16, 2027-2033.

Kuhn, HG; Winkler, J; Kempermann, G; Thal, LJ; Gage, FH. Epidermal growth factor and fibroblast growth factor-2 have different effects on neural progenitors in the adult rat brain. *J. Neurosci.,* 1997, 17, 5820-5829.

Kukekov, VG; Laywell, ED; Suslov, O; *et al.* Multipotent stem/progenitor cells with similar properties arise from two neurogenic regions of adult human brain. *Exp. Neurol.,* 1999, 156, 333-344.

Kulkarni, VA; Jha, S; Vaidya, VA. Depletion of norepinephrine decreases the proliferation, but does not influence the survival and differentiation, of granule cell progenitors in the adult rat hippocampus. *Eur. J. Neurosci.,* 2002, 16, 2008-2012.

Laforet, GA; Sapp, E; Chase, K; *et al.* Changes in cortical and striatal neurons predict behavioral and electrophysiological abnormalities in a transgenic murine model of Huntington's disease. *J. Neurosci.,* 2001, 21, 9112-9123.

Lai, K; Kaspar, BK; Gage, FH; Schaffer, DV. Sonic hedgehog regulates adult neural progenitor proliferation in vitro and in vivo. *Nat. Neurosci.*, 2003, 6, 21-27.

Lawrence, AD; Hodges, JR; Rosser, AE; *et al.* Evidence for specific cognitive deficits in preclinical Huntington's disease. *Brain.*, 1998, 121, 1329-1341.

Lawrence, AD; Watkins, LH; Sahakian, BJ; Hodges, JR; Robbins, TW. Visual object and visuospatial cognition in Huntington's disease: implications for information processing in corticostriatal circuits. *Brain.*, 2000, 123, 1349-1364.

Lazic, SE; Grote, H; Armstrong, RJ; *et al.* Decreased hippocampal cell proliferation in R6/1 Huntington's mice. *Neuroreport.*, 2004, 15, 811-813.

Lazic, SE; Grote, HE; Blakemore, C; *et al.* Neurogenesis in the R6/1 transgenic mouse model of Huntington's disease: effects of environmental enrichment. *Eur. J. Neurosci.*, 2006, 23, 1829-1838.

Lee, J; Duan, W; Long, JM; Ingram, DK; Mattson, MP. Dietary restriction increases the number of newly generated neural cells, and induces BDNF expression, in the dentate gyrus of rats. *J. Mol. Neurosci.*, 2000, 15, 99-108.

Lee, J; Duan, W; Mattson, MP. Evidence that brain-derived neurotrophic factor is required for basal neurogenesis and mediates, in part, the enhancement of neurogenesis by dietary restriction in the hippocampus of adult mice. *J. Neurochem.*, 2002, 82, 1367-1375.

Lee, ST; Chu, K; Park, JE; *et al.* Intravenous administration of human neural stem cells induces functional recovery in Huntington's disease rat model. *Neurosci. Res.*, 2005, 52, 243-249.

Lemkine, GF; Raj, A; Alfama, G; *et al.* Adult neural stem cell cycling in vivo requires thyroid hormone and its alpha receptor. *FASEB J.*, 2005, 19, 863-865.

Levine, MS; Klapstein, GJ; Koppel, A; *et al.* Enhanced sensitivity to N-methyl-D-aspartate receptor activation in transgenic and knockin mouse models of Huntington's disease. *J. Neurosci. Res.*, 1999, 58, 515-532.

Li, H; Li, SH; Johnston, H; Shelbourne, PF; Li, XJ. Amino-terminal fragments of mutant huntingtin show selective accumulation in striatal neurons and synaptic toxicity. *Nat. Genet.*, 2000, 25, 385-389.

Li, H; Li, SH; Yu, ZX; Shelbourne, P; Li, XJ. Huntingtin aggregate-associated axonal degeneration is an early pathological event in Huntington's disease mice. *J. Neurosci.*, 2001, 21, 8473-8481.

Lichtenwalner, RJ; Forbes, ME; Bennett, SA; Lynch, CD; Sonntag, WE; Riddle, DR. Intracerebroventricular infusion of insulin-like growth factor-I ameliorates the age-related decline in hippocampal neurogenesis. *Neuroscience.*, 2001, 107, 603-613.

Lie, DC; Colamarino, SA; Song, HJ; *et al.* Wnt signalling regulates adult hippocampal neurogenesis. *Nature.*, 2005, 437, 1370-1375.

Lie, DC; Dziewczapolski, G; Willhoite, AR; Kaspar, BK; Shults, CW; Gage, FH. The adult substantia nigra contains progenitor cells with neurogenic potential. *J. Neurosci.*, 2002, 22, 6639-6649.

Lievens, JC; Woodman, B; Mahal, A; *et al.* Impaired glutamate uptake in the R6 Huntington's disease transgenic mice. *Neurobiol. Dis.*, 2001, 8, 807-821.

Lim, DA; Tramontin, AD; Trevejo, JM; Herrera, DG; Garcia-Verdugo, JM; varez-Buylla, A. Noggin antagonizes BMP signaling to create a niche for adult neurogenesis. *Neuron.*, 2000, 28, 713-726.

Limke, TL; Rao, MS. Neural stem cells in aging and disease. *J. Cell Mol. Med.,* 2002, 6, 475-496.

Lin, CH; Tallaksen-Greene, S; Chien, WM; *et al.* Neurological abnormalities in a knock-in mouse model of Huntington's disease. *Hum. Mol. Genet.,* 2001, 10, 137-144.

Lione, LA; Carter, RJ; Hunt, MJ; Bates, GP; Morton, AJ; Dunnett, SB. Selective discrimination learning impairments in mice expressing the human Huntington's disease mutation. *J. Neurosci.,* 1999, 19, 10428-10437.

Liu, B; Hong, JS. Role of microglia in inflammation-mediated neurodegenerative diseases: mechanisms and strategies for therapeutic intervention. *J. Pharmacol. Exp. Ther.,* 2003, 304, 1-7.

Lobo, A; Launer, LJ; Fratiglioni, L; *et al.* Prevalence of dementia and major subtypes in Europe: A collaborative study of population-based cohorts. Neurologic Diseases in the Elderly Research Group. *Neurology.,* 2000, 54, S4-S9.

Lowenstein, DH; Arsenault, L. The effects of growth factors on the survival and differentiation of cultured dentate gyrus neurons. *J. Neurosci.,* 1996, 16, 1759-1769.

Lu, M; Grove, EA; Miller, RJ. Abnormal development of the hippocampal dentate gyrus in mice lacking the CXCR4 chemokine receptor. *Proc. Natl. Acad. Sci., U. S. A.* 2002, 99, 7090-7095.

Lunkes, A; Lindenberg, KS; Ben-Haiem, L; *et al.* Proteases acting on mutant huntingtin generate cleaved products that differentially build up cytoplasmic and nuclear inclusions. *Mol. Cell.,* 2002, 10, 259-269.

Luskin, MB. Restricted proliferation and migration of postnatally generated neurons derived from the forebrain subventricular zone. *Neuron.,* 1993, 11, 173-189.

Luthi-Carter, R; Strand, A; Peters, NL; *et al.* Decreased expression of striatal signaling genes in a mouse model of Huntington's disease. *Hum. Mol. Genet.,* 2000, 9, 1259-1271.

Ma, L; Morton, AJ; Nicholson, LF. Microglia density decreases with age in a mouse model of Huntington's disease. *Glia.,* 2003, 43, 274-280.

MacDonald, ME; Barnes, G; Srinidhi, J; *et al.* Gametic but not somatic instability of CAG repeat length in Huntington's disease. *J. Med. Genet.,* 1993, 30, 982-986.

MacGibbon, GA; Hamilton, LC; Crocker, SF; *et al.* Immediate-early gene response to methamphetamine, haloperidol, and quinolinic acid is not impaired in Huntington's disease transgenic mice. *J. Neurosci. Res.,* 2002, 67, 372-378.

Machold, R; Hayashi, S; Rutlin, M; *et al.* Sonic hedgehog is required for progenitor cell maintenance in telencephalic stem cell niches. *Neuron.,* 2003, 39, 937-950.

MacQueen, GM; Campbell, S; McEwen, BS; *et al.* Course of illness, hippocampal function, and hippocampal volume in major depression. *Proc. Natl. Acad. Sci., U. S. A.* 2003, 100, 1387-1392.

Magavi, SS; Leavitt, BR; Macklis, JD. Induction of neurogenesis in the neocortex of adult mice. *Nature.,* 2000, 405, 951-955.

Malberg, JE; Duman, RS. Cell proliferation in adult hippocampus is decreased by inescapable stress: reversal by fluoxetine treatment. *Neuropsychopharmacology.,* 2003, 28, 1562-1571.

Malberg, JE; Eisch, AJ; Nestler, EJ; Duman, RS. Chronic antidepressant treatment increases neurogenesis in adult rat hippocampus. *J. Neurosci.,* 2000, 20, 9104-9110.

Mangiarini, L; Sathasivam, K; Seller, M; *et al.* Exon 1 of the HD gene with an expanded CAG repeat is sufficient to cause a progressive neurological phenotype in transgenic mice. *Cell.,* 1996, 87, 493-506.

Manley, K; Shirley, TL; Flaherty, L; Messer, A. Msh2 deficiency prevents in vivo somatic instability of the CAG repeat in Huntington disease transgenic mice. *Nat. Genet.,* 1999, 23, 471-473.

Martindale, D; Hackam, A; Wieczorek, A; *et al.* Length of huntingtin and its polyglutamine tract influences localization and frequency of intracellular aggregates. *Nat. Genet.,* 1998, 18, 150-154.

Mason, HA; Ito, S; Corfas, G. Extracellular signals that regulate the tangential migration of olfactory bulb neuronal precursors: inducers, inhibitors, and repellents. *J. Neurosci.,* 2001, 21, 7654-7663.

Mason, ST; Fibiger, HC. Kainic acid lesions of the striatum: behavioural sequalae similar to Huntington's chorea. *Brain Res.,* 1978, 155, 313-329.

Mattson, MP; Duan, W; Wan, R; Guo, Z. Prophylactic activation of neuroprotective stress response pathways by dietary and behavioral manipulations. *NeuroRx.,* 2004, 1, 111-116.

Mende-Mueller, LM; Toneff, T; Hwang, SR; Chesselet, MF; Hook, VY. Tissue-specific proteolysis of Huntingtin (htt) in human brain: evidence of enhanced levels of N- and C-terminal htt fragments in Huntington's disease striatum. *J. Neurosci.,* 2001, 21, 1830-1837.

Metzler, M; Chen, N; Helgason, CD; *et al.* Life without huntingtin: normal differentiation into functional neurons. *J. Neurochem.,* 1999, 72, 1009-1018.

Metzler, M; Helgason, CD; Dragatsis, I; *et al.* Huntingtin is required for normal hematopoiesis. *Hum. Mol. Genet.,* 2000, 9, 387-394.

Michell, AW; Phillips, W; Barker, RA. Can endogenous stem cells be stimulated to repair the degenerating brain? *J. Pharm. Pharmacol.,* 2004, 56, 1201-1210.

Miller, MW; Nowakowski, RS. Use of bromodeoxyuridine-immunohistochemistry to examine the proliferation; migration and time of origin of cells in the central nervous system. *Brain Res.,* 1988, 457, 44-52.

Mitchell, IJ; Cooper, AJ; Griffiths, MR. The selective vulnerability of striatopallidal neurons. *Prog. Neurobiol.,* 1999, 59, 691-719.

Miyata, T; Maeda, T; Lee, JE. NeuroD is required for differentiation of the granule cells in the cerebellum and hippocampus. *Genes Dev.,* 1999, 13, 1647-1652.

Moberg, PJ; Pearlson, GD; Speedie, LJ; Lipsey, JR; Strauss, ME; Folstein, SE. Olfactory recognition: differential impairments in early and late Huntington's and Alzheimer's diseases. *J. Clin. Exp. Neuropsychol.,* 1987, 9, 650-664.

Moe, MC; Varghese, M; Danilov, AI; *et al.* Multipotent progenitor cells from the adult human brain: neurophysiological differentiation to mature neurons. *Brain.,* 2005, 128, 2189-2199.

Mohapel, P; Frielingsdorf, H; Haggblad, J; Zachrisson, O; Brundin, P. Platelet-derived growth factor (PDGF-BB) and brain-derived neurotrophic factor (BDNF) induce striatal neurogenesis in adult rats with 6-hydroxydopamine lesions. *Neuroscience.,* 2005, 132, 767-776.

Monje, ML; Mizumatsu, S; Fike, JR; Palmer, TD. Irradiation induces neural precursor-cell dysfunction. *Nat. Med.,* 2002, 8, 955-962.

Monje, ML; Toda, H; Palmer, TD. Inflammatory blockade restores adult hippocampal neurogenesis. *Science.*, 2003, 302, 1760-1765.

Montaron, MF; Petry, KG; Rodriguez, JJ; et al. Adrenalectomy increases neurogenesis but not PSA-NCAM expression in aged dentate gyrus. *Eur. J. Neurosci.*, 1999, 11, 1479-1485.

Morshead, CM; van der Kooy, D. Disguising adult neural stem cells. *Curr. Opin. Neurobiol.*, 2004, 14, 125-131.

Morton, AJ; Hunt, MJ; Hodges, AK; et al. A combination drug therapy improves cognition and reverses gene expression changes in a mouse model of Huntington's disease. *Eur. J. Neurosci.*, 2005a, 21, 855-870.

Morton, AJ; Lagan, MA; Skepper, JN; Dunnett, SB. Progressive formation of inclusions in the striatum and hippocampus of mice transgenic for the human Huntington's disease mutation. *J. Neurocytol.*, 2000, 29, 679-702.

Morton, AJ; Leavens, W. Mice transgenic for the human Huntington's disease mutation have reduced sensitivity to kainic acid toxicity. *Brain Res. Bull.*, 2000, 52, 51-59.

Morton, AJ; Wood, NI; Hastings, MH; Hurelbrink, C; Barker, RA; Maywood, ES. Disintegration of the sleep-wake cycle and circadian timing in Huntington's disease. *J. Neurosci.*, 2005b, 25, 157-163.

Murphy, KP; Carter, RJ; Lione, LA; et al. Abnormal synaptic plasticity and impaired spatial cognition in mice transgenic for exon 1 of the human Huntington's disease mutation. *J. Neurosci.*, 2000, 20, 5115-5123.

Myers, RH; Vonsattel, JP; Paskevich, PA; et al. Decreased neuronal and increased oligodendroglial densities in Huntington's disease caudate nucleus. *J. Neuropathol. Exp. Neurol.*, 1991, 50, 729-742.

Nacher, J; onso-Llosa, G; Rosell, DR; McEwen, BS. NMDA receptor antagonist treatment increases the production of new neurons in the aged rat hippocampus. *Neurobiol. Aging.*, 2003, 24, 273-284.

Nacher, J; Rosell, DR; McEwen, BS. Widespread expression of rat collapsin response-mediated protein 4 in the telencephalon and other areas of the adult rat central nervous system. *J. Comp. Neurol.*, 2000, 424, 628-639.

Nakatomi, H; Kuriu, T; Okabe, S; et al. Regeneration of hippocampal pyramidal neurons after ischemic brain injury by recruitment of endogenous neural progenitors. *Cell.*, 2002, 110, 429-441.

Nguyen, L; Malgrange, B; Breuskin, I; et al. Autocrine/paracrine activation of the GABA(A) receptor inhibits the proliferation of neurogenic polysialylated neural cell adhesion molecule-positive (PSA-NCAM+) precursor cells from postnatal striatum. *J. Neurosci.*, 2003, 23, 3278-3294.

Nichols, NR; Zieba, M; Bye, N. Do glucocorticoids contribute to brain aging? *Brain Res. Brain Res. Rev.*, 2001, 37, 273-286.

Nicniocaill, B; Haraldsson, B; Hansson, O; O'Connor, WT; Brundin, P. Altered striatal amino acid neurotransmitter release monitored using microdialysis in R6/1 Huntington transgenic mice. *Eur. J. Neurosci.*, 2001, 13, 206-210.

Nunes, MC; Roy, NS; Keyoung, HM; et al. Identification and isolation of multipotential neural progenitor cells from the subcortical white matter of the adult human brain. *Nat. Med.*, 2003, 9, 439-447.

Ona, VO; Li, M; Vonsattel, JP; *et al*. Inhibition of caspase-1 slows disease progression in a mouse model of Huntington's disease. *Nature.*, 1999, %20, 399, 263-267.

Packer, MA; Stasiv, Y; Benraiss, A; *et al*. Nitric oxide negatively regulates mammalian adult neurogenesis. *Proc. Natl. Acad. Sci., U. S. A.* 2003, 100, 9566-9571.

Palmer, TD; Markakis, EA; Willhoite, AR; Safar, F; Gage, FH. Fibroblast growth factor-2 activates a latent neurogenic program in neural stem cells from diverse regions of the adult CNS. *J. Neurosci.*, 1999, 19, 8487-8497.

Palmer, TD; Willhoite, AR; Gage, FH. Vascular niche for adult hippocampal neurogenesis. *J. Comp. Neurol.*, 2000, 425, 479-494.

Panov, AV; Gutekunst, CA; Leavitt, BR; *et al*. Early mitochondrial calcium defects in Huntington's disease are a direct effect of polyglutamines. *Nat. Neurosci.*, 2002, 5, 731-736.

Parent, JM. Injury-induced neurogenesis in the adult mammalian brain. *Neuroscientist.*, 2003, 9, 261-272.

Parent, JM; Valentin, VV; Lowenstein, DH. Prolonged seizures increase proliferating neuroblasts in the adult rat subventricular zone-olfactory bulb pathway. *J. Neurosci.*, 2002a, 22, 3174-3188.

Parent, JM; Vexler, ZS; Gong, C; Derugin, N; Ferriero, DM. Rat forebrain neurogenesis and striatal neuron replacement after focal stroke. *Ann. Neurol.*, 2002b, 52, 802-813.

Parent, JM; von dem Bussche, N; Lowenstein, DH. Prolonged seizures recruit caudal subventricular zone glial progenitors into the injured hippocampus. *Hippocampus.*, 2006, 22, 3174-3188.

Parent, JM; Yu, TW; Leibowitz, RT; Geschwind, DH; Sloviter, RS; Lowenstein, DH. Dentate granule cell neurogenesis is increased by seizures and contributes to aberrant network reorganization in the adult rat hippocampus. *J. Neurosci.*, 1997, 17, 3727-3738.

Pavese, N; Andrews, TC; Brooks, DJ; *et al*. Progressive striatal and cortical dopamine receptor dysfunction in Huntington's disease: a PET study. *Brain.*, 2003, 126, 1127-1135.

Pearson, AG; Curtis, MA; Waldvogel, HJ; Faull, RL; Dragunow, M. Activating transcription factor 2 expression in the adult human brain: association with both neurodegeneration and neurogenesis. *Neuroscience.*, 2005, 133, 437-451.

Pekcec, A; Loscher, W; Potschka, H. Neurogenesis in the adult rat piriform cortex. *Neuroreport.*, 2006, 17, 571-574.

Pencea, V; Bingaman, KD; Wiegand, SJ; Luskin, MB. Infusion of brain-derived neurotrophic factor into the lateral ventricle of the adult rat leads to new neurons in the parenchyma of the striatum, septum, thalamus, and hypothalamus. *J. Neurosci.*, 2001, 21, 6706-6717.

Perutz, MF; Johnson, T; Suzuki, M; Finch, JT. Glutamine repeats as polar zippers: their possible role in inherited neurodegenerative diseases. *Proc. Natl. Acad. Sci., U. S. A.* 1994, 91, 5355-5358.

Petersen, A; Chase, K; Puschban, Z; DiFiglia, M; Brundin, P; Aronin, N. Maintenance of susceptibility to neurodegeneration following intrastriatal injections of quinolinic acid in a new transgenic mouse model of Huntington's disease. *Exp. Neurol.*, 2002, 175, 297-300.

Petersen, A; Hansson, O; Puschban, Z; *et al*. Mice transgenic for exon 1 of the Huntington's disease gene display reduced striatal sensitivity to neurotoxicity induced by dopamine and 6-hydroxydopamine. *Eur. J. Neurosci.*, 2001, 14, 1425-1435.

Phillips, HS; Hains, JM; Armanini, M; Laramee, GR; Johnson, SA; Winslow, JW. BDNF mRNA is decreased in the hippocampus of individuals with Alzheimer's disease. *Neuron.*, 1991, 7, 695-702.

Phillips, W; Barker, R. Brain repair: do it yourself. *The Biochemist.*, 2006, 28, 15-18.

Phillips, W; Jennifer, MA; Barker, RA. Limbic neurogenesis/plasticity in the R6/2 mouse model of Huntington's disease. *Neuroreport.*, 2006, 17, 1623-1627.

Phillips, W; Morton, AJ; Barker, RA. Abnormalities of neurogenesis in the R6/2 mouse model of Huntington's disease are attributable to the in vivo microenvironment. *J. Neurosci.*, 2005, 25, 11564-11576.

Popoli, P; Pintor, A; Domenici, MR; *et al.* Blockade of striatal adenosine A2A receptor reduces, through a presynaptic mechanism, quinolinic acid-induced excitotoxicity, possible relevance to neuroprotective interventions in neurodegenerative diseases of the striatum. *J. Neurosci.*, 2002, 22, 1967-1975.

Portera-Cailliau, C; Hedreen, JC; Price, DL; Koliatsos, VE. Evidence for apoptotic cell death in Huntington disease and excitotoxic animal models. *J Neurosci*, 1995, 15, 3775-3787.

Pratley, RE; Salbe, AD; Ravussin, E; Caviness, JN. Higher sedentary energy expenditure in patients with Huntington's disease. *Ann. Neurol.*, 2000, 47, 64-70.

Prickaerts, J; Koopmans, G; Blokland, A; Scheepens, A. Learning and adult neurogenesis: survival with or without proliferation? *Neurobiol. Learn Mem.*, 2004, 81, 1-11.

Rakic, P. Adult neurogenesis in mammals: an identity crisis. *J. Neurosci.*, 2002, 22, 614-618.

Ravikumar, B; Vacher, C; Berger, Z; *et al.* Inhibition of mTOR induces autophagy and reduces toxicity of polyglutamine expansions in fly and mouse models of Huntington disease. *Nat. Genet.*, 2004, 36, 585-595.

Reddy, PH; Williams, M; Charles, V; *et al.* Behavioural abnormalities and selective neuronal loss in HD transgenic mice expressing mutated full-length HD cDNA. *Nat. Genet.*, 1998, 20, 198-202.

Reynolds, DS; Carter, RJ; Morton, AJ. Dopamine modulates the susceptibility of striatal neurons to 3-nitropropionic acid in the rat model of Huntington's disease. *J. Neurosci.*, 1998, 18, 10116-10127.

Reynolds, GP; Dalton, CF; Tillery, CL; Mangiarini, L; Davies, SW; Bates, GP. Brain neurotransmitter deficits in mice transgenic for the Huntington's disease mutation. *J. Neurochem.*, 1999, 72, 1773-1776.

Reynolds, GP; Garrett, NJ. Striatal dopamine and homovanillic acid in Huntington's disease. *J. Neural. Transm.*, 1986, 65, 151-155.

Reynolds, GP; Pearson, SJ. Decreased glutamic acid and increased 5-hydroxytryptamine in Huntington's disease brain. *Neurosci. Lett.*, 1987, 78, 233-238.

Ribchester, RR; Thomson, D; Wood, NI; *et al.* Progressive abnormalities in skeletal muscle and neuromuscular junctions of transgenic mice expressing the Huntington's disease mutation. *Eur. J. Neurosci.*, 2004, 20, 3092-3114.

Ridley, RM; Frith, CD; Crow, TJ; Conneally, PM. Anticipation in Huntington's disease is inherited through the male line but may originate in the female. *J. Med. Genet.*, 1988, 25, 589-595.

Rigamonti, D; Sipione, S; Goffredo, D; Zuccato, C; Fossale, E; Cattaneo, E. Huntingtin's neuroprotective activity occurs via inhibition of procaspase-9 processing. *J. Biol. Chem.*, 2001, 276, 14545-14548.

Rochefort, C; Gheusi, G; Vincent, JD; Lledo, PM. Enriched odor exposure increases the number of newborn neurons in the adult olfactory bulb and improves odor memory. *J. Neurosci.*, 2002, 22, 2679-2689.

Rosas, HD; Feigin, AS; Hersch, SM. Using advances in neuroimaging to detect, understand, and monitor disease progression in Huntington's disease. *NeuroRx.*, 2004, 1, 263-272.

Rosas, HD; Koroshetz, WJ; Chen, YI; *et al.* Evidence for more widespread cerebral pathology in early HD: an MRI-based morphometric analysis. *Neurology.*, 2003, 60, 1615-1620.

Rosas, HD; Liu, AK; Hersch, S; *et al.* Regional and progressive thinning of the cortical ribbon in Huntington's disease. *Neurology.*, 2002, 58, 695-701.

Ross, CA. Huntington's disease: new paths to pathogenesis. *Cell.*, 2004, 118, 4-7.

Roy, NS; Wang, S; Jiang, L; *et al.* In vitro neurogenesis by progenitor cells isolated from the adult human hippocampus. *Nat. Med.*, 2000, 6, 271-277.

Rubinsztein, DC. Lessons from animal models of Huntington's disease. *Trends Genet.*, 2002, 18, 202-209.

Rubinsztein, DC; Leggo, J; Chiano, M; *et al.* Homozygotes and heterozygotes for ciliary neurotrophic factor null alleles do not show earlier onset of Huntington's disease. *Neurology.*, 1997, 49, 890-892.

Rubinsztein, DC; Leggo, J; Coles, R; *et al.* Phenotypic characterization of individuals with 30-40 CAG repeats in the Huntington disease (HD) gene reveals HD cases with 36 repeats and apparently normal elderly individuals with 36-39 repeats. *Am. J. Hum. Genet.*, 1996, 59, 16-22.

Sanai, N; Tramontin, AD; Quinones-Hinojosa, A; *et al.* Unique astrocyte ribbon in adult human brain contains neural stem cells but lacks chain migration. *Nature.*, 2004, 19; 427, 740-744.

Sanberg, PR; Calderon, SF; Giordano, M; Tew, JM; Norman, AB. The quinolinic acid model of Huntington's disease: locomotor abnormalities. *Exp. Neurol.*, 1989, 105, 45-53.

Sanchez, I; Mahlke, C; Yuan, J. Pivotal role of oligomerization in expanded polyglutamine neurodegenerative disorders. *Nature.*, 2003, 421, 373-379.

Santarelli, L; Saxe, M; Gross, C; *et al.* Requirement of hippocampal neurogenesis for the behavioral effects of antidepressants. *Science.*, 2003, 301, 805-809.

Sapolsky, RM. Stress hormones: good and bad. *Neurobiol. Dis.*, 2000, 7, 540-542.

Sapolsky, RM. Is impaired neurogenesis relevant to the affective symptoms of depression? *Biol. Psychiatry.*, 2004, 56, 137-139.

Sapp, E; Kegel, KB; Aronin, N; *et al.* Early and progressive accumulation of reactive microglia in the Huntington disease brain. *J. Neuropathol. Exp. Neurol.*, 2001, 60, 161-172.

Sapp, E; Schwarz, C; Chase, K; *et al.* Huntingtin localization in brains of normal and Huntington's disease patients. *Ann. Neurol.*, 1997, 42, 604-612.

Saudou, F; Finkbeiner, S; Devys, D; Greenberg, ME. Huntingtin acts in the nucleus to induce apoptosis but death does not correlate with the formation of intranuclear inclusions. *Cell.*, 1998, 95, 55-66.

Schauwecker, PE. Modulation of cell death by mouse genotype: differential vulnerability to excitatory amino acid-induced lesions. *Exp. Neurol.*, 2002, 178, 219-235.

Schauwecker, PE; Steward, O. Genetic determinants of susceptibility to excitotoxic cell death: implications for gene targeting approaches. *Proc. Natl. Acad. Sci., U. S. A.* 1997, 94, 4103-4108.

Scherzinger, E; Lurz, R; Turmaine, M; *et al.* Huntingtin-encoded polyglutamine expansions form amyloid-like protein aggregates in vitro and in vivo. *Cell.,* 1997, 90, 549-558.

Schiefer, J; Alberty, A; Dose, T; Oliva, S; Noth, J; Kosinski, CM. Huntington's disease transgenic mice are resistant to global cerebral ischemia. *Neurosci. Lett.,* 2002, 334, 99-102.

Schilling, G; Savonenko, AV; Klevytska, A; *et al.* Nuclear-targeting of mutant huntingtin fragments produces Huntington's disease-like phenotypes in transgenic mice. *Hum. Mol. Genet.,* 2004, 13, 1599-1610.

Schmidt-Hieber, C; Jonas, P; Bischofberger, J. Enhanced synaptic plasticity in newly generated granule cells of the adult hippocampus. *Nature.,* 2004, 429, 184-187.

Seaberg, RM; van der Kooy, D. Adult rodent neurogenic regions: the ventricular subependyma contains neural stem cells, but the dentate gyrus contains restricted progenitors. *J. Neurosci.,* 2002, 22, 1784-1793.

Seri, B; Garcia-Verdugo, JM; McEwen, BS; varez-Buylla, A. Astrocytes give rise to new neurons in the adult mammalian hippocampus. *J. Neurosci.,* 2001, 21, 7153-7160.

Sharp, AH; Loev, SJ; Schilling, G; *et al.* Widespread expression of Huntington's disease gene (IT15) protein product. *Neuron.,* 1995, 14, 1065-1074.

Sheline, YI; Gado, MH; Kraemer, HC. Untreated depression and hippocampal volume loss. *Am. J. Psychiatry.,* 2003, 160, 1516-1518.

Shen, Q; Goderie, SK; Jin, L; *et al.* Endothelial cells stimulate self-renewal and expand neurogenesis of neural stem cells. *Science.,* 2004, 304, 1338-1340.

Shetty, AK; Rao, MS; Hattiangady, B; Zaman, V; Shetty, GA. Hippocampal neurotrophin levels after injury, Relationship to the age of the hippocampus at the time of injury. *J. Neurosci. Res.,* 2004, 78, 520-532.

Shingo, T; Gregg, C; Enwere, E; *et al.* Pregnancy-stimulated neurogenesis in the adult female forebrain mediated by prolactin. *Science.,* 2003, 299, 117-120.

Shingo, T; Sorokan, ST; Shimazaki, T; Weiss, S. Erythropoietin regulates the in vitro and in vivo production of neuronal progenitors by mammalian forebrain neural stem cells. *J. Neurosci.,* 2001, 21, 9733-9743.

Shors, TJ; Miesegaes, G; Beylin, A; Zhao, M; Rydel, T; Gould, E. Neurogenesis in the adult is involved in the formation of trace memories. *Nature.,* 2001, 410, 372-376.

Shors, TJ; Townsend, DA; Zhao, M; Kozorovitskiy, Y; Gould, E. Neurogenesis may relate to some but not all types of hippocampal-dependent learning. *Hippocampus.,* 2002, 12, 578-584.

Shuttleworth, CW; Connor, JA. Strain-dependent differences in calcium signaling predict excitotoxicity in murine hippocampal neurons. *J. Neurosci.,* 2001, 21, 4225-4236.

Siesling, S; Vegter-van, DV; Roos, RA. Juvenile Huntington disease in the Netherlands. *Pediatr. Neurol.,* 1997, 17, 37-43.

Silvestri, R; Raffaele, M; De, DP; *et al.* Sleep features in Tourette's syndrome, neuroacanthocytosis and Huntington's chorea. *Neurophysiol. Clin.,* 1995, 25, 66-77.

Slow, EJ; Graham, RK; Osmand, AP; *et al.* Absence of behavioral abnormalities and neurodegeneration in vivo despite widespread neuronal huntingtin inclusions. *Proc. Natl. Acad. Sci., U. S. A.* 2005, 102, 11402-11407.

Smith, MT; Pencea, V; Wang, Z; Luskin, MB; Insel, TR. Increased number of BrdU-labeled neurons in the rostral migratory stream of the estrous prairie vole. *Horm. Behav.,* 2001, 39, 11-21.

Snell, RG; MacMillan, JC; Cheadle, JP; *et al.* Relationship between trinucleotide repeat expansion and phenotypic variation in Huntington's disease. *Nat. Genet.,* 1993, 4, 393-397.

Song, H; Stevens, CF; Gage, FH. Astroglia induce neurogenesis from adult neural stem cells. *Nature.,* 2002, 417, 39-44.

Sonino, N; Fava, GA. Psychiatric disorders associated with Cushing's syndrome. Epidemiology, pathophysiology and treatment. *CNS Drugs.,* 2001, 15, 361-373.

Sorensen, SA; Fenger, K. Causes of death in patients with Huntington's disease and in unaffected first degree relatives. *J. Med. Genet.,* 1992, 29, 911-914.

Spires, TL; Grote, HE; Garry, S; *et al.* Dendritic spine pathology and deficits in experience-dependent dendritic plasticity in R6/1 Huntington's disease transgenic mice. *Eur. J. Neurosci.,* 2004a, 19, 2799-2807.

Spires, TL; Grote, HE; Varshney, NK; *et al.* Environmental enrichment rescues protein deficits in a mouse model of Huntington's disease, indicating a possible disease mechanism. *J. Neurosci.,* 2004b, 24, 2270-2276.

Sprengelmeyer, R; Schroeder, U; Young, AW; Epplen, JT. Disgust in pre-clinical Huntington's disease: A longitudinal study. *Neuropsychologia.,* 2006, 44, 518-533.

Squitieri, F; Gellera, C; Cannella, M; *et al.* Homozygosity for CAG mutation in Huntington disease is associated with a more severe clinical course. *Brain.,* 2003, 126, 946-955.

Steffan, JS; Kazantsev, A; Spasic-Boskovic, O; *et al.* The Huntington's disease protein interacts with p53 and CREB-binding protein and represses transcription. *Proc. Natl. Acad. Sci., U. S. A.,* 2000, 97, 6763-6768.

Stemmelin, J; Lazarus, C; Cassel, S; Kelche, C; Cassel, JC. Immunohistochemical and neurochemical correlates of learning deficits in aged rats. *Neuroscience.,* 2000, 96, 275-289.

Sun, L; Lee, J; Fine, HA. Neuronally expressed stem cell factor induces neural stem cell migration to areas of brain injury. *J. Clin. Invest.,* 2004, 113, 1364-1374.

Sun, Y; Savanenin, A; Reddy, PH; Liu, YF. Polyglutamine-expanded huntingtin promotes sensitization of N-methyl-D-aspartate receptors via post-synaptic density 95. *J. Biol. Chem.,* 2001, 276, 24713-24718.

Tabrizi, SJ; Workman, J; Hart, PE; *et al.* Mitochondrial dysfunction and free radical damage in the Huntington R6/2 transgenic mouse. *Ann. Neurol.,* 2000, 47, 80-86.

Takemura, NU. Evidence for neurogenesis within the white matter beneath the temporal neocortex of the adult rat brain. *Neuroscience.,* 2005, 134, 121-132.

Tanapat, P; Hastings, NB; Reeves, AJ; Gould, E. Estrogen stimulates a transient increase in the number of new neurones in the dentate gyrus of the adult female rat. *J. Neurosci.,* 1999, 19, 5792-5801.

Tarditi, A; Camurri, A; Varani, K; *et al.* Early and transient alteration of adenosine A(2A) receptor signaling in a mouse model of Huntington disease. *Neurobiol. Dis.,* 2006, 23, 44-53.

Tattersfield, AS; Croon, RJ; Liu, YW; Kells, AP; Faull, RL; Connor, B. Neurogenesis in the striatum of the quinolinic acid lesion model of Huntington's disease. *Neuroscience.,* 2004, 127, 319-332.

Tekin, S; Cummings, JL. Frontal-subcortical neuronal circuits and clinical neuropsychiatry: an update. *J. Psychosom. Res.*, 2002, 53, 647-654.

Telenius, H; Kremer, B; Goldberg, YP; *et al.* Somatic and gonadal mosaicism of the Huntington disease gene CAG repeat in brain and sperm. *Nat. Genet.* 1994, 6, 409-414.

Telenius, H; Kremer, HP; Theilmann, J; *et al.* Molecular analysis of juvenile Huntington disease: the major influence on (CAG)n repeat length is the sex of the affected parent. *Hum. Mol. Genet.*, 1993, 2, 1535-1540.

Thieben, MJ; Duggins, AJ; Good, CD; *et al.* The distribution of structural neuropathology in pre-clinical Huntington's disease. *Brain.*, 2002, 125, 1815-1828.

Tooyama, I; Kremer, HP; Hayden, MR; Kimura, H; McGeer, EG; McGeer, PL. Acidic and basic fibroblast growth factor-like immunoreactivity in the striatum and midbrain in Huntington's disease. *Brain Res.*, 1993, 610, 1-7.

Tozuka, Y; Fukuda, S; Namba, T; Seki, T; Hisatsune, T. GABAergic excitation promotes neuronal differentiation in adult hippocampal progenitor cells. *Neuron.* 2005, 47, 803-815.

Tropepe, V; Coles, BL; Chiasson, BJ; *et al.* Retinal stem cells in the adult mammalian eye. *Science.*, 2000, 287, 2032-2036.

Tropepe, V; Craig, CG; Morshead, CM; van der, KD. Transforming growth factor-alpha null and senescent mice show decreased neural progenitor cell proliferation in the forebrain subependyma. *J. Neurosci.*, 1997, 17, 7850-7859.

Turmaine, M; Raza, A; Mahal, A; Mangiarini, L; Bates, GP; Davies, SW. Nonapoptotic neurodegeneration in a transgenic mouse model of Huntington's disease. *Proc. Natl. Acad. Sci., U. S. A.* 2000, 97, 8093-8097.

Ueki, T; Tanaka, M; Yamashita, K; *et al.* A novel secretory factor, Neurogenesin-1, provides neurogenic environmental cues for neural stem cells in the adult hippocampus. *J. Neurosci.*, 2003, 23, 11732-11740.

Ullian, EM; Sapperstein, SK; Christopherson, KS; Barres, BA. Control of synapse number by glia. *Science.*, 2001, 291, 657-661.

van Dellen, A; Grote, HE; Hannan, AJ. Gene-environment interactions, neuronal dysfunction and pathological plasticity in Huntington's disease. *Clin. Exp. Pharmacol. Physiol.*, 2005, 32, 1007-1019.

Van Kampen, JM; Hagg, T; Robertson, HA. Induction of neurogenesis in the adult rat subventricular zone and neostriatum following dopamine D receptor stimulation. *Eur. J. Neurosci.*, 2004, 19, 2377-2387.

van Praag, H; Kempermann, G; Gage, FH. Neural consequences of environmental enrichment. *Nat. Rev. Neurosci.*, 2000, 1, 191-198.

van Praag, H; Schinder, AF; Christie, BR; Toni, N; Palmer, TD; Gage, FH. Functional neurogenesis in the adult hippocampus. *Nature.*, 2002, 415, 1030-1034.

van Praag, H; Shubert, T; Zhao, C; Gage, FH. Exercise enhances learning and hippocampal neurogenesis in aged mice. *J. Neurosci.*, 2005, 25, 8680-8685.

Veizovic, T; Beech, JS; Stroemer, RP; Watson, WP; Hodges, H. Resolution of stroke deficits following contralateral grafts of conditionally immortal neuroepithelial stem cells. *Stroke.*, 2001, 32, 1012-1019.

Vetter, JM; Jehle, T; Heinemeyer, J; *et al.* Mice transgenic for exon 1 of Huntington's disease: properties of cholinergic and dopaminergic pre-synaptic function in the striatum. *J. Neurochem.*, 2003, 85, 1054-1063.

Vollmayr, B; Simonis, C; Weber, S; Gass, P; Henn, F. Reduced cell proliferation in the dentate gyrus is not correlated with the development of learned helplessness. *Biol Psychiatry.*, 2003, 54, 1035-1040.

Vonsattel, JP; Myers RH; Stevens, TJ; Ferrante, RJ; Bird, ED; Richardson, EP; Jr. Neuropathological classification of Huntington's disease. *J. Neuropathol. Exp. Neurol.* 1985, 44, 559-577.

Vythilingam, M; Heim, C; Newport, J; *et al.* Childhood trauma associated with smaller hippocampal volume in women with major depression. *Am. J. Psychiatry.*, 2002, 159, 2072-2080.

Waeber, C; Palacios, JM. Serotonin-1 receptor binding sites in the human basal ganglia are decreased in Huntington's chorea but not in Parkinson's disease: a quantitative in vitro autoradiography study. *Neuroscience.*, 1989, 32, 337-347.

Wang, R; Dineley, KT; Sweatt, JD; Zheng, H. Presenilin 1 familial Alzheimer's disease mutation leads to defective associative learning and impaired adult neurogenesis. *Neuroscience.*, 2004, 126, 305-312.

Wang, S; Scott, BW; Wojtowicz, JM. Heterogenous properties of dentate granule neurons in the adult rat. *J. Neurobiol.*, 2000, 42, 248-257.

Weiss, S; Dunne, C; Hewson, J; *et al.* Multipotent CNS stem cells are present in the adult mammalian spinal cord and ventricular neuroaxis. *J. Neurosci.*, 1996, 16, 7599-7609.

Wellington, CL; Ellerby, LM; Gutekunst, CA; *et al.* Caspase cleavage of mutant huntingtin precedes neurodegeneration in Huntington's disease. *J. Neurosci.*, 2002, 22, 7862-7872.

Wellington, CL; Leavitt, BR; Hayden, MR. Huntington disease: new insights on the role of huntingtin cleavage. *J. Neural. Transm. Suppl.*, 2000, 1-17.

Wen, PH; Hof, PR; Chen, X; *et al.* The presenilin-1 familial Alzheimer disease mutant P117L impairs neurogenesis in the hippocampus of adult mice. *Exp. Neurol.*, 2004, 188, 224-237.

Wexler, NS; Lorimer, J; Porter, J; *et al.* Venezuelan kindreds reveal that genetic and environmental factors modulate Huntington's disease age of onset. *Proc. Natl. Acad. Sci., U. S. A.* 2004, 101, 3498-3503.

Wheeler, VC; White, JK; Gutekunst, CA; *et al.* Long glutamine tracts cause nuclear localization of a novel form of huntingtin in medium spiny striatal neurons in HdhQ92 and HdhQ111 knock-in mice. *Hum. Mol. Genet.*, 2000, 9, 503-513.

White, JK; Auerbach, W; Duyao, MP; *et al.* Huntingtin is required for neurogenesis and is not impaired by the Huntington's disease CAG expansion. *Nat. Genet.*, 1997, 17, 404-410.

Wichterle, H; Garcia-Verdugo, JM; varez-Buylla, A. Direct evidence for homotypic, glia-independent neuronal migration. *Neuron.*, 1997, 18, 779-791.

Wilkinson, FL; Nguyen, TM; Manilal, SB; *et al.* Localization of rabbit huntingtin using a new panel of monoclonal antibodies. *Brain Res. Mol. Brain Res.*, 1999, 69, 10-20.

Winocur, G; Wojtowicz, JM; Sekeres, M; Snyder, JS; Wang, S. Inhibition of neurogenesis interferes with hippocampus-dependent memory function. *Hippocampus.*, 2006, 16, 296-304.

Wong, EY; Herbert, J. The corticoid environment, a determining factor for neural progenitors' survival in the adult hippocampus. *Eur. J. Neurosci.*, 2004, 20, 2491-2498.

Yamamoto, A; Lucas, J; Hen, R. Reversal of neuropathology and motor dysfunction in a conditional model of Huntington's disease. *Cell.*, 2000, 101, 57-66.

Yang, W; Dunlap, JR; Andrews, RB; Wetzel, R. Aggregated polyglutamine peptides delivered to nuclei are toxic to mammalian cells. *Hum. Mol. Genet.*, 2002, 11, 2905-2917.

Yoshimi, K; Ren, YR; Seki, T; *et al.* Possibility for neurogenesis in substantia nigra of parkinsonian brain. *Ann. Neurol.*, 2005, 58, 31-40.

Yoshimizu, T; Chaki, S. Increased cell proliferation in the adult mouse hippocampus following chronic administration of group II metabotropic glutamate receptor antagonist, MGS0039. *Biochem. Biophys. Res. Commun.*, 2004, 315, 493-496.

Zeron, MM; Fernandes, HB; Krebs, C; *et al.* Potentiation of NMDA receptor-mediated excitotoxicity linked with intrinsic apoptotic pathway in YAC transgenic mouse model of Huntington's disease. *Mol. Cell Neurosci.*, 2004a, 25, 469-479.

Zhao, M; Momma, S; Delfani, K; *et al.* Evidence for neurogenesis in the adult mammalian substantia nigra. *Proc. Natl. Acad. Sci.*, U. S. A. 2003, 100, 7925-7930.

Zhou, H; Cao, F; Wang, Z; *et al.* Huntingtin forms toxic NH2-terminal fragment complexes that are promoted by the age-dependent decrease in proteasome activity. *J. Cell Biol.*, 2003, 163, 109-118.

Ziu, M; Schmidt, NO; Cargioli, TG; Aboody, KS; Black, PM; Carroll, RS. Glioma-produced extracellular matrix influences brain tumor tropism of human neural stem cells. *J. Neurooncol.*, 2006, 79, 125-133.

Zuccato, C; Ciammola, A; Rigamonti, D; *et al.* Loss of huntingtin-mediated BDNF gene transcription in Huntington's disease. *Science.*, 2001, 20, 293, 493-498.

Zuccato, C; Liber, D; Ramos, C; *et al.* Progressive loss of BDNF in a mouse model of Huntington's disease and rescue by BDNF delivery. *Pharmacol. Res.*, 2005, 52, 133-139.

Index

A

abatement, 334
ABC, 593
abduction, 312
ablations, 83, 85, 109, 183
abnormalities, 1, 2, 4, 6, 8, 9, 10, 11, 17, 25, 28, 30, 33, 34, 37, 40, 43, 45, 49, 50, 195, 260, 306, 314, 342, 343, 347, 361, 398, 413, 416, 420, 428, 431, 470, 477, 518, 519, 529, 532, 538, 548, 558, 578, 579, 581, 613, 751, 791, 795
Aboriginal, 596
abortion, 759, 771, 772, 773
Abraham, 500, 514, 554, 671
absorption, 248
abstinence, 68
abstract, 179, 196
abstraction, 16, 164, 241, 409
abuse, xiii, 13, 25, 26, 143, 308, 340, 350, 351, 575, 597, 604, 730, 805
abusive, 604, 780
AC, 35, 43, 66, 87, 175, 230, 231, 233, 234, 235, 236, 241, 248, 296, 367, 439, 470, 559, 819, 842
academic, xiv, 408, 415, 428, 469, 470, 471, 508, 509, 511, 597
academic learning, 508
academic performance, 471
ACC, 312, 341, 342, 347
access, 80, 81, 82, 83, 102, 110, 229, 264, 292, 383, 591, 662, 753, 754, 761, 762, 771, 780, 782
accessibility, 285
accidental, 309
accidents, 786, 796

accounting, 361, 680, 777, 825
accuracy, 229, 237, 238, 239, 248, 253, 254, 263, 265, 286, 291, 292, 451, 462, 466, 468, 469, 752, 759
acetate, 103
acetylation, 640
acetylcholine, 47, 51, 54, 56, 63, 64, 65, 66, 67, 69, 70, 137, 196, 392, 423, 529, 550, 551, 667, 669, 822
acetylcholinesterase, 51, 486, 488, 544, 545, 612, 840
acetylcholinesterase inhibitor, 51, 486, 488
achievement, xiv, 408, 470, 593, 775
acid, 5, 28, 29, 34, 38, 44, 45, 53, 103, 106, 118, 137, 259, 281, 322, 331, 334, 335, 355, 398, 408, 412, 421, 425, 438, 441, 480, 488, 492, 535, 545, 635, 644, 647, 648, 674, 685, 802, 818, 833, 834, 835, 837, 838, 839, 842, 846, 847, 848, 849, 850, 851, 853
acidic, 481, 531, 540, 559, 801, 834
acoustic, 70, 118, 164, 289, 313
actin, 529, 537
action potential, 86, 285, 421
activity level, 162, 438
actuality, 235
actuation, 563, 564
acute, 31, 41, 59, 139, 282, 299, 304, 312, 332, 389, 399, 400, 401, 417, 429, 478, 525
acute stress, 139
AD pathological changes, 479
Adams, 63, 142, 149, 297, 537, 555, 570
adaptability, 255
adaptation, x, xvii, 75, 77, 116, 128, 159, 515, 587, 589, 590, 594, 595, 596, 597, 698, 710, 735
adaptations, 70, 462, 593
adaptive functioning, 580

addiction, 138, 146
additives, 612
adducts, 525
adenine, 678, 680, 681, 807, 813, 833, 834
adenosine, 423, 828, 833, 841, 850, 853
ADHD, vii, xiv, 146, 149, 183, 189, 190, 195, 282, 443, 444, 445, 446, 447, 448, 449, 450, 451, 452, 453, 454, 455, 456, 457, 458, 459, 460, 461, 462, 463, 464, 465, 466, 467, 468, 469, 470, 471, 472, 473, 474, 500, 513, 517, 580, 584
ADHD-C, 458, 461, 473, 474
ADHD-CT, 458, 461, 473, 474
adhesion, 35, 477, 526, 529, 536, 537, 583, 834, 835, 848
adhesions, 666
adjunctive therapy, 70, 328, 420, 436
adjustment, 142, 436
administration, xii, 21, 36, 43, 47, 55, 56, 77, 111, 183, 185, 187, 228, 255, 398, 420, 423, 429, 440, 445, 448, 457, 462, 469, 484, 487, 493, 494, 561, 567, 568, 569, 587, 588, 589, 591, 592, 594, 804, 818, 845, 856
administrative, 591, 594
administrators, 587, 592
adolescence, ix, 2, 5, 6, 9, 32, 37, 215, 223, 409, 424, 440, 500, 519
adolescent, 193
adolescents, 143, 148, 150, 195, 319, 396, 414, 427, 433, 473
ADP, 643
adrenoceptors, 177, 181, 182, 184, 187, 188, 191, 193, 196
adrenoceptors (ARs), 177, 181, 182, 184, 187, 188, 191, 193, 196
adulthood, ix, xx, 2, 8, 9, 41, 142, 223, 224, 619, 753, 758, 759, 799, 800, 809, 811
adults, xv, xvii, 144, 146, 151, 179, 180, 187, 189, 191, 216, 217, 222, 267, 297, 325, 353, 355, 358, 362, 396, 415, 420, 421, 425, 430, 433, 437, 457, 460, 471, 482, 484, 487, 499, 500, 501, 503, 504, 507, 508, 513, 514, 516, 517, 518, 519, 561, 563, 565, 570, 600, 601, 602, 603, 604, 605, 606, 649, 714
advanced glycation end products, 493, 551
advancement, 172, 711
advancements, xii, xv, 256, 260, 265, 271, 291, 293, 499
adverse effects, 21, 278, 320, 368, 381, 400, 414, 422, 429, 484, 695, 735
adverse event, 11, 237, 306, 309, 314, 326, 328, 419, 484, 485, 486, 488, 496
aerosol, 570

aerosols, 569, 570
aetiology, xvii, 574
affective disorder, 25, 57, 401, 439, 733
affective meaning, 157
Africa, xvii, 587, 590, 595, 596, 719, 731
African American, 511
African-American, 490, 510, 556
ageing, 538, 556, 606, 687
agent, 21, 486, 491
agents, xvii, 8, 20, 21, 22, 47, 51, 53, 55, 56, 64, 66, 70, 105, 137, 182, 188, 278, 342, 396, 427, 441, 474, 482, 490, 609, 612, 614
aggregates, 483, 530, 620, 634, 637, 638, 639, 640, 641, 647, 648, 651, 652, 656, 658, 659, 660, 661, 662, 663, 666, 667, 673, 739, 806, 810, 811, 817, 839, 842, 844, 847, 852
aggregation, 481, 482, 483, 486, 487, 488, 493, 494, 496, 544, 618, 620, 634, 650, 667, 671, 810, 811
aggression, xviii, 81, 82, 85, 106, 376, 509, 510, 656, 733
aggressiveness, xix, 692, 697
aging, xv, 480, 508, 516, 517, 518, 522, 533, 536, 538, 540, 548, 550, 555, 556, 558, 607, 843, 846, 848
aging process, 685
agonist, 51, 53, 55, 56, 62, 64, 65, 66, 69, 70, 172, 182, 183, 189, 190, 191, 193, 195, 196, 306, 818, 827
aid, 462, 522, 601, 789
aiding, 458, 746, 789
AIDS, 723, 740
air, 375, 612
airway responsiveness, 570
airways, xvi, 562, 563, 565
akathisia, 693
akinesia, 274, 277, 305, 310, 311, 313, 314, 317, 366, 373, 696
AKT, 43
alanine, 619
alcohol, xvii, 11, 25, 31, 35, 308, 460, 472, 574, 575, 576, 577, 578, 579, 580, 581, 582, 583, 584, 585, 707, 712
alcohol abuse, 575
alcohol consumption, xvii, 574, 578, 579, 581, 582
alcohol dependence, 35
alcoholism, 32, 576, 734
alcohols, 530
aldolase, 528
alertness, 99
algorithm, 252, 292, 327, 334, 547
alienation, 69

Index

alkaline, 524, 525
alkaline phosphatase, 524, 525
alkylation, 680
allele, xvi, 10, 513, 522, 525, 526, 535, 614, 620, 630, 631, 656, 661, 679, 684, 686, 689, 730, 807
alleles, 191
allosteric, 53, 55, 66, 421, 424, 439, 494
alpha, 7, 38, 67, 69, 70, 188, 189, 190, 191, 192, 193, 195, 196, 451, 463, 489, 491, 507, 537, 539, 540, 554, 593, 594
ALS, 313
alternative, xiii, xvi, 135, 143, 240, 255, 256, 257, 258, 259, 260, 273, 274, 276, 278, 280, 283, 287, 319, 340, 346, 377, 428, 471, 478, 508, 526, 562, 565, 572, 593, 751, 753, 755, 756, 820
alternative behaviors, 143, 508
alternatives, 281, 602, 758, 764, 775
alters, 42, 104, 216, 440, 484, 493, 530, 549, 656, 812
aluminium, 324
aluminum, 610, 611
Alzheimer disease, 175, 491, 492, 493, 494, 495, 533, 536, 538, 553, 554, 555, 557, 605, 606
Alzheimer's disease, xv, xvi, xvii, 475, 485, 486, 490, 491, 492, 493, 494, 495, 521, 522, 541, 543, 544, 551, 562, 605, 606, 610, 805, 833
Alzheimer's disease (AD), 805
Alzheimer's, 840, 842, 843, 847, 850, 855
Alzheimer's disease, 489, 490, 491, 492, 493, 495, 533, 534, 535, 536, 537, 538, 539, 540, 542, 543, 546, 547, 552, 553, 554, 555, 556, 557, 558, 787, 840, 842, 843, 847, 850, 855
ambivalence, 777
ambivalent, 756
amelioration, 311, 346, 367, 508
American Psychiatric Association, 402, 444, 469, 502
amino, 53, 358, 408, 421, 423, 424, 545, 624, 635, 636, 638, 639, 652, 802, 809, 818, 834, 848, 851
amino acid, 53, 421, 423, 424, 545, 635, 636, 638, 809, 818, 848, 851
amino acids, 53, 423, 424, 635, 636, 638, 809, 818
amnesia, 415, 423, 434, 435
AMPA, 408, 422, 423, 427, 429, 438, 803, 827, 836
amphetamine, 37, 50, 177, 187, 190
amphetamines, 187
amplitude, 87, 238, 239, 240, 241, 242, 246, 292, 308, 375, 382, 383, 414, 696, 828

Amsterdam, 106, 113, 168, 253, 300, 379, 438, 515, 793
amygdala, 12, 16, 19, 20, 23, 25, 26, 27, 35, 79, 82, 83, 84, 85, 106, 109, 110, 111, 142, 145, 147, 148, 149, 151, 184, 185, 186, 191, 192, 194, 261, 269, 324, 325, 326, 331, 336, 341, 342, 376, 398, 409, 413, 414, 423, 636, 801, 812, 814, 815, 816, 829, 837
amyloid, xvi, 34, 44, 377, 476, 478, 479, 482, 483, 484, 485, 488, 489, 490, 491, 492, 493, 494, 495, 522, 526, 532, 534, 536, 539, 540, 543, 544, 545, 546, 549, 550, 553, 554, 555, 556, 557, 558
Amyloid, 475, 476, 482, 483, 489, 490, 493, 545, 549, 555, 839
amyloid angiopathy, 377, 493, 544, 545, 554, 555, 556, 557, 558
amyloid beta, 490, 491, 493, 494, 534, 540, 550, 553, 557, 842
amyloid deposits, 479, 544, 545, 549, 554
amyloid fibril formation, 526
amyloid fibrils, 536
amyloid plaques, 479, 482, 488, 550
amyloid precursor protein, 489, 490, 494, 532, 539, 540, 544, 545, 550, 554, 557
amyloidosis, 493, 549, 550, 554, 557, 558
amyotrophic lateral sclerosis, 313, 317, 522
anaesthesia, 235, 236, 314, 353, 397
analgesia, 272, 289
analgesics, 236
analog, 87, 91, 175, 292
analysis of variance, 309, 463, 523
anatomy, 32, 103, 211, 230, 231, 351, 387, 390, 392, 431, 829, 836, 844
anchoring, 391
androgen, 82, 104, 113, 640
androgen receptors, 82
anemia, 486
anger, 430, 622
angiogenesis, 550, 802, 804, 805, 824
anhedonia, 376
anhydrase, 422
animal models, 55, 69, 79, 173, 262, 279, 320, 323, 325, 326, 327, 382, 425, 483, 487, 536, 541, 548, 549, 550, 551, 552, 554, 807, 812, 850, 851
animal studies, 6, 20, 173, 320, 321, 323, 419, 420, 578
animals, x, xi, xii, 20, 21, 74, 85, 86, 115, 116, 118, 119, 120, 122, 124, 125, 126, 127, 128, 130, 132, 134, 135, 154, 156, 164, 178, 181, 187, 188, 216, 218, 327, 347, 366, 381, 382,

383, 384, 385, 386, 389, 417, 423, 425, 426, 429, 477, 483, 548, 549, 551, 579, 802, 805
Animals, 117, 120, 383
anode, 307, 309, 310
ANOVA, 89, 91, 94, 96, 98, 100, 218, 220, 221, 222, 449, 463
anoxia, 361
anoxic, 353, 358, 359, 422, 424
ANS, 287, 288
antagonism, 109, 429, 442
antagonist, 32, 54, 56, 63, 70, 174, 182, 183, 193, 195, 427
antagonistic, 281
antagonists, ix, 48, 50, 51, 53, 55, 67, 139, 182, 195, 438
anterior, 192, 194
anterior cingulate cortex, 18, 29, 35, 42, 61, 135, 340, 341, 342, 344, 348, 351, 371, 372, 388, 398, 413, 515
anthropological, 211, 763
anthropology, 763
anti-apoptotic, 20, 479, 827
antibiotic, 250
antibody, 483, 484, 524, 525, 688
anticholinergic, 55, 697
anticholinergic effect, 697
anticoagulant, 250
anticonvulsant, 270, 278, 296, 321, 331, 336, 397, 401, 419, 422, 430, 435, 439, 441, 442
anticonvulsant treatment, 397
anticonvulsants, 401, 426
antidepressant, xiv, 38, 261, 283, 296, 342, 396, 397, 399, 400, 401, 402, 697, 733, 846
antidepressant medication, 733
antidepressants, 340, 342, 376, 482, 697, 707, 851
antiepileptic drugs, xiii, 278, 280, 281, 302, 319, 324, 407, 408, 431, 432, 436, 438, 439, 441
antigen, 483, 681, 683, 688, 824, 828, 834, 835
anti-inflammatory agents, 490
anti-inflammatory drugs, 478, 485, 494
antimony, 610
antioxidant, 38, 480, 481, 487, 522, 530, 532, 535, 538
antipsychotic, ix, 13, 18, 20, 21, 22, 31, 32, 34, 36, 38, 39, 41, 42, 43, 44, 47, 48, 50, 51, 52, 53, 58, 61, 62, 63, 65, 196, 695, 697
antipsychotic drugs, ix, 21, 31, 32, 34, 38, 41, 44, 47, 48, 50, 51, 52, 53, 58, 61, 62, 63, 65
Antipsychotic drugs, 61
antipsychotic effect, 21, 53
antipsychotics, 3, 13, 14, 20, 21, 22, 23, 31, 33, 40, 44, 47, 48, 51, 53, 57, 61, 64

anxiety, x, xiii, xviii, 55, 99, 105, 110, 111, 119, 137, 141, 142, 144, 145, 146, 147, 150, 152, 228, 236, 261, 262, 269, 275, 283, 301, 306, 309, 340, 341, 344, 346, 348, 350, 351, 456, 461, 464, 467, 473, 474, 551, 612, 656, 693, 694, 697, 709, 724, 726, 730, 732, 734, 756, 760, 778, 786, 808
Anxiety, 141, 148, 150, 151, 456, 461, 464, 465, 466, 474
anxiety disorder, x, xiii, 141, 144, 145, 146, 150, 269, 301, 340, 344, 346, 348, 350, 461, 709
anxiolytic, 103, 474
anxious/depressed, 149
AP, 87, 91, 95, 96, 97, 103, 106, 107, 110, 317, 331, 852, 853
apathy, ix, xix, 2, 274, 275, 288, 346, 376, 612, 695, 705, 732
aphasia, 285, 368, 369, 378, 380, 424, 487, 495, 543
aplasia, 379
APOE, 533
Apolipoprotein E, 554
apoptosis, xviii, 9, 15, 20, 32, 38, 41, 44, 422, 425, 426, 427, 429, 441, 526, 527, 529, 530, 531, 532, 539, 610, 611, 613, 643, 648, 651, 652, 653, 656, 661, 662, 666, 667, 671, 673, 678, 801, 804, 826, 843, 851
apoptotic, 9, 15, 20, 33, 426, 427, 432, 441, 442, 534, 538, 578, 818, 827, 850, 856
Apoptotic, 10, 33, 35, 584
apoptotic effect, 426, 427
apoptotic mechanisms, 10, 15, 33
apoptotic pathway, 856
APP, xv, 476, 477, 478, 483, 485, 490, 531, 532, 544, 545, 546, 550
appetite, 78, 695
application, xv, 42, 104, 111, 125, 183, 204, 205, 218, 238, 253, 256, 258, 259, 260, 261, 262, 263, 272, 277, 280, 282, 286, 289, 292, 324, 325, 336, 346, 412, 444, 445, 458, 537, 556, 589, 789
applications, 777, 792
appointments, 709, 710
appropriate technology, 204
AR, 845, 849
Argentina, 617
argument, 204, 428, 611
arithmetic, 411, 502, 517, 519, 547
arousal, x, xi, 74, 78, 83, 84, 87, 89, 99, 101, 104, 112, 123, 125, 129, 134, 135, 137, 177, 178, 180, 190, 192, 509, 510
arrest, 120, 301, 313, 652, 833
arrests, 423, 494

Index

arsenic, 610
arteries, 482, 545, 546, 547, 549, 551
arterioles, 545, 547
artery, 383, 543, 551, 556
arthritis, 564
ascorbic, 480
ascorbic acid, 480
ASD, xvii, 610, 611
aseptic, 483
Asia, 272, 555
ASL, 313
aspartate, 30, 408, 435, 628, 652, 653, 812, 845, 853
aspiration, 699, 706, 709, 721, 723
aspiration pneumonia, 706
assessment, xiv, xvii, 11, 12, 21, 22, 39, 187, 237, 238, 239, 243, 248, 252, 293, 298, 304, 309, 316, 344, 367, 376, 377, 409, 412, 428, 430, 433, 443, 449, 453, 456, 457, 458, 459, 460, 462, 463, 465, 466, 468, 473, 483, 503, 541, 547, 570, 571, 572, 574, 582, 590, 594, 596, 597, 605, 625, 674, 716, 725, 730, 737, 748, 780, 785, 788, 836
assessment models, 409
assessment procedures, 594
assignment, 489
association, 179, 184, 185, 195
associations, 371
assumptions, 588, 775
asthma, xvi, 561, 562, 563, 565, 570, 571
astrocyte, 36, 542
Astrocyte, 830
astrocytes, 21, 479, 481, 491, 532, 542, 545, 557, 558, 611, 802, 826, 828, 839
astrogliosis, 611, 637, 812, 819
astronomy, 229
asymmetry, 8, 31, 104, 528, 529, 538
asymptomatic, 629, 687, 688, 723, 815
ataxia, 257, 276, 622, 635, 653, 671, 807, 818, 835, 844
Athens, 255, 271, 595
atherosclerosis, 482, 493, 544
athetosis, 312
Atlas, 241, 253, 254
atmosphere, 510
ATP, 363, 528, 536, 628, 643, 667, 673
atrial fibrillation, 482
atrioventricular node, 397
atrophy, xiii, xvi, 7, 19, 30, 44, 173, 174, 228, 257, 263, 276, 287, 304, 305, 306, 314, 315, 361, 413, 419, 423, 433, 439, 483, 522, 533, 548, 549, 555, 556, 622, 628, 632, 635, 636, 640, 649, 650, 737, 806, 807, 812, 815, 819, 833, 835, 836
atropine, 51
attachment, 511
attacks, 272
attention, xi, 177, 178, 180, 181, 183, 184, 185, 186, 188, 189, 190, 191, 192, 193, 194, 195, 196, 197, 341, 351, 372, 375, 808
Attention Deficit Disorder, 470, 471
attention deficit hyperactivity disorder, 191, 193, 196
Attention Deficit Hyperactivity Disorder, xi, 178, 188, 189, 195, 469, 470, 471, 473, 474, 579, 580
attention problems, 511
attentional bias, 144, 150
attitudes, 144, 584, 717, 719, 724, 726, 727, 735, 736, 738, 751, 757, 759, 764, 773, 774, 780, 788, 790, 791, 793, 795, 796
attribution, 17, 752
attribution theory, 752
atypical, 21, 22, 31, 32, 38, 39, 40, 41, 47, 48, 51, 58, 61, 62, 65, 69
auditory cortex, 8, 16, 43, 117, 118, 119, 120, 138, 139, 140, 155, 164, 165, 167, 168, 169, 170, 184, 191
auditory hallucinations, 18, 28, 33, 293, 296
auditory stimuli, 68, 119, 153, 159, 160, 162, 164, 166
Australia, 47, 171, 199, 474, 601, 717, 720, 737, 738, 795
authorities, 603
authority, 263
authors, 627, 631, 684, 693, 720, 721, 723, 726, 729, 730, 732, 733, 734, 747, 800
autism, xvii, 573, 583, 609, 610, 611, 612, 613, 614, 615
autistic spectrum disorders, 580
autonomic nerve, 144
autonomic nervous system, 77
autonomy, 735, 741
autophagy, 811, 850
autopsy, 483, 631
autosomal dominant, xviii, 547, 620, 656, 661, 678, 691, 692, 699, 787, 790, 806
autosomal recessive, 363
availability, 51, 63, 260, 273, 285, 289, 382, 421, 588, 600, 601, 695, 721, 773, 777
aversion, x, 141, 151
avoidance, 27, 35, 115, 116, 118, 119, 120, 121, 122, 123, 124, 125, 126, 127, 128, 129, 130, 131, 132, 133, 134, 136, 138, 139, 140, 143, 418, 419, 423, 736, 780

avoidance behavior, 122
awareness, xiv, xv, xvii, xix, 265, 293, 408, 411, 428, 443, 462, 510, 573, 579, 584, 609, 693, 696, 713, 746, 757, 764, 765, 766, 767, 768, 769, 770, 772, 773, 777, 779, 783, 784
axon, 39, 43, 45, 205, 212
axon terminals, 39, 43, 45, 810
axonal, 3, 5, 7, 8, 342, 529, 531, 538
axonal degeneration, 638, 648, 845
axons, 5, 10, 42, 181, 188, 243, 476, 477, 636, 637, 642, 802, 809, 810
Aβ, 488, 531, 532, 544, 545, 549, 550

B

B cell, 484, 801, 828
B cells, 484, 801, 828
babies, 754, 773
background, 664, 700, 723, 728, 764, 826
background information, 764
bacteria, 706, 707, 712, 714
Badia, 596
bandwidth, 235, 248
barbiturates, 418
barrier, 485, 545, 546, 550, 552, 554, 555, 559, 564, 565, 569, 776
barriers, 561, 564, 723, 748, 767
basal forebrain, 180, 430, 544
basal ganglia, xiii, xiv, 21, 22, 23, 36, 60, 75, 77, 173, 175, 257, 258, 266, 269, 274, 277, 278, 289, 309, 320, 323, 324, 329, 330, 334, 336, 342, 350, 354, 361, 365, 370, 371, 372, 374, 377, 379, 382, 388, 390, 392, 393, 410, 412, 413, 431, 579, 584, 612, 622, 632, 669, 814, 815, 833, 836, 841, 855
Basal ganglia, 323, 329, 377, 390, 836
basal lamina, xx, 799, 800, 802
base, 172, 230, 231, 237, 241, 254, 292, 300, 363, 364, 428, 431, 448, 459, 503, 517, 518, 524, 555, 565, 568, 573, 629, 651, 677, 678, 679, 680, 686, 687, 699, 726, 747, 846, 851
base pair, 629
basement membrane, 557
basic fibroblast growth factor, 841, 854
basic research, 47, 188, 594
basket cells, 5
batteries, 264, 400, 446, 472, 588
battery, 57, 218, 228, 250, 256, 259, 262, 275, 283, 289, 291, 307, 362, 368, 373, 400, 411, 445, 448, 462, 469, 503, 596
Bax, 44
BBB, 542, 548, 550
BDNF, 5, 10, 20, 28, 40, 408, 426, 427

Beck Depression Inventory, 729, 735
bedding, 74, 84, 98, 99, 100, 101
behavior therapy, 148
behavioral assessment, 309
behavioral change, 184, 281, 345, 408, 551, 613
behavioral effects, 38, 435, 851
behavioral problems, 419, 509
behaviors, x, xiii, 73, 74, 75, 77, 78, 79, 80, 81, 82, 83, 84, 85, 86, 87, 91, 93, 94, 95, 96, 97, 101, 102, 106, 142, 143, 179, 340, 410, 411, 442, 501, 508, 510, 513, 593, 594, 604, 612
behavioural disorders, 60
behaviours, 54, 55, 157, 461, 573
Belgium, 601, 739
beliefs, 726, 747, 751, 760, 790
beneficial effect, xi, 20, 141, 146, 171, 177, 181, 182, 188, 247, 262, 273, 275, 276, 278, 282, 313, 323, 326, 346, 366, 422, 423, 436, 483, 487, 488, 503, 696, 697, 811
benefits, xii, 146, 256, 258, 259, 261, 262, 263, 265, 275, 276, 277, 282, 284, 288, 289, 307, 322, 347, 389, 428, 429, 512, 599, 603, 709, 714, 723, 724, 725, 726, 730, 738, 747, 751, 818
benign, 416, 434, 435, 437
benzodiazepine, 103, 421, 697
benzodiazepines, 237, 355, 418
beryllium, 610
beverages, 707
bias, x, 12, 141, 143, 144, 150, 290, 458, 588, 597, 618, 752, 777
bilateral, 366, 371, 378, 379, 380
bilingual, 379, 591
bilirubin, 480
binding, 29, 50, 52, 53, 54, 56, 62, 63, 64, 65, 66, 69, 71, 175, 310, 363, 421, 424, 477, 478, 481, 482, 487, 489, 491, 497, 514, 517, 524, 525, 526, 529, 535, 640, 641, 648, 649, 652, 665, 673, 674, 810, 811, 812, 813, 815, 833, 853, 855
binge drinking, 573, 574
biochemistry, 224
bioenergy, 628
biogenesis, 477
bioinformatics, 658
biological processes, 612, 655, 657, 664, 666
biomarker, 495, 535
biomarkers, 522, 526, 532, 535, 537
biopsies, 355
biosynthesis, 674
biotechnology, xiii, 270, 302, 320, 793, 796
biotin, 812, 835

bipolar, xiii, 25, 28, 29, 32, 33, 34, 38, 39, 240, 242, 243, 286, 287, 304, 305, 307, 313, 314, 325, 366, 400, 401, 402, 436, 439, 500, 514, 518
bipolar disorder, 25, 28, 29, 32, 33, 34, 38, 39, 401, 439, 500, 514, 518, 697
birth, xvii, 81, 224, 435, 477, 574, 580, 581, 582, 583, 613, 619, 751, 753, 754, 758, 763, 786, 789
birth control, 754, 763
births, 578, 625, 757, 760
black hole, 618
bladder, 658
blame, 750, 753, 767, 768, 769
bleeding, 243, 249, 250, 287, 368, 373, 386, 484, 709
blends, 154
blindness, 262, 283, 461, 581, 623
blocking, 177, 183, 188
blocks, 145, 187, 335, 393, 422, 423, 439, 543, 708
blood, 44, 49, 50, 58, 59, 60, 149, 172, 173, 175, 180, 183, 189, 355, 371, 381, 383, 384, 386, 390, 392, 393, 397, 398, 438, 472, 480, 485, 492, 525, 542, 543, 544, 545, 546, 550, 551, 552, 554, 555, 559, 575, 611, 613, 658, 679, 684, 685, 687, 771, 826
blood flow, 44, 49, 50, 58, 59, 60, 149, 172, 173, 175, 180, 183, 189, 371, 381, 386, 390, 392, 393, 398, 542, 543, 544, 546, 555
blood pressure, 384, 397, 492, 543, 546, 551
blood supply, 542
blood vessels, 543, 544, 545, 546, 550, 551, 826
blood-brain barrier, 545, 546, 550, 552, 554, 555, 559
blot, 525
blurring, 790
BMA, 583, 635, 807
body, 181, 340, 342, 354, 362
body mass index, 627
body temperature, 384
body weight, 362
BOLD, 61, 180
bolus, 236, 237
bone, 526, 536, 658, 801, 833
bone marrow, 658
bone remodeling, 536
bones, 757
borderline, 27, 41, 144, 261, 565
borderline personality disorder, 27, 41
Boston, 167, 412, 473, 791
bottleneck, 501
botulinum, 312, 363

bowel, 760, 787
boys, 193, 473, 474, 513, 610
brachial plexus, 272
bradykinesia, 274, 275, 305, 373, 375, 389, 806, 808
bradykinetic, 706, 808, 814
brain abnormalities, 30, 361, 736, 809
brain activity, 61, 107, 151, 262, 291, 292, 293, 392, 410, 414, 421, 501
brain asymmetry, 8
brain damage, 170, 581
brain development, 10, 44, 426, 429, 809
brain functioning, 74, 542, 613
brain functions, xi, 111, 154, 177, 568
brain growth, 425, 441
brain injury, 423
brain size, 32, 204
brain stem, 387, 398
brain structure, 10, 11, 28, 38, 83, 84, 86, 87, 97, 102, 199, 205, 279, 342, 382, 411, 414, 423, 430, 439, 509
brain tumor, 856
brainstem, 75, 276, 279, 281, 289, 342, 360, 398, 612, 815, 837
brainstem nuclei, 398
branching, 30, 206, 822
breakdown, 546, 612, 614, 730, 775
breast cancer, 795
breast feeding, 104
breeding, 550
Britain, 574, 584, 757
Brno, 1
broad spectrum, 57
Broca's area, 422
bromodeoxyuridine, 833, 847
bronchodilator, xvi, 562, 569, 570, 571
bronchopneumonia, 376
bruxism, 696, 706, 714
buffer, 707
bulbs, 82
bundle branch block, 228
bundling, 537
Bureau of Labor Statistics, 606
burn, 735
burning, 707
burnout, 599, 603
businesses, 783
Butcher, 66, 67, 70
buttons, 87
butyric, 408
bypass, 771

C

cables, 87
cadmium, 610
Cairo, 409
calcitonin, 550
calcium, 6, 412, 421, 422, 423, 439, 610, 618, 643, 650, 652, 813, 817, 822, 824, 827, 842, 849, 852
calmodulin, 417
calorie, 694
Canada, 239, 272, 344, 601, 718, 720, 724, 738, 739, 740, 775, 790
cancer, 272, 613, 687, 723, 741, 781, 787, 795, 801
candidates, 228, 259, 260, 273, 277, 278, 280, 282, 367, 378, 396, 402, 725, 738
cannabis, 26
capacity, 4, 10, 37, 50, 77, 83, 193, 194, 208, 210, 216, 217, 412, 420, 425, 472, 508, 514
Capacity, 480
capillary, 545
capsule, 261, 266, 268, 281, 283, 296, 299, 340, 343, 344, 346, 347, 375, 564, 637
carbamazepine, 408, 435, 436, 437, 439, 697
carbohydrates, 524, 711
carbon, 265, 267, 270, 292, 297, 302, 721, 722
Carbon, 292
carbon monoxide, 721, 722
Carbonyl, 530
carbonyl groups, 524
card sorting test, 17, 571
cardiac, 186
cardiac pacemaker, 249, 250
cardiomyopathy, 622
cardiovascular disease, 488, 525, 543, 553
cardiovascular disorders, 546
carefulness, 144
caregiver, 81, 600, 605, 606
caregivers, 18, 484, 486, 510, 599, 600, 601, 603, 604, 605, 695, 697, 698, 709, 710, 711
caregiving, xvii, 599, 600, 602, 605, 606, 607
cargoes, 786
caries, xix, 705, 707, 710, 711, 712, 715, 716
carnivores, 210
carotid arteries, 551
carrier, 525, 528, 693, 717, 718, 724, 725, 726, 727, 728, 730, 731, 735, 740, 742, 786, 787, 792, 793, 794
Cartesian coordinates, 229
case examples, 736
case studies, 779, 782, 790
case study, 380, 597, 791

Caspase-8, 651, 674
caspases, 811, 841
castration, 104, 105, 111
catalase, 480
catatonia, 695
catechol, 40, 148
catecholamine, 61, 180, 182, 187
catecholamines, 179, 180, 187
Catecholamines, 191
categorization, 16, 216, 217, 222, 223, 224
category a, 219, 454
category b, 723
category d, 216
catheter, 236, 251
catheterization, 251
cathode, 309, 310
cats, 4, 84, 103, 110, 111, 113, 320
Caucasian, 614
Caucasian population, xviii, 625, 656
Caucasians, 625
causal relationship, 611, 805
causality, 747
cDNA, 539, 650, 819, 821, 827, 850
CDR, 485, 486
CEA, 319
ceiling effect, 457, 459, 593
cell adhesion, 35, 477, 526, 529, 537, 583, 834, 835, 848
cell biology, 554
cell body, 206, 207
cell culture, 423, 578
cell cycle, 526, 531, 532, 539, 666, 817
cell death, 7, 412, 427, 432, 439, 441, 478, 528, 530, 613, 620, 644, 648, 651, 667, 674, 688, 804, 805, 810, 823, 826, 827, 828, 829, 842, 850, 851, 852
cell differentiation, 493, 530
cell division, 679, 801
cell growth, 478
cell line, 643, 647, 684, 802, 842
cell lines, 643, 647, 684, 842
cell signaling, 9, 38, 530, 532, 552, 642
cell size, 3, 6
cell surface, 477
cells, 180, 184, 185, 187
cellular phone, 164
cement, 383
central executive, 106, 119, 129, 138, 212
central nervous system, 47, 67, 457, 479, 481, 483, 491, 551, 552, 576, 581, 611, 614, 707, 847, 848
centromere, 10
cerebellar ataxia, 257, 276

cerebellum, 13, 14, 15, 18, 19, 26, 153, 155, 165, 169, 185, 258, 279, 312, 320, 330, 389, 410, 412, 413, 417, 431, 440, 537, 579, 612, 633, 636, 683, 687, 847
cerebral amyloid angiopathy, 493, 544, 554, 555, 556, 557, 558
cerebral amyloidosis, 549
cerebral asymmetry, 31
cerebral autosomal dominant arteriopathy, 547
cerebral blood flow, 44, 49, 50, 58, 59, 60, 149, 172, 173, 183, 189, 371, 381, 382, 386, 390, 393, 398, 544, 555
cerebral blood flow (CBF), 544
cerebral cortex, xiv, 32, 33, 37, 38, 42, 101, 108, 112, 175, 193, 194, 199, 204, 205, 206, 211, 212, 213, 253, 289, 290, 302, 322, 330, 365, 410, 516, 549, 551, 632, 633
cerebral function, 60, 102
cerebral hemisphere, 285, 411, 546
cerebral hemorrhage, 550, 554, 558
cerebral hypoperfusion, 550, 551, 556, 559
cerebral ischemia, 556, 559, 843, 852
cerebral metabolism, 480
cerebral palsy, 353, 355, 358, 359, 361, 611
cerebrospinal fluid, xvi, 36, 228, 236, 250, 287, 398, 481, 522, 535, 834
cerebrovascular, 282, 541, 542, 544, 546, 548, 549, 550, 552, 553, 555, 558, 603
cerebrovascular disease, 282, 542, 544, 548, 603
cerebrovascular diseases, 544
cerebrum, 200, 203, 540
cervix, 658
C-fibres, 397
c-fos, 84, 85, 95, 111
c-Fos, 139, 140
CGC, 622
challenges, 59, 484, 493, 573, 582, 711, 751, 789, 794
channels, 54, 155, 191, 242, 296, 297, 408, 421, 422, 431, 650, 817
chaos, 789
chaperones, 639, 663, 666, 673
Chaperones, 540
character, 722, 732, 797
charge density, 321
chemical, 32, 74, 172, 178, 186, 292, 412, 425, 540, 551, 610, 615, 645, 669, 712
chemical characteristics, 32
chemicals, 84, 609, 610, 612, 613, 614
chemokine, 480, 491
chemokine receptor, 491, 846
chemokine synthesis, 491
chemokines, 479, 542, 545, 825

chemotactic cytokines, 479
chemotaxis, 526
chemotherapeutic agent, 710
chemotherapy, 646, 840
Chicago, 70, 348, 705
Child Behavior Checklist, 447
child development, xvii, 587, 590, 594, 596
childbearing, 429, 575, 748, 758, 791, 797
childhood, 8, 9, 141, 144, 179, 190, 215, 223, 300, 355, 409, 416, 427, 430, 434, 437, 442, 499, 501, 510, 512, 515, 516, 519, 520, 573, 588, 610, 619, 764, 782, 789
childhood disorders, 190, 610
childless, 755, 756, 773, 789, 792
chimpanzee, 204
China, xviii, 171, 656
Chi-square, 450, 452
chloride, 103, 486
chlorpromazine, 53, 62
chocolate, 151
cholesterol, 426, 475, 476, 481, 488, 489, 538, 674
choline, 51, 423
cholinergic, ix, 47, 48, 51, 52, 54, 57, 63, 64, 67, 69, 70, 71, 75, 136, 176, 423, 439, 529, 532, 537, 538, 544, 554
cholinergic neurons, 423, 805
cholinesterase, 51, 55, 377, 482, 489
cholinesterase inhibitors, 51, 377, 482
chorea, xviii, 617, 620, 622, 625, 692, 695, 696, 698, 706, 714, 715, 737, 740, 741, 742, 743, 788, 789, 791, 793, 808, 814, 839, 841, 847, 852, 855
choreoathetosis, 267, 297, 623
CHRM2, 52
chromatography, 524, 531
chromosomal abnormalities, 613
chromosome, xviii, 10, 34, 68, 324, 329, 476, 544, 624, 629, 650, 656, 678, 718, 807, 820, 835
chromosome 20, 324, 329
chromosomes, 476
chronic, 195, 366, 373, 378, 380
chronic illness, 13
chronic obstructive pulmonary disease, xvi, 561, 562, 571, 714
chronic pain, 260, 272, 273, 283, 287, 294, 297, 298, 299, 300
chronic stress, 439
chronically ill, 15
cigarette smoking, 55, 68, 69
cilia, 801, 830
cilium, 830

cingulated, 76, 83, 84, 112, 379, 380, 389
circadian, 53, 65
circadian rhythm, 53, 65
circadian rhythms, 53, 65
circadian timing, 848
circulation, 550
CIS, 447, 448, 451, 452, 453, 461, 462, 464, 465, 466
citizens, 601
Civil Rights, 596
clarity, 564, 589, 591, 594, 622
classes, 78, 509, 691
classical, 20, 21, 22, 69, 80, 116, 147, 237, 372, 397, 544, 550, 579, 789
classical conditioning, 69, 116, 579
classification, 78, 79, 97, 216, 217, 218, 221, 222, 223, 251, 280, 354, 363, 445, 450, 452, 454, 455, 457, 652, 664, 734, 855
classroom, 471, 510, 518, 582
classrooms, 509, 510, 582
cleaning, 698, 711
cleavage, 476, 477, 485, 490, 618, 638, 639, 640, 652, 680, 682, 811, 841, 855
cleavages, 476, 639
clients, 603, 759, 760
clinical, 366, 376, 377, 378, 380
clinical application, 272, 290, 458
clinical approach, 583
clinical assessment, 243, 263, 287, 309
clinical diagnosis, 228, 378, 548, 578, 693, 733, 738, 740
clinical examination, 355
clinical heterogeneity, 2
clinical judgment, 444
clinical presentation, 459, 625, 628
clinical symptoms, 18, 250, 251, 327, 734
clinical syndrome, xvi, 19, 541
clinical trial, 58, 258, 273, 280, 289, 294, 320, 322, 328, 331, 395, 396, 398, 400, 401, 402, 460, 483, 484, 488, 489, 496, 510, 533, 547
clinical trials, xix, 58, 258, 273, 280, 289, 320, 322, 395, 396, 398, 400, 401, 402, 460, 483, 496, 533, 547, 692, 736
clinically significant, 418, 456
clinician, 443, 446, 571
clinicopathologic, 380
clinics, 503, 760, 768
clone, 629
clonidine, 182, 183, 190, 191, 192
cloning, 807
closed-loop, xiii, 259, 265, 266, 280, 291, 320, 328, 331
closure, 292

clozapine, 20, 21, 22, 34, 41, 43, 50, 51, 55, 58, 61, 62, 63, 69, 196, 697
cluster headache, 228, 249, 253, 272, 294, 299, 300, 301
cluster headaches, 249
clustering, 812
clusters, 5, 6, 22, 23, 307, 772
CNS, 47, 52, 54, 65, 67, 70, 149, 176, 265, 292, 424, 429, 438, 479, 491, 514, 617, 628, 807, 834, 840, 849, 853, 855
Co, 87, 91, 192, 253, 254, 489, 550
CO2, 236
coagulation, 236, 525
cobalt, 320, 333
cocaine, 26, 33, 42, 109
cocaine use, 42
Cochrane, 440, 488, 495, 606
codes, xviii, 656, 767, 797, 807
coding, xviii, 10, 137, 167, 264, 290, 614, 617, 619, 621, 634
co-existence, 600
cognitive abilities, xix, 10, 19, 50, 187, 204, 379, 486, 509, 513, 518, 595, 698, 705
cognitive ability, 210, 411, 503
cognitive activity, 173, 373, 375, 416
cognitive biases, 777
cognitive deficit, ix, xi, 29, 47, 48, 49, 51, 52, 53, 54, 55, 56, 57, 68, 70, 171, 173, 190, 371, 413, 414, 415, 416, 419, 420, 422, 492, 541, 544, 613, 623, 627, 633, 698, 734, 742, 845
cognitive deficits, ix, xi, 29, 47, 48, 49, 51, 52, 53, 54, 55, 57, 68, 171, 173, 190, 413, 414, 415, 416, 419, 420, 422, 492, 541, 544, 613, 623, 627, 633, 698, 734, 742, 845
cognitive development, xii, 216, 221, 222, 223, 430, 511, 516, 583
cognitive disorders, 604
cognitive dissonance, 752, 790
cognitive domains, 16, 409, 413, 548
cognitive dysfunction, ix, xiv, 2, 16, 17, 47, 50, 57, 58, 171, 172, 173, 175, 288, 407, 408, 412, 414, 419, 437, 458, 468, 470, 500, 558, 559, 628, 815
cognitive flexibility, 49, 116, 123, 172, 380, 627, 698
cognitive functions, 179, 181
cognitive load, 28
cognitive models, 792, 796
cognitive performance, xii, 29, 31, 55, 60, 117, 128, 135, 175, 196, 216, 223, 293, 370, 372, 415, 416, 434, 435, 436, 445, 446, 447, 449, 450, 452, 455, 456, 457, 458, 466, 468, 483, 508, 597

cognitive process, 74, 116, 183, 190, 211, 371, 372, 376, 415, 420, 423, 564, 612, 748, 752
cognitive processes, 183, 190
cognitive processing, 211, 376, 415, 420, 748, 752
cognitive profile, 367, 368, 373, 419
cognitive set, 372
cognitive skills, 588, 589
cognitive style, 144
cognitive system, 119
cognitive tasks, 448, 457, 459, 547
cognitive test, 411, 412, 420, 444, 445, 446, 455, 460, 470, 471, 503, 511, 547, 548, 595, 596
cognitive testing, 420, 595
cognitive therapy, 147
cognitive-behavioral therapies, 342, 349
cognitive-behavioral therapy, 342
coherence, 9, 111
cohesins, 650
cohort, 30, 32, 52, 59, 322, 413, 419, 429, 478, 553, 680
coil, 280, 283
collaboration, 228, 236, 286, 514, 781, 807
collagen, 477, 802
collateral, 5, 210
college students, 179, 187
Colombia, 724
colon, 658
colonization, 706
colostrum, 483, 493
commercial, 146, 487
commissure, 229, 230, 232
common symptoms, xviii, 656
communication, 33, 129, 146, 237, 285, 292, 293, 306, 315, 502, 510, 528, 531, 579, 588, 611, 632, 698, 708, 717, 750, 767
communication skills, 502, 510
communication strategies, 588
communities, 349, 510
community, xvii, 260, 300, 428, 431, 502, 503, 507, 563, 565, 570, 590, 592, 600, 601, 602, 603, 604, 611, 613, 618, 700, 711, 715, 726, 832
co-morbidities, 263, 287
comorbidity, 256, 275, 411, 474
compatibility, 780
compensation, 6, 60, 698, 815
competence, 442, 507, 510, 513, 515, 567, 595, 735, 748, 749, 750, 782
complement, xx, 478, 526, 544, 545, 554, 683, 708, 799, 802, 827
complement components, 545
complement pathway, 544

complex partial seizure, 258, 279, 280, 322, 414
complex regional pain syndrome, 312
complex systems, 429
complexity, xvi, 148, 199, 208, 210, 211, 224, 225, 229, 290, 362, 562, 670, 777, 783
compliance, 401, 402, 457, 579, 695
complications, xii, xix, 8, 249, 250, 256, 263, 273, 275, 287, 291, 299, 306, 308, 315, 321, 324, 340, 343, 344, 345, 355, 361, 400, 484, 555, 559, 610, 698, 705, 706, 707, 709, 733
components, x, 31, 47, 48, 49, 77, 78, 79, 80, 86, 91, 92, 97, 107, 116, 126, 162, 167, 172, 275, 286, 322, 433, 477, 508, 510, 544, 545, 752, 759, 765, 802, 804, 823, 831
composition, 559, 645
compounds, 69, 172, 182, 185, 186, 475, 484, 487, 530, 612, 613, 614, 615, 645
comprehension, 117, 124, 125, 368, 369, 370, 373, 375, 431, 547, 580, 612
compression, 564
compulsion, 144
compulsive behavior, xviii, 656
computation, 264, 425, 451
computed tomography, 59, 228, 252, 264, 411
computer, 34, 264, 292, 293, 298, 302, 462, 507, 513
computers, 795
computing, 447, 449, 450, 723
concentration, 12, 13, 14, 17, 18, 22, 26, 33, 39, 51, 56, 178, 187, 346, 398, 414, 416, 419, 421, 422, 472, 479, 488, 545, 610, 669, 708, 712, 813
concentric annuli, 206, 207
concept, 195
conception, xix, 55, 144, 746
conceptual model, 471
conceptualization, 502
conceptualizations, 589
concordance, 633, 636
concrete, xii, 117, 215, 216, 217, 218, 219, 220, 221, 222, 223
concussion, 445, 455, 470
condensation, 818
conditioned response, 62, 120, 125, 128, 130, 131, 132, 133, 134, 135
conditioned stimulus, 120, 124
conditioning, 69, 102, 103, 116, 117, 119, 120, 122, 124, 128, 129, 130, 131, 133, 134, 136, 138, 149, 164, 579
conductance, 143, 145, 148, 265, 290, 424, 438
conduction, 622
conductivity, 263, 265, 292
conductor, 186

confidence, 594, 627, 759, 776, 781
confidence interval, 627
confidentiality, 693, 761
configuration, 242, 307, 421
confinement, xix, 692
conflict, 31, 61, 105, 115, 116, 117, 119, 123, 125, 127, 128, 133, 135, 143, 341, 372, 380, 511, 512, 797
confusion, 203, 204, 251, 343, 344, 346, 376, 419, 428
congress, 596
conjugation, 530
Connecticut, 177
connectionist, 58
connective tissue, 622
connectivity, 4, 5, 6, 12, 17, 38, 61, 68, 205, 210, 212, 264, 417
consciousness, 116, 123, 129, 250, 384, 410
consensus, 234, 260, 282, 457, 552, 709, 722, 754
consent, 228, 308, 369, 461, 484, 709, 764
consolidation, 185, 414, 508, 749, 752, 796
constraints, 164, 239, 591, 747, 749, 751, 759
construction, xii, 216, 264, 291, 411, 500, 593, 751
consultants, 759, 760
consumers, 605
consumption, xvii, 136, 327, 363, 382, 387, 388, 389, 393, 574, 576, 578, 579, 580, 581, 582, 611
contaminated food, 612
contamination, 511
context, 346
contiguity, 119
contingency, 118, 122, 123, 125, 132, 138
continuity, 603, 604, 741
contour, 264, 291
contracture, 694
contrast sensitivity, 138
control condition, 503, 509
control group, 369, 370, 424, 445, 446, 447, 448, 449, 450, 451, 452, 460, 482, 502, 503, 507, 508, 510, 511, 512, 513, 592, 610, 611
controlled studies, 195, 259, 275, 279, 283, 285, 328, 414
controlled study, 195
controlled trials, 257, 262, 274, 277, 281, 402, 437, 438, 459, 460, 487, 515, 602
controversial, 237, 366, 420, 637, 801
controversies, 791
contusion, 288
convention, 466
convergence, 290

conversion, 15, 522
conviction, 365
cooking, 698, 711
cooperation, 287, 683, 713
coordination, 6, 28, 169, 192, 374, 383, 419, 508, 563, 565, 566, 683, 684
COPD, xvi, 562, 563, 565, 571
coping strategies, 766
coping strategy, 726, 792
copper, 477, 492, 495, 497
copulation, 79, 85, 101, 104, 105, 107, 109, 112
corepressor, 648
corpus callosum, 579, 585, 825
correlation, 9, 17, 18, 19, 22, 51, 53, 54, 73, 87, 89, 90, 91, 93, 100, 102, 106, 116, 120, 122, 123, 129, 130, 133, 172, 315, 391, 450, 470, 479, 481, 546, 550, 567, 593, 620, 626, 628, 630, 651, 685
correlation analysis, 315
correlation coefficient, 87, 450
correlations, x, 34, 116, 173, 550, 593, 625, 628, 685
cortical, 367, 371, 372, 373, 374, 376, 379
cortical asymmetry, 8
cortical functions, 175
cortical information processing, 173
cortical inhibition, 415
cortical neurons, 169, 180, 185, 196, 279, 325, 821
cortical pathway, 207, 389
cortical processing, 181, 367, 371
cortical systems, 212, 335
corticobasal degeneration, 478
corticospinal, 258
corticosteroids, 416
corticosterone, 136, 138, 140, 423, 439
cortisol, 805
cosmetic, 291
cost, 234, 239, 262, 265, 276, 284, 288, 293, 362, 382, 402, 509, 581, 602, 776
cost-effective, 262, 265, 276, 293
costs, 143, 238, 372, 599, 600, 726, 736, 738, 751, 788
cotton, 712
cough, 400, 708
coughing, 706
counseling, 429, 693, 694, 789, 792, 795, 797
counseling psychology, 795
couples, 751, 755, 756, 760, 762, 769, 775, 776, 777, 778, 781, 783, 786, 789
coupling, 60, 112, 825, 839
courtship, 109
covariate, 431

covering, 391
CPT, 187, 459
cracking, 546
cranial nerve, 397
craniotomy, 287, 291, 314
cranium, 264, 291
craving, 401
creatine, 355, 537
creatine kinase, 537
creative abilities, 613
credibility, 759
CRH, 440
criminal justice, 581
criminal justice system, 581
criminals, 500
crises, 559
critical period, 9, 427
criticism, 260, 771
crops, 613
cross sectional study, 611
cross-cultural, 588, 592, 595, 596, 597, 614
cross-cultural comparison, 592
cross-fertilization, 266
cross-sectional, 22, 412, 413, 415
cross-sectional study, 687
crying, 85
cryogenic, 28
CSF, xvi, 287, 355, 375, 398, 413, 479, 480, 481, 482, 483, 484, 485, 488, 491, 522, 523, 525, 526, 530, 532, 535, 539, 557, 830, 834
CT scan, 19, 235, 247, 250, 311, 375
C-terminal, 476, 526
Cuba, 239
cues, 74, 80, 83, 85, 103, 109, 170, 179, 519, 854
cultural differences, 588
cultural practices, 588, 591
culture, 251, 410, 512, 573, 574, 587, 588, 590, 595, 596, 597, 679, 788, 801, 802, 805
curcumin, 487
cure, xviii, 656, 691, 724, 725, 726, 735, 736, 759, 785, 786
curiosity, 749
current limit, 256
curricula, 510
curriculum, 509, 510, 515, 516, 594, 768
Cushing's syndrome, 853
CXC, 479
cycles, 416, 786
cycling, 97, 104, 280, 401, 402, 837, 845
cyclooxygenase, 494
cyclotron, 385
cysteine, 479, 538, 807, 833
cysteine residues, 479

cystic fibrosis, 779, 786, 794
cytoarchitecture, xi, 154, 165, 166
cytochrome, 480, 492, 643, 843
cytochrome oxidase, 480
cytokine, 425, 441, 479, 487, 490, 526, 550
cytokines, 478, 479, 491, 542, 802, 823, 825, 843
cytomegalovirus, 833
cytoplasm, 531, 618, 634, 637, 638, 639, 641, 662, 666, 809, 810, 811, 817
cytoskeleton, 4, 10, 21, 526, 539, 642, 665, 666
cytosol, 427
Czech Republic, 1, 28

D

daily care, 710
daily living, xv, 306, 367, 485, 486, 503, 522, 601, 694, 708
damping, 54
dance, 625, 808
danger, 142, 382, 793
data analysis, 448
data collection, 594, 727, 764
data distribution, 445, 448
data generation, 457
data set, 199, 203, 235, 449, 457, 460, 674
database, 218, 524, 645, 658
deacetylation, 640
deafness, 788
death, xiii, xviii, 7, 9, 52, 304, 346, 423, 426, 440, 441, 477, 529, 545, 563, 612, 620, 622, 638, 644, 645, 648, 651, 656, 657, 667, 669, 672, 673, 674, 677, 678, 688, 706, 717, 718, 719, 720, 721, 722, 723, 724, 738, 739, 741, 742, 743, 758, 770, 771, 782, 792, 804, 805, 810, 812, 818, 824, 826, 827, 828, 829, 836, 842, 844, 850, 851, 852, 853
death rate, 563
deaths, 345, 720, 721, 723, 734, 738
debt, 758
decay, 710, 713
decision, 181
decision makers, 751, 765
decision making, xix, 116, 117, 124, 125, 129, 132, 134, 135, 136, 137, 144, 148, 149, 151, 152, 282, 348, 350, 443, 698, 745, 746, 749, 759, 771, 772, 789, 790, 792, 793, 794, 796, 797, 798
decision trees, 764, 765
decision-making, 181
decision-making process, 753, 759, 765, 777, 780, 781

decisions, xiv, xix, 124, 143, 146, 239, 428, 443, 444, 445, 446, 455, 693, 745, 746, 747, 748, 749, 750, 751, 752, 753, 754, 756, 757, 758, 759, 760, 761, 763, 765, 767, 768, 769, 770, 771, 772, 773, 775, 776, 777, 781, 782, 783, 784, 789, 792, 796
declarative memory, 60, 410, 414, 423, 432
decoding, 292, 410
decortication, 105, 189
deep brain stimulation, vi, ix, xii, xiii, xiv, 227, 252, 253, 254, 255, 256, 266, 267, 268, 269, 270, 271, 291, 293, 294, 296, 297, 298, 299, 300, 301, 302, 304, 320, 331, 334, 336, 337, 340, 346, 351, 354, 355, 364, 366, 367, 371, 373, 376, 378, 379, 380, 381, 382, 390, 391, 392, 393, 396
deep venous thrombosis, 250
deep-sea, 278
Deer, 436
defecation, 81
defects, 393, 435, 547, 571, 581, 613, 622, 629, 643, 650, 796, 849
defence, 136, 480
defense, 530, 532
deficiencies, 576, 686
deficiency, 175, 180, 355, 440, 492, 544, 551, 614, 686, 688, 847
definition, x, xi, 74, 86, 131, 200, 201, 280, 295, 374, 411, 463, 574, 589, 590, 614, 751, 770
deformities, 361
degenerate, 547, 632, 812, 813, 814
degenerative disease, 353, 541
degradation, 526, 532, 536, 632, 656, 663, 674, 810
degrading, 546
degree, 369
dehydrate, 707
dehydrogenase, 528
delivery, xvi, 287, 327, 342, 427, 428, 562, 565, 568, 571, 820, 856
Delphi, 552
Delta, 65
delusions, ix, 1, 48, 515
demand, 119, 128, 132, 133, 134, 135, 788
demographic characteristics, 727
dendrites, 3, 5, 6, 10, 30, 42, 207, 210, 637, 821, 824
dendritic cell, 537
dendritic spines, 3, 4, 5, 6, 8, 9, 11, 21, 27, 32, 43, 182, 206, 207, 208, 529, 821, 828
denervation, 32, 393, 554
denial, 732, 770
Denmark, 239, 381, 383, 390, 719, 720

density, 3, 4, 5, 6, 8, 10, 11, 12, 18, 19, 21, 24, 25, 26, 27, 29, 30, 31, 32, 33, 35, 36, 38, 39, 40, 42, 43, 45, 52, 53, 54, 63, 65, 182, 206, 207, 261, 376, 417, 525, 633, 637, 652, 812, 821, 824, 835, 846, 853
dental care, 709, 715
dental caries, xix, 705, 707, 711, 716
dental implants, 710
dental offices, 713
dental plaque, 708, 712, 713
dental restorations, xix, 705
dentate gyrus, xx, 412, 417, 435, 799, 800, 834, 836, 837, 838, 839, 841, 843, 844, 845, 846, 848, 852, 853, 855
dentate gyrus (DG), 800
dentin, 715
dentist, xix, 699, 705, 706, 708, 709, 710, 711, 712, 713
dentures, 710, 713
Department of Labor, 602, 603
dependency ratio, 602
dependent variable, 448, 462, 463
dephosphorylation, 539
depolarization, 415, 421, 422, 423, 424, 643
deposition, xvi, 478, 482, 487, 488, 495, 496, 534, 543, 544, 545, 550, 551, 554, 557, 562, 570
deposits, 476, 479, 488, 544, 545, 549, 554, 557
depressed, xiv, 146, 149, 150, 261, 281, 282, 296, 396, 398, 399, 401, 414, 415
depressive disorder, 603
depressive symptoms, 309, 399, 400, 401, 402
deprivation, 11, 43, 830
depth, 185, 280, 332, 590, 670, 727, 734, 746, 764, 768
deregulation, 656, 660, 661
derivatives, 494, 535
desensitization, 71, 113
desert, 116, 596
desipramine, 186, 195
desire, 78, 750
desires, 768
desorption, 523, 526
despair, 736
destruction, 101, 553, 707
detectable, 5, 479
detection, x, xiii, xvii, 11, 70, 74, 78, 79, 84, 87, 111, 118, 119, 120, 122, 128, 129, 130, 131, 132, 133, 134, 194, 195, 259, 280, 291, 292, 320, 327, 334, 341, 445, 448, 452, 458, 459, 464, 466, 467, 468, 469, 524, 539, 548, 574, 632, 742, 751, 795, 842
determinism, 757, 785

Index

detoxification, 530, 532, 544, 610
Detoxification, 530
devaluation, 150
developed countries, 561, 601, 602, 603, 727
developing brain, 26, 425, 429, 440, 441, 442, 444, 578
developing countries, xvii, 574, 575, 595
developmental change, 6
developmental disabilities, 611, 612, 615
developmental disorder, xvii, 9, 32, 609, 610, 611, 614, 619
developmental process, 9, 429
developmental psychology, 594
developmental psychopathology, 473
developmental theories, 9
deviation, 80, 233, 445, 449, 451
diabetes, 228, 522, 541, 543, 551, 555, 559, 622, 723, 843
diabetes mellitus, 522, 555
diagnosis, 187
diagnostic, 367, 369
Diagnostic and Statistical Manual of Mental Disorders, 469, 502
diagnostic criteria, 367, 369, 555, 577, 579
Diamond, 11, 31, 142, 148, 223, 224, 500, 501, 509, 511, 515
diarrhoea, 486
diathesis-stress model, 610
dichotomy, 181, 760, 773
diet, 440, 694, 699, 706
dietary, 482
differential diagnosis, 2, 25, 578, 580
differentiation, 49, 471, 484, 493, 530, 749, 752, 819, 821, 823, 844, 846, 847, 854
diffusion, 60, 548, 584
diffusivity, 50
digestion, 523
dioxin, 612, 615
dioxin-like compounds, 612, 615
diplopia, 273
direct observation, 20, 234, 590
directionality, 643
disabilities, 191
disability, xvi, xvii, 59, 284, 298, 353, 356, 358, 366, 371, 395, 547, 562, 572, 573, 574, 576, 579, 582, 600, 601, 604, 606
disabled, 292
disaster, 748
discharges, 323, 324, 325, 326, 328, 332, 333, 366, 407, 412, 414, 415, 416, 433, 434
discomfort, 91, 236, 248, 288, 706, 707, 710, 711
Discovery, 66

discrimination, 116, 124, 125, 126, 127, 128, 140, 156, 195, 196, 224, 437, 580, 685, 693, 724, 800, 841, 846
discrimination learning, 116, 126, 437, 846
discrimination tasks, 156
discrimination training, 125, 126, 127
discriminatory, 289
disease gene, 620, 646, 647, 650, 651, 653, 687, 836, 838, 840, 843, 849, 852, 854
disease model, 174
disease progression, 313, 494, 532, 544, 548, 619, 627, 628, 629, 637, 639, 650, 821, 842, 849, 851
diseases, x, xiv, xviii, 30, 48, 255, 316, 320, 353, 354, 355, 358, 361, 365, 415, 480, 481, 489, 541, 542, 543, 544, 552, 604, 611, 617, 620, 625, 632, 635, 637, 643, 645, 651, 660, 670, 671, 678, 686, 710, 713, 723, 742, 787, 807, 817, 832, 837, 841, 844, 846, 847, 849, 850
disequilibrium, 68, 624, 629
disgust, 808
disinhibition, 373, 808
displacement, 234, 241, 361, 375, 380
dissociation, 217, 223, 442, 642
dissonance, 752, 790
distinctness, 127
distortions, 263, 286
distraction, 518
distress, 603, 725, 726, 730, 734, 776, 782, 783, 787, 788
distribution, xviii, 21, 22, 34, 35, 39, 43, 83, 170, 172, 285, 354, 425, 445, 512, 536, 546, 548, 555, 570, 587, 593, 617, 637, 649, 756, 817, 839, 840, 854
disulfide, 536
diurnal, 376
divergence, 290
divergent thinking, 411
diversity, 18, 64, 213, 233, 517, 688, 761
division, 122, 130, 133, 134, 217, 679
divorce, 187, 730, 748, 752
dizygotic, 24, 35
dizygotic twins, 35
dizziness, 486
DNA, viii, ix, xviii, 426, 480, 492, 522, 535, 617, 624, 629, 630, 634, 637, 641, 647, 661, 671, 677, 678, 679, 680, 681, 682, 683, 684, 685, 686, 687, 688, 689, 725, 730, 732, 739, 741, 742, 743, 790, 795, 796, 807, 840, 841
DNA damage, 480, 535, 677, 679, 680, 681, 682, 683, 686, 687
DNA lesions, 682, 683
DNA ligase, 681, 683, 685, 688

DNA polymerase, 681
DNA repair, ix, 480, 492, 677, 678, 680, 682, 685, 686, 688, 689, 807
DNA testing, 725, 730, 742, 790, 796
DNAs, 684
doctors, 547, 601
dogs, 549, 557
dominance, 223, 629, 794
donor, 756
donors, 685
doors, 159
dopamine agonist, 172, 274, 306
dopamine antagonists, 139, 195
dopaminergic, xi, 4, 5, 6, 7, 8, 15, 27, 50, 54, 63, 67, 70, 77, 95, 118, 119, 136, 137, 138, 139, 156, 171, 172, 173, 174, 175, 176, 274, 275, 324, 350, 351, 366, 376, 391, 393, 398, 674, 737, 802, 806, 814, 823, 841, 854
dopaminergic modulation, 136
dopaminergic neurons, xi, 171, 350, 351, 841
dorsolateral prefrontal cortex, 3, 4, 5, 6, 7, 13, 14, 15, 16, 17, 21, 25, 26, 27, 28, 39, 47, 58, 59, 60, 63, 109, 142, 154, 169, 170, 172, 173, 188, 189, 193, 215, 217, 222, 223, 224, 281, 305, 307, 341, 370, 372, 374, 388, 410, 500
dorsomedial nucleus, 410
dosage, 306, 367, 375, 420, 462, 576, 707
dose-response relationship, 460
dosing, 821
DOT, 382, 383, 385
double-blind, 195
double-blind trial, 336
Down syndrome, 544
Down's Syndrome, xvii, 574
down-regulation, 426
draft, 591
drawing, 145, 385, 547, 561, 568, 572, 749, 757, 764, 774, 779
drinking, 77, 81, 573, 574, 576, 577, 581, 582, 584
drinking pattern, 582
drinking patterns, 582
dropouts, 400
Drosophila, 490, 640, 642, 647, 652, 671, 672, 673
DRS, 399
drug, 185, 188, 190
drug abuse, 308, 351
drug action, 188
drug addict, 500, 519
drug addiction, 519
drug delivery, 569, 820
drug dependence, 228

drug exposure, 42
drug targets, 48, 553
drug therapy, 314, 365, 695, 848
drug treatment, xvi, 48, 51, 54, 63, 299, 429, 562
drug use, 53
drug-resistant, 258, 273, 278, 280, 283, 328, 334, 436
drugs, ix, xiii, xvi, 21, 31, 32, 34, 38, 41, 44, 47, 48, 50, 51, 52, 53, 55, 56, 57, 58, 61, 62, 63, 65, 70, 77, 185, 236, 259, 274, 278, 280, 302, 319, 324, 325, 342, 355, 407, 408, 416, 424, 426, 428, 431, 432, 436, 438, 439, 440, 441, 472, 475, 482, 485, 494, 496, 556, 562, 691, 692, 695, 696, 697, 796
DSM, 447, 461, 473, 474, 502
DSM-IV, 447, 461, 473, 474
dual task, 508
duplication, 619
duration, 18, 21, 22, 37, 42, 43, 54, 81, 84, 105, 119, 134, 237, 242, 265, 280, 285, 293, 314, 354, 383, 400, 402, 412, 413, 414, 415, 416, 419, 431, 457, 461, 485, 488, 565, 575, 628, 692, 699, 808, 837
duration of untreated psychosis, 18, 37
dust, 610
duties, 758
dynamics, 775, 797
dysarthria, xix, 361, 705, 808
dyskinesia, xiii, 51, 62, 306, 353, 354, 355, 358, 360, 361, 363
dysphagia, 311, 694, 699, 706, 714, 808
dysphoria, 732, 733
dysplasia, 280, 324, 329, 333, 477, 490
dysregulation, 1, 6, 9, 27, 282, 341, 350, 519
dysthymia, 423
dystonia, ix, xiv, xviii, 227, 228, 236, 251, 256, 257, 258, 264, 267, 268, 269, 277, 278, 291, 293, 294, 295, 297, 298, 301, 312, 316, 329, 353, 354, 355, 358, 359, 360, 361, 362, 363, 364, 366, 389, 623, 625, 656, 693, 694, 696, 698, 808

E

ears, 163, 164, 167, 367, 581, 812
eating, 77, 151, 401, 698, 706, 707, 808
eating disorders, 401
economic indicator, 489
economic status, 717, 731
edema, 243, 248, 545, 546
Eden, 75, 76, 77, 106, 112
editors, 69, 168, 169, 268, 294, 298, 300, 302, 391

education, xv, xvii, 142, 146, 216, 419, 428, 499, 510, 514, 547, 573, 574, 581, 582, 583, 590, 596, 600, 708, 718, 792
Education, 28, 582, 583, 585, 596, 597, 742, 789, 797
educational attainment, 568
educational research, 582
educational system, 582
educators, 516, 573
EEG, xiii, 9, 73, 74, 86, 87, 88, 89, 91, 92, 93, 94, 95, 97, 99, 101, 102, 104, 106, 107, 110, 111, 112, 136, 137, 238, 253, 259, 260, 280, 301, 308, 309, 320, 321, 322, 323, 326, 329, 333, 334, 335, 343, 404, 408, 411, 414, 415, 416, 419, 420, 425, 433, 434, 437, 470
EEG activity, 73, 87, 99, 104, 259, 280, 301, 335, 415, 425, 434
EEG patterns, 74, 87, 97
effective, 177
efficacy, 182, 190, 369
EGB, 91
Egypt, 407, 409
ejaculation, 79, 84, 87, 91, 92, 104, 106, 113
elaboration, 23, 101
elderly, xv, xvi, xvii, 276, 287, 295, 408, 475, 487, 494, 507, 515, 517, 518, 519, 541, 542, 548, 555, 561, 562, 563, 564, 565, 567, 568, 569, 570, 571, 572, 600, 601, 602, 603, 604, 630, 714, 715, 805, 851
elderly population, 295
elders, 490, 599, 600, 605
election, 239, 266, 294, 351, 372
electric current, 289
electrical, 366, 367, 368, 376, 379, 380
electrical conductivity, 263, 265, 292
electricity, xii, 256
Electroconvulsive Therapy, 402, 405
electrode surface, 314
electrodes, 87, 227, 239, 243, 246, 247, 248, 249, 253, 263, 264, 265, 271, 273, 281, 287, 288, 290, 291, 292, 300, 311, 322, 325, 326, 327, 356, 361, 367, 373, 375, 380, 384, 386, 388, 389, 400
electroencephalogram, 86, 228
electroencephalography, 411, 434
electrographic seizures, xiii, 320, 328
electrolyte, 236
electrolyte imbalance, 236
electromyography, 238
electron, 21, 42, 481, 528
electron microscopy, 21
electrophoresis, 523, 535
electrophysiological properties, 801

electrophysiological study, 167, 212, 301
electrophysiology, 106, 142, 248
elongation, 529, 531, 539, 673
elucidation, 295, 382, 389
embolus, 543
embryo, 575, 786
embryonic, 800, 802, 809, 834
embryonic development, 802
embryos, 786
emergency, 250, 695
EMG, 238, 239, 240, 242, 244, 265, 291, 292
emission, 49, 59, 60, 86, 93, 172, 175, 191, 377, 379, 381, 382, 383, 391, 411, 417, 434, 513, 528, 548, 736, 815, 835
emotion, 103, 142, 147, 151, 164, 177, 192, 210, 260, 341, 348, 509, 510, 511, 517, 783, 801, 816, 832
emotion regulation, 509
emotional, ix, x, xv, xvii, 2, 17, 75, 77, 117, 123, 135, 137, 141, 145, 148, 185, 228, 341, 342, 343, 347, 348, 397, 409, 412, 433, 499, 508, 509, 510, 513, 515, 579, 600, 613, 747, 749, 752, 767, 776, 783, 784
emotional distress, 734
emotional processes, 75, 77, 347
emotional reactions, 767
emotional state, 123, 783
emotional valence, 17
emotions, 112, 142, 500, 510, 748, 755, 792
employment, 146, 415, 427, 581, 724, 730, 751, 775
employment opportunities, 427
EMS, 235
enamel, 581
enantiomer, 485
enantiomers, 495
encapsulated, 782
encephalitis, 483
encoding, xv, xviii, 5, 129, 136, 150, 152, 153, 156, 289, 290, 364, 371, 407, 416, 476, 478, 548, 580, 619, 664, 677, 678, 679, 818
Encoding, 149, 151, 379, 405
encouragement, 788
endocrine, 77, 82, 103, 112, 612, 613
endocrine system, 77, 612
endocytosis, 811
endonuclease, 680
endophenotypes, 24, 30, 470
endoplasmic reticulum, 477, 532
Endothelial, 802, 838, 852
endothelin-1, 550
endothelium, 542, 543, 550

energy, 110, 117, 344, 363, 480, 528, 532, 536, 544, 547, 643, 644, 648, 649, 651, 699, 812, 813, 841, 843, 844, 850
energy consumption, 363
energy expenditure, 699, 850
engagement, 153, 165, 510, 597, 752, 784, 794
engineering, 292, 478
England, 32, 239, 308, 376, 570, 578, 584, 604, 605, 606, 626, 721, 725, 740, 741, 743, 745, 789, 791, 799
enlargement, 213, 483
enolase, 528, 537
enthusiasm, 289
entorhinal cortex, xvi, 321, 377, 413, 479, 522, 533
entropy, 236
environment, x, 77, 78, 116, 128, 142, 178, 239, 252, 372, 408, 579, 582, 591, 592, 612, 688, 697, 698, 712, 800, 838, 854, 855
environmental audit, 153, 159, 164
environmental change, 137
environmental conditions, 186
environmental factors, 543, 678, 855
environmental influences, 11, 808
environmental stimuli, 153, 159, 160, 165, 186, 436
enzymatic, 522, 528, 530, 542, 614
enzymatic activity, 522, 528, 614
enzyme, 5, 51, 180, 421, 476, 480, 489, 492, 529, 530, 546, 613, 614, 677, 679, 682, 684, 685
enzymes, 407, 477, 480, 481, 491, 522, 528, 535, 538, 679, 680, 681, 683
ependymal, xx, 799, 800
ependymal cell, xx, 799, 800
epidemic, 604
epidemics, 611
epidemiologic studies, 490
epidemiology, 347, 349, 351, 645, 647, 672, 791
epidermal growth factor, 834
epileptic seizures, xiii, 279, 302, 309, 320, 323, 324, 325, 331, 333, 336
epileptogenesis, 258, 407, 417, 429, 440
episodic memory, xv, 50, 414, 503, 522
epithelial cell, 537
epithelial cells, 537
epitopes, 483
epoxy, 87
EPS, 62
equal opportunity, 583
equilibrium, 693
equipment, 238, 239, 244, 254, 263, 276, 286, 288, 382, 821
ERK1, 491

erosion, 322, 326
ERPs, 155, 512
error detection, 195, 341
ESI, 523, 524
essential tremor, 228, 257, 266, 269, 276, 299, 301, 311, 312, 316, 353, 362
EST, 658
esterase, 51, 423
estimating, 204, 446
estradiol, 109
estrogen, 99, 103, 105, 111
Estrogen, 82, 110, 803, 853
estrogens, 99
etanercept, 487, 495
ethanol, 423
Ethanol, 583, 584, 585
ethical concerns, 382
ethical issues, 594, 786
ethical standards, 262
ethics, 461, 788, 794
ethnic groups, 625
ethology, 112
etiology, xiv, xvii, 27, 182, 353, 354, 355, 358, 414, 609, 611, 714
eugenics, 790, 796
eukaryotic, 539, 540
Europe, 272, 482, 486, 587, 604, 605, 712, 846
European, 379, 380
euthanasia, 739
event-related potential, 242, 399, 417
event-related potentials, 242, 399, 417
everyday life, 143, 708
evoked potential, 228, 238, 240, 241, 243, 285, 286, 360
evolution, xii, 67, 77, 112, 200, 211, 216, 357, 358, 434, 475, 480, 481
ewe, 111
examinations, 3, 512, 611, 613
excision, 268, 280, 300, 677, 678, 679, 680, 683, 686, 687
excitability, 97, 144, 302, 322, 327, 330, 407, 421, 422, 427, 438
excitation, 289, 325, 421, 422, 423, 427, 697, 825, 854
excitatory synapses, 4, 206, 208
excitotoxic, 29, 32, 33, 37, 111, 412, 419, 441
excitotoxicity, 10, 15, 393, 422, 423, 528, 529, 532, 636, 653, 656, 812, 813, 828, 831, 837, 842, 843, 850, 852, 856
excitotoxins, 812, 826, 827, 828
exclusion, xix, 203, 228, 308, 447, 468, 692, 769, 775, 796

execution, 77, 81, 82, 83, 91, 123, 124, 157, 162, 342, 370, 372
executive function, ix, xvi, 2, 16, 17, 19, 41, 119, 138, 139, 149, 178, 183, 187, 218, 225, 275, 288, 309, 366, 371, 377, 378, 407, 409, 410, 411, 412, 413, 414, 431, 445, 448, 457, 459, 469, 470, 472, 473, 474, 499, 501, 502, 508, 513, 514, 515, 516, 517, 519, 520, 561, 562, 567, 568, 572, 579, 815
executive functioning, 16, 41, 367, 414, 579
executive functions, ix, 2, 16, 17, 19, 119, 139, 178, 183, 187, 218, 275, 288, 309, 371, 377, 407, 409, 411, 412, 414, 431, 473, 508, 513, 514, 516, 517, 519
exercise, 81, 512, 697, 698, 758, 830, 840
exons, xviii, 478, 656
exonuclease, 685
expectation, 341, 349
experimental condition, 80, 468
experimental design, 449, 456, 457, 460, 461, 463
expert, 588
expertise, 237, 239, 587, 590, 591, 700, 749, 781, 788
explicit memory, 584
exploitation, 143
exposure, xvii, 3, 11, 13, 18, 20, 21, 22, 32, 36, 42, 83, 84, 85, 86, 136, 234, 342, 399, 418, 419, 426, 427, 430, 435, 436, 437, 438, 441, 510, 568, 575, 576, 578, 579, 581, 583, 584, 585, 590, 591, 609, 610, 611, 612, 613, 614, 615, 708, 712, 756, 851
Exposure, 187, 193, 584, 585, 610, 611, 612, 613, 614, 615
expressed sequence tag, 658
expressivity, 618, 619
extensor, 239, 242
extensor digitorum, 239, 242
external environment, 78, 698
external locus of control, 773
extinction, 134, 150, 172, 192
extracellular, 190, 193
extracellular matrix, 477, 545, 549, 554, 825, 856
extracranial, 396
extraction, 185
extracts, 124, 487, 658, 667, 682, 683, 684, 767
extravasation, 546
extroversion, 512
eye movement, 67, 153, 157, 158, 159, 165, 166, 167, 168, 169, 180, 625, 808
eyelid, 292
eyes, 95, 98, 581

F

fabrication, 708
face validity, 590
facial expression, 239
facial muscles, 625
facial pain, 313
factor analysis, 508
factorial, 107, 371, 543
factors, 193
factual knowledge, 781
failure, ix, 2, 143, 146, 238, 260, 276, 278, 284, 291, 307, 482, 542, 563, 564, 568, 623, 686, 723, 813, 814, 820, 824, 828
faith, 736
false belief, 515
false negative, 14, 450
false positive, 14, 22, 449, 450, 452, 458, 463, 658
familial, 476, 478, 492, 546, 550, 787
family history, xix, 724, 729, 746, 760, 787
family life, 793, 794
family members, 489, 490, 692, 693, 700, 709, 720, 721, 722, 724, 727, 732, 736, 745, 746, 753, 755, 757, 759, 761, 762, 764, 768, 769, 770, 771, 775, 777, 778, 780, 781, 782, 783, 807
family physician, xvii, 562
family planning, 694
family structure, 763
family support, 764
family system, 760, 796
family therapy, 752, 763
FAS, 575, 576, 577, 578, 579, 581, 582, 583, 584, 667, 669
fasciculation, 579
FASD, 573, 576, 577, 578, 579, 580, 581, 582, 584, 585
fat, 131
fatigue, 460
fatty acid, 38
fatty acids, 480, 487, 699
faults, 786
FDA, 275, 366, 395, 396, 640
FDG, 417, 548
fear, 103, 106, 110, 139, 142, 143, 144, 147, 150, 151, 260, 262, 283, 724, 725, 770, 793
fears, 729, 782
febrile seizure, 418, 437
February, 297
feces, 346
feedback, 49, 115, 116, 117, 124, 129, 134, 135, 172, 334, 410, 411, 579, 592, 708, 749, 765

feeding, 104, 110, 601
feelings, 142, 510, 722, 725, 729, 733, 734, 736, 756, 772, 781, 782
feet, 159
female rat, 70, 73, 80, 83, 87, 88, 91, 93, 94, 97, 98, 100, 101, 102, 103, 104, 105, 106, 108, 109, 110, 111, 113, 853
females, 74, 81, 83, 84, 88, 90, 93, 98, 105, 721, 730, 731, 733, 787
fermentable carbohydrates, 711
ferritin, 544
fertility, 623, 765, 770, 774, 789, 791, 794, 818
fertilization, 266, 786
fetal, 426, 435, 582, 583, 584, 585, 751, 758, 790, 791, 793, 795
fetal abnormalities, 751, 791, 795
fetal alcohol syndrome, 582, 584
fetal tissue, 832
fetus, 408, 426, 758, 769, 776
fetuses, 758, 759, 776
FFT, 87
fiber, 4, 185, 192, 194, 343, 440, 837, 839
fiber bundles, 4, 343
fibers, 4, 290, 314, 340, 366, 628
fibrillar, 478
fibrillation, 482, 554
fibrils, 476, 810
fibrin, 525
fibrinogen, 525, 551
fibroblast growth factor, 802, 834, 841, 844, 854
fibroblasts, 677, 679, 680, 685
fibrosis, 314, 544, 779, 786, 794
filament, 532
fillers, 611
film, 243
films, 243, 246, 247
filters, 238
filtration, 575
financial, 486, 599, 600, 604, 670, 736, 756, 768, 776, 832
financial support, 670
fire, 179, 180
firearms, 721
fires, 177
first degree relative, 742, 853
first dimension, 523
first generation, ix, 22, 48, 50
fish, 512, 611, 612
fitness, 698
fixation, 32, 153, 157, 159, 162, 163, 165, 166, 167, 233, 236, 252, 386
flame, 610, 612
flame retardants, 610, 612

flaws, 591
flexibility, 16, 49, 116, 123, 172, 178, 189, 362, 380, 414, 508, 546, 627, 698, 782, 808
flow, 44, 49, 50, 58, 59, 60, 149, 172, 173, 175, 180, 183, 189, 313, 371, 381, 386, 390, 392, 393, 398, 417, 542, 543, 546, 555, 764
fluctuations, xiv, 86, 228, 306, 308, 314, 367, 375, 376, 381, 382
fluid, xvi, 36, 228, 236, 248, 250, 287, 373, 398, 481, 483, 508, 512, 522, 525, 535, 699, 834
fluorine, 191
fluoxetine, 36, 344, 697, 805, 819, 841, 846
fMRI, 49, 50, 60, 61, 148, 149, 155, 172, 191, 224, 281, 285, 286, 289, 366, 379, 398, 517
FMRI, 286
focal seizure, xiii, 319, 320, 325, 428
focusing, 366, 395, 509, 511, 523, 580, 718, 747, 751, 755, 807
folate, 488
Folate, 497
folding, 531, 539
folding catalyst, 539
folic acid, 488
follicles, 610
food, 78, 136, 142, 383, 401, 601, 611, 613, 699, 706, 712, 713, 770
Food and Drug Administration, 272
food chain, 611
food intake, 401, 699
force, 79, 248, 281, 351
Ford, 9, 33, 212, 355, 363
forebrain, 44, 70, 75, 180, 386, 430, 544, 612, 837, 840, 846, 849, 852, 854
forgetfulness, 178, 375
forgetting, 410, 434
formation, 10, 39, 49, 63, 85, 115, 116, 117, 119, 120, 122, 124, 125, 128, 129, 130, 133, 134, 135, 139, 140, 155, 285, 287, 329, 385, 386, 417, 425, 476, 479, 482, 483, 488, 492, 526, 530, 538, 539, 544, 551, 557, 628, 634, 637, 641, 645, 647, 648, 651, 663, 666, 667, 677, 680, 682, 683, 685, 686, 750, 752, 800, 810, 826, 836, 837, 848, 851, 852
fornix, 321, 413
foundations, 457, 582, 792
Fourier, 87, 319
Fox, 37, 431, 442, 492, 493, 556, 596, 754, 791
fracture, 250, 357, 361, 362
fragmentation, 840
fragments, 476, 478, 485, 486, 634, 638, 639, 648, 651, 809, 811, 813, 826, 827, 841, 845, 847, 852
framing, 152, 777

France, 108, 319, 324, 327, 339, 353, 395, 600, 601, 602, 606, 719
free radical, 422, 424, 480, 482, 530
free radicals, 422, 424, 482, 530, 813
free recall, 376
freedom, 239, 428, 749, 754, 758, 770
freedom of choice, 754, 758
freezing, 81, 275, 305, 306, 308, 314, 373, 374
frontal, 189, 191, 192, 193, 194, 195, 196
frontal cortex, xi, 3, 16, 18, 23, 28, 48, 49, 50, 52, 56, 63, 64, 84, 103, 108, 135, 136, 138, 149, 154, 156, 157, 158, 160, 161, 162, 163, 167, 168, 170, 173, 175, 189, 192, 195, 204, 210, 213, 296, 398, 410, 413, 414, 502, 571, 633, 734, 815
frontal lobe, xi, 3, 12, 14, 16, 19, 29, 30, 45, 50, 60, 75, 105, 108, 111, 112, 142, 148, 149, 151, 155, 168, 169, 170, 173, 191, 193, 196, 200, 201, 203, 204, 210, 211, 216, 217, 224, 343, 345, 346, 349, 350, 374, 409, 410, 411, 412, 413, 417, 422, 425, 433, 500, 567, 571, 579
frontal lobes, 108, 168, 170, 193, 203, 216, 217, 224, 343, 346, 410, 412, 417, 422, 500, 579
frontal-subcortical circuits, 341, 378, 548
frontotemporal dementia, 543, 555
Frontotemporal Lobar Degeneration, 479, 491
frustration, 708, 781
FSH, 362
FTLD, 479
fulfillment, 772
functional, 180, 183, 193, 194, 195, 196
functional activation, 50
functional analysis, 223
functional architecture, 106, 212
functional changes, 9, 172
functional imaging, 165, 168, 260, 262, 273, 282, 366
functional magnetic resonance imaging, 49, 170, 194, 285, 378
functional MRI, 193, 195, 225, 380, 433, 517
funding, 204, 599, 600, 601, 788
furniture, xii, 216
fusiform, 19, 20, 25, 417
fusion, 238, 647, 842
futures, 751

G

G protein, 65, 66
GABA, 3, 4, 5, 6, 7, 15, 21, 27, 32, 34, 38, 39, 42, 43, 44, 259, 296, 325, 326, 330, 331, 392, 394, 398, 408, 421, 422, 423, 424, 425, 426, 438, 439, 440, 441, 579, 641, 802, 803, 813, 814, 816, 834, 841, 848
$GABA_B$, 426, 435, 438
GABAergic, 31, 43, 186, 189, 279, 323, 324, 325, 366, 416, 417, 421, 422, 854
gait, xix, 228, 275, 285, 305, 306, 311, 314, 315, 366, 373, 374, 556, 579, 625, 694, 705
gambling, xi, 141, 143, 145, 146, 148, 149, 151
games, 462, 511, 512, 513
gamete, 769, 774, 785
gametes, 757, 773, 776
gamma-aminobutyric acid, 44, 45, 438
ganglia, xiii, xiv, 21, 22, 23, 36, 60, 75, 77, 173, 175, 257, 258, 266, 269, 274, 277, 278, 289, 309, 320, 323, 324, 329, 330, 334, 336, 342, 350, 354, 361, 365, 370, 372, 374, 377, 379, 388, 390, 392, 393, 397, 410, 412, 413, 431, 579, 584, 612
Ganglia, 382, 814
gastrointestinal, 484, 485, 486, 487
gauge, 91
GC, 28, 167, 329, 552, 555
GE, 28, 29, 37, 235, 390, 491, 523, 524, 526, 531, 553
gel, 333, 523, 524, 525, 535, 658, 712, 713
gels, 523, 524, 525
gender, x, 54, 74, 447, 461, 467, 472, 571, 626, 627, 760, 762, 774
gene expression, 32, 39, 427, 441, 539, 628, 633, 634, 640, 644, 648, 649, 661, 671, 672, 811, 843, 848
gene promoter, 638, 671, 818
gene targeting, 852
general anaesthesia, 236, 353, 397
general anesthesia, 354, 356, 361, 709
General Electric, 230
General Health Questionnaire, 734, 735, 739
general practitioner, 806
generalization, 224, 335, 373, 512
generalized anxiety disorder, 146, 150
generalized seizures, xiv, 279, 280, 322, 325, 336, 408, 413, 414
generalized tonic-clonic seizure, 279, 322, 433
generation, ix, 20, 21, 22, 33, 48, 50, 51, 140, 215, 217, 218, 219, 220, 224, 258, 277, 279, 299, 320, 323, 326, 350, 351, 372, 374, 410, 476, 481, 508, 549, 630, 636, 640, 699, 761, 762, 763, 767, 768, 769, 773, 779, 780, 781, 782, 785, 806, 842
generators, 414, 750
genes, xv, xix, 10, 40, 52, 64, 70, 74, 86, 180, 191, 476, 477, 479, 492, 554, 612, 618, 620, 629, 635, 640, 641, 648, 658, 661, 662, 664,

666, 667, 674, 675, 681, 746, 779, 787, 817, 827, 846
genetic abnormalities, 614
genetic alteration, 180, 426
genetic background, 664
genetic counselling, 632, 757, 760
genetic defect, xviii, 617, 634
genetic disease, 670, 723, 795
genetic disorders, 756, 787, 790
genetic endowment, 576
genetic factors, 627, 728, 746, 784, 808
genetic information, 732, 760, 785
genetic linkage, 55, 624, 629
genetic marker, 629
genetic predisposition, xvii, 610
genetic screening, 740
genetic testing, xix, 717, 718, 725, 726, 727, 729, 731, 732, 737, 738, 739, 746, 748, 751, 759, 762, 763, 764, 769, 771, 773, 774, 775, 776, 781, 782, 786, 787, 790, 793, 796
genetics, xix, 414, 519, 618, 620, 672, 741, 745, 746, 757, 759, 762, 763, 765, 781, 793, 795, 796, 797, 806
Geneva, 585
genome, 24, 655, 664, 666, 667, 682, 809, 818
genomic, 539
genotype, 24, 28, 54, 148, 191, 533, 625, 630, 664, 685, 821, 851
genotyping, 512
Ger, 295
geriatric, 315, 606
geriatricians, xvii, 561, 562
Germany, 23, 113, 115, 141, 147, 149, 152, 319, 601, 602, 757
Gestalt, 594
gestation, 98, 613
GFAP, 531, 532
GHQ, 734, 735
gift, 792
gingival, 707
gingivitis, 707, 712
Ginkgo biloba, 487, 495
gland, 658
glass, 98, 188
glia, 3, 425, 545, 618, 637, 667, 679, 800, 801, 813, 823, 854, 855
glial, xx, 3, 20, 21, 25, 41, 42, 479, 531, 532, 540, 542, 559, 799, 800, 801, 802, 804, 813, 814, 834, 849
Glial, 42, 812
glial cells, xx, 3, 4, 20, 21, 479, 799, 800, 813
glial fibrillary acidic protein, 531, 540, 559
glial fibrillary acidic protein (GFAP), 531

glioma, 674, 825
gliosis, 8, 263, 287, 376, 540, 828
globus, xiv, 14, 18, 238, 254, 256, 267, 268, 269, 274, 297, 300, 353, 362, 363, 364, 365, 366, 371, 375, 378, 388, 390, 394, 410, 633, 636, 669, 814, 834
glucocorticoids, 351, 848
glucose, 34, 393, 417, 431, 434, 435, 528, 531, 536, 542, 545, 548, 550, 556, 558, 667, 673, 713
glucose metabolism, 393, 417, 431, 434, 435, 528, 550, 556
glucose oxidase, 713
glutamate, 15, 34, 56, 71, 175, 392, 394, 421, 422, 423, 426, 427, 429, 441, 529, 542, 545, 546, 611, 636, 650, 667, 673, 805, 812, 813, 814, 817, 818, 827, 837, 845, 856
glutamate decarboxylase, 421
glutamatergic, 10, 70, 330, 423, 605
glutamic acid, 5, 28, 34, 44, 137, 421, 850
glutamine, 529, 537, 635, 638, 648, 653, 807, 835, 837, 855
glutathione, 480, 530, 538
glutathione peroxidase, 480
glycerin, 713
glycine, 424, 439
glycogen, 530, 538
glycogen synthase kinase, 530, 538
glycolysis, 528, 667, 673
glycoprotein, 477, 525, 526
glycoproteins, 524
glycosaminoglycans, 544
glycosylation, 522, 524, 546
goal-directed, 77, 78, 94, 115, 116, 117, 123, 125, 126, 128, 133, 135, 411, 509
goal-directed behavior, 77, 94, 115, 123, 125, 411
goals, 78, 178, 290, 411, 713, 748, 749, 752, 754, 772, 774, 780, 782
goblet cells, 484
God, 597
gold, 11, 234, 258, 277, 328, 447
gold standard, 11, 234, 258, 277, 328, 447
Gore, 379
Gori, 315
government, xvii, 574, 602, 603
G-protein, 52, 423, 817
grades, 632, 637
grading, 571, 633, 734
grain, 478
grants, 489, 533, 552, 601, 737
granule cells, 847, 852
gravitational force, 248

gravity, 95
gray matter, 1, 5, 10, 11, 12, 13, 14, 15, 16, 17, 18, 19, 20, 22, 23, 24, 25, 26, 27, 28, 29, 30, 31, 32, 33, 35, 36, 38, 39, 41, 42, 43, 45, 413, 500, 547, 628
Greece, 255, 271
green tea, 487
grey matter, 31, 35, 36, 40, 43, 44, 50, 261, 579, 815
grounding, 761
growth, 33, 38, 78, 223, 361, 425, 426, 440, 441, 476, 478, 529, 539, 550, 557, 558, 576, 578, 641, 649, 651, 707, 801, 802, 804, 805, 820, 822, 823, 824, 826, 829, 830, 832, 834, 835, 837, 841, 843, 844, 845, 846, 847, 849, 854
growth factor, 426, 550, 557, 558, 641, 649, 651, 801, 802, 804, 805, 820, 823, 825, 826, 829, 830, 832, 834, 835, 837, 841, 843, 844, 845, 846, 847, 849, 854
growth factors, 426, 801, 802, 804, 805, 820, 823, 825, 826, 830, 832, 843, 846
GSK-3, 494, 530, 531
GSR, 250
GST, 530, 538
Guangdong, 171
Guangzhou, 171
guanine, 677, 680, 682, 685, 807, 833
guardian, 462
guidance, 251, 252, 573, 667, 788
guidelines, 235, 266, 281, 282, 295, 429, 565, 571, 574, 591, 600, 605, 611, 693, 696, 718, 763, 764, 793, 795
guilt, 500, 724, 725, 768, 775
guilt feelings, 725
guilty, 725
gustatory, 77
gynecomastia, 623
gyri, 11, 17, 18, 19, 20, 25
gyrus, 2, 12, 13, 14, 15, 16, 17, 18, 19, 20, 22, 23, 24, 25, 26, 34, 35, 50, 155, 207, 208, 211, 261, 268, 281, 300, 326, 371, 376, 389, 417, 548
Gyrus, 397

H

habituation, 69, 118, 291, 314, 418, 436, 551, 612
hair, 485, 610, 611, 612, 615, 757
hair follicle, 610
hairpins, 682, 810
half-life, 460
hallucinations, ix, 1, 7, 18, 28, 33, 40, 42, 48, 282, 288, 293, 296, 304, 376

haloperidol, 20, 21, 22, 29, 32, 34, 36, 40, 41, 43, 50, 53, 54, 56, 58, 61, 62
handedness, 369
handicapped, 361
handling, 136, 416, 492, 570
hands, 420, 564, 569, 581, 710
hanging, 124, 763
haploinsufficiency, 809
haplotypes, 10, 31, 629, 649
haptoglobin, 525, 535
harm, 27, 35, 574, 576, 581
Harvard, 596, 609, 789, 797
Hawaii, 65
hazards, 792
HDAC, 640
head, 373
head and neck cancer, 801
head injury, 276, 460
Head Start, 510, 514, 515
headache, 227, 228, 253, 272, 294, 299, 300, 301, 346, 612
headache,, 346
healing, 526
health, xix, 21, 263, 442, 511, 516, 517, 552, 575, 581, 582, 587, 589, 590, 596, 599, 603, 605, 606, 611, 612, 613, 692, 699, 705, 706, 707, 708, 709, 710, 711, 713, 714, 715, 718, 726, 727, 728, 732, 735, 747, 751, 752, 758, 769, 781, 782, 784, 787, 791, 792, 793, 795, 797
health care, xix, 599, 603, 699, 705, 706, 709, 711, 713, 718, 758, 769, 781
health care professionals, 699, 713, 758, 769, 781
health care system, 718
health care workers, 603
health education, 708
health problems, 581, 713, 751
health psychology, 747, 784, 795
health services, 795
health status, xix, 596, 705, 792
healthcare, 603
hearing, 289, 292, 461, 501, 564, 581
hearing impairment, 461
hearing loss, 581
heart, 148, 292, 341, 397, 476, 522, 541, 581
Heart, 557, 559
heart disease, 522, 541, 721
heart rate, 148, 292, 397
heat, 250, 480, 530, 537, 538, 673
heat shock protein, 480, 530, 538, 673
heating, 384
heavy metal, 609, 610, 612, 613, 614
heavy metals, 609, 610, 613, 614
hedonic, 350

height, 362, 757
helix, 65, 793, 795, 796
helplessness, 855
hematocrit, 551
hematoma, 288, 324
hematopoiesis, 847
hemiparesis, 250, 297, 343
hemisphere, 8, 17, 180, 284, 306, 309, 321, 411, 416
hemispheric asymmetry, 104
hemodynamic, 74, 86, 544
hemoglobin, 525
hemorrhage, 263, 272, 273, 288, 289, 314, 315, 322, 343, 346, 354, 361, 545, 546, 549, 550, 554, 555, 558
hemorrhages, 550, 554
hemorrhagic stroke, 550
heredity, 790, 806
heritability, 513
heterogeneity, 2, 17, 18, 19, 22, 25, 76, 212, 361, 646, 840
heterogeneous, ix, 1, 20, 27, 259, 273, 358, 420, 458
heterozygote, 740
heterozygotes, 851
heuristic, 797
high blood pressure, 543, 550
high pressure, 546
high resolution, 59, 307
high risk, 23, 35, 147, 152, 419, 574, 581, 599, 603, 759, 760, 769, 775, 776
high school, 409
high tech, 792
high-frequency, xiv, 282, 304, 322, 324, 329, 331, 336, 381, 382, 390, 391
high-level, 172, 174
high-risk, 15, 39, 762
high-tech, 263, 266, 286, 293
hip, 698
hippocampal, xv, 7, 8, 20, 24, 26, 30, 37, 38, 39, 43, 45, 63, 129, 136, 138, 139, 140, 195, 290, 302, 323, 326, 327, 329, 330, 333, 336, 412, 413, 417, 418, 420, 421, 422, 423, 424, 425, 426, 427, 430, 431, 432, 435, 439, 440, 441, 489, 492, 522, 527, 528, 531, 533, 534, 554, 556, 559, 578
Hippocampus, 138, 258, 439, 440, 849, 852, 855
hips, 146
HIS, 159, 160, 161, 162, 163
Hispanic, 510, 511
histochemical, 67
histogram, 159, 160, 161, 162, 163, 240, 241, 244
histological, 25, 29, 326, 388, 393, 440, 554

histone, 640, 647, 817
histone deacetylase, 640, 647, 817
histopathology, 2, 27, 537
history, xix, 18, 51, 228, 275, 300, 308, 355, 373, 402, 410, 442, 460, 461, 543, 547, 576, 580, 613, 693, 724, 729, 730, 731, 733, 734, 742, 746, 757, 760, 763, 785, 787, 791, 797
HIV, 740, 777, 788, 789
HIV/AIDS, 740
HLA, 41
HNE, 521, 522, 524, 527, 528, 530, 534, 538
Holland, 142, 150, 222, 224, 515
home care services, 601, 602, 603
home-care, 601, 603, 606
homeostasis, 77, 529, 532, 551, 842
homes, xix, 590, 604, 605, 692, 763, 764
homework, 507
homicide, 727
homocysteine, 480, 488, 492
Homocysteine, 488
homogeneity, 259, 280, 457
homogenous, ix, 2, 793
homologous genes, 658, 662
homology, 205
homovanillic acid, 118, 398, 850
homozygote, 809
hopelessness, 733, 734, 736, 738
Hops, 147, 150
hormone, 81, 82, 612, 673, 803, 839, 845
hormones, 79, 81, 82, 97, 105, 106, 107, 113, 612, 851
hospital, 250, 263, 356, 503, 563, 572, 574, 584, 607, 759, 763
hospitalization, 737
hospitalized, 18, 237, 399
hostility, 733, 783
hourly wage, 603
household, 591
housing, 503
HPLC, 122
Hubble, 252
human behavior, 348
human body, xii, 255
human brain, 28, 40, 67, 138, 150, 151, 172, 175, 176, 194, 203, 204, 225, 256, 265, 270, 275, 289, 292, 295, 431, 674, 801, 836, 839, 844, 847, 848, 849, 851
human cerebral cortex, 42, 212, 516
human genome, 655, 664, 666, 667, 682
Human Genome Project, 792, 797
human immunodeficiency virus, 33
human remains, 552
human right, 262, 284

human rights, 262, 284
human subjects, 6, 193
human welfare, xv, 499, 514
humans, xi, 9, 20, 55, 56, 71, 76, 82, 86, 112, 129, 135, 149, 155, 163, 174, 175, 179, 181, 182, 185, 188, 191, 192, 193, 200, 201, 203, 204, 210, 213, 216, 258, 292, 302, 320, 323, 326, 327, 334, 382, 417, 426, 430, 483, 500, 502, 541, 548, 549, 550, 611, 612, 613
humidity, 383
Hungary, 720, 731
Hunter, 335, 539, 701
Huntington disease, 41, 789, 790, 792, 794, 796, 797, 838, 840, 841, 847, 850, 851, 852, 853, 854, 855
Huntington's disease, xx, 355, 379, 799, 801, 834
Huntington's chorea, 788, 789, 791, 793, 839, 847, 852, 855
Huntington's disease, 769, 788, 790, 791, 793, 795, 796, 797, 836, 837, 838, 839, 840, 841, 842, 843, 844, 845, 846, 847, 848, 849, 850, 851, 852, 853, 854, 855, 856
husband, 722, 769, 778
husbandry, 804
hybrid, 292, 641, 658, 824
hybridization, 67, 633
hydrolyzed, 488
hydroperoxides, 530
hydrophobic, 524, 535
hydroxide, 324
hydroxyl, 424
hygiene, 698, 699, 706, 707, 708, 709, 710, 715
hyperactivity, 49, 63, 94, 178, 179, 183, 184, 188, 189, 190, 191, 192, 193, 194, 195, 196, 197, 418, 419, 462, 470, 471, 472, 473, 474, 519, 551, 558, 583, 613
hypercholesterolemia, 543
hyperemia, 557
hyperhomocysteinemia, 480, 482
hyperphosphorylation, 530, 531
hypertension, 482, 493, 522, 541, 546, 547, 550, 555, 558
Hypertension, 482, 543, 546, 550, 555
hypertensive, 138, 550, 551, 558, 559
hypertrophy, 144
hypoglossal nerve, 376
hypoperfusion, 482, 547
hypotensive, 189
hypothalamic, 104, 105, 106, 110, 272, 299, 301, 326, 332
hypothalamus, 75, 82, 83, 104, 107, 110, 111, 239, 249, 281, 341, 398, 410, 636, 641, 801, 844, 849

hypothesis, xvii, 7, 9, 17, 22, 29, 30, 32, 36, 38, 42, 50, 51, 53, 61, 120, 123, 125, 128, 129, 139, 148, 149, 150, 151, 185, 189, 284, 321, 366, 373, 446, 457, 460, 461, 476, 477, 480, 481, 484, 489, 492, 538, 609, 636, 638, 640, 641, 644, 660, 663, 664, 680, 684, 693, 727, 729, 814
hypoxia, 420, 438, 667, 674

I

ice, 183
ideal, 181, 289, 565, 629, 699, 710, 749
identification, x, 2, 57, 74, 204, 216, 231, 233, 237, 240, 249, 286, 297, 310, 332, 355, 372, 383, 395, 427, 445, 448, 452, 458, 464, 467, 468, 503, 521, 523, 534, 535, 537, 538, 539, 540, 554, 594, 607, 615, 636, 661, 673, 691, 692, 719, 759, 776, 782
identity, 155, 300, 547, 750, 756, 757, 759, 772, 777, 779, 850
idiopathic, ix, 228, 257, 269, 276, 277, 300, 308, 354, 363, 367, 369, 378, 407, 409, 415, 434, 840
IFN, 838
IGF, 362
IL-1, 479, 491
IL-6, 479, 491
IL-8, 479
Illinois, 170, 296
illusion, 760
image, 11, 27, 231, 234, 235, 236, 239, 248, 251, 252, 385, 745, 747, 751, 752, 784
imagery, 348
images, 11, 12, 86, 229, 231, 233, 234, 235, 247, 248, 287, 386, 391, 525, 751, 752, 756
imagination, 146
imaging systems, 263, 286
imaging techniques, 199, 205, 252
imbalances, 236
immortality, 782
immune regulation, 536
immune response, 484, 550
immunity, 526
immunization, 483, 484, 493
immunocompetent cells, 491
immunocytochemistry, 35, 840
immunodeficiency, 33
immunohistochemical, 393, 537, 555
immunohistochemistry, 392, 479, 847
immunoprecipitation, 658
immunoreactivity, 21, 33, 103, 113, 479, 483, 550, 554, 667, 828, 832, 854

impaired energy metabolism, 480, 812
impairments, ix, xv, 6, 8, 48, 50, 51, 52, 56, 64, 149, 284, 285, 286, 367, 408, 412, 414, 417, 422, 427, 428, 438, 446, 468, 494, 499, 517, 528, 544, 612, 613, 623, 686, 698, 806, 817, 833, 846, 847
implants, 242, 270, 289, 292, 302, 710
implementation, 290, 713, 716, 724, 748, 751, 780, 781, 782
improvements, 2, 47, 257, 285, 288, 305, 322, 328, 399, 424, 459, 467, 468, 483, 501, 512, 563
impulses, x, 141, 314, 397, 579
impulsive, 143, 145, 146, 458
impulsiveness, x, 141, 143, 144, 444
impulsivity, 143, 148, 178, 183, 184, 188, 461, 509, 511
in situ, 142, 341, 371, 512, 601, 804, 830
in situ hybridization, 633
in utero, 430
in vitro, 56, 212, 292, 329, 479, 480, 487, 490, 491, 526, 643, 658, 662, 666, 682, 683, 685, 686, 710, 786, 800, 801, 810, 841, 843, 845, 852, 855
in vitro fertilization, 786
in vitro fertilization (IVF), 786
in vivo, xiv, 11, 53, 54, 69, 135, 138, 327, 408, 440, 481, 488, 495, 532, 628, 638, 641, 649, 651, 671, 673, 678, 684, 688, 710, 801, 802, 805, 810, 832, 843, 845, 847, 850, 852
inactivation, 85, 393, 531
inactive, 94, 97
inattention, 188, 416, 421, 444
incentive, x, 73, 74, 79, 80, 83, 84, 85, 86, 90, 101, 102, 107, 109, 150, 288, 767, 785
incentives, 73, 79, 83, 102, 104, 146
incidence, xiv, xvii, 285, 331, 340, 345, 408, 427, 444, 487, 515, 541, 543, 544, 546, 549, 553, 578, 583, 609, 611, 614, 707, 725, 736, 737, 757
inclusion, 308, 314, 382, 461, 468, 593, 620, 645, 810, 837
income, 510
incompatibility, 228
independence, 284, 502, 516, 694, 735, 756
independent living, 581
independent variable, 218
indexing, 727
India, 655, 670, 671, 673
Indians, xviii, 656
indication, 120, 258, 293, 680, 681, 684, 735, 768, 773, 775, 777
indicators, 599, 605, 707

indices, 93, 522
individual differences, 474, 726, 770, 772, 788
Individuals with Disabilities Education Improvement Act, 581
indomethacin, 495
induction, 7, 103, 104, 111, 112, 145, 274, 417, 427, 438, 442, 480, 481, 492, 542, 824
industrial, 288, 486
industry, 288
ineffectiveness, 55, 286
inequity, 596
infancy, 409, 424, 426, 584, 758
infants, 424, 585, 592
infarction, 228, 343, 487, 551
infection, xiii, 249, 250, 263, 287, 304, 308, 314, 324, 328, 361, 368, 613, 706, 713
infections, 355, 384, 486, 581
infectious, 8, 250
infectious episodes, 250
inferences, 216, 444, 458, 460, 463
inferior frontal gyrus, 13, 14, 15, 17, 18, 19, 20, 22, 24, 25
infertile, 755, 756, 771, 785
infertility, 754, 756, 768, 793, 794
inflammation, 386, 475, 476, 479, 481, 525, 526, 532, 536, 546, 551, 706, 823, 846
inflammatory, 376, 388, 478, 479, 485, 487, 490, 494, 496, 542, 544, 545, 564, 817, 823
inflammatory mediators, 478
inflammatory response, 479, 823
inflammatory responses, 479
influence, xi, 177, 178, 180, 181, 184, 193, 195
influenza, 8, 568
informal sector, 600
information exchange, 265, 293
information processing, ix, 7, 48, 56, 69, 71, 73, 99, 116, 119, 124, 125, 128, 129, 130, 132, 133, 134, 144, 173, 379, 407, 418, 420, 434, 471, 508, 774, 845
information seeking, 772, 774, 794
informed consent, 228, 308, 369, 461
infusions, 182, 183
ingestion, 612, 711
inhalation, 172, 324, 325, 383, 393, 563, 569, 570, 571
inhaled therapy, 561, 565, 568
inhaler, xvi, 561, 562, 563, 564, 565, 566, 567, 568, 569, 570, 571, 572
inheritance, 620, 629, 699, 757, 785, 786
inherited, 257, 276, 355, 363, 476, 760, 769, 787
inherited disorder, 645
inhibition, x, xiv, 7, 9, 29, 37, 49, 55, 62, 66, 68, 69, 70, 141, 142, 143, 144, 147, 160, 178, 179,

183, 188, 189, 190, 194, 195, 258, 279, 289, 314, 323, 324, 325, 331, 332, 336, 337, 343, 365, 372, 380, 393, 412, 415, 417, 421, 422, 424, 427, 435, 439, 459, 470, 473, 474, 485, 494, 508, 509, 512, 513, 515, 519, 640, 644, 646, 710, 814, 840, 850
inhibitor, 51, 55, 421, 438, 484, 487, 489, 493, 494, 554, 647, 669, 801, 802, 804, 837
inhibitors, 51, 56, 351, 377, 378, 423, 482, 486, 488, 489, 493, 494, 496
inhibitory, 21, 34, 37, 44, 61, 68, 69, 102, 136, 153, 155, 162, 165, 166, 179, 185, 189, 193, 284, 320, 324, 330, 337, 388, 389, 408, 415, 417, 421, 422, 479, 483, 508, 512, 513
inhibitory effect, 153, 162, 165, 337, 422, 483
initiation, 78, 119, 178, 311, 341, 342, 382, 388, 418, 435, 539
injection, 156, 193, 211, 324, 383, 393, 532, 540
injections, 56, 139, 195, 322, 355, 393, 483, 818, 824, 825, 849
innate immunity, 526
innervation, 28, 95, 181, 185, 194, 423, 537
innovation, 268, 299
inorganic, 611
iNOS, 542, 545
inositol, 423, 483, 496, 641, 652
input, 776
inputs, 180, 186, 196
insanity, xix, 718, 806
insemination, 756
insertion, 125, 248, 264, 291, 322, 619, 809, 818
insight, x, 18, 19, 74, 522, 533, 535, 604, 655
inspection, 455
Inspection, 463
instability, 197, 275, 307, 619, 630, 631, 647, 650, 677, 678, 679, 680, 681, 683, 684, 686, 687, 688, 807, 813, 841, 844, 846, 847
institutions, 324
instruction, 218, 568, 595, 710
instructional techniques, 581
instruments, 457, 513, 589, 591, 596, 597, 709
insulin, 536, 541, 546, 551, 561, 569, 622, 802, 834, 845
insulin resistance, 541, 551, 622
insults, 355, 412, 424, 531, 536
insurance, 723, 724, 730
integration, 73, 74, 77, 83, 96, 97, 99, 101, 103, 106, 155, 164, 167, 211, 224, 258, 266, 277, 292, 342, 417, 424, 458, 510
integrin, 477
integrins, 477, 490
integrity, 10, 237, 372, 529, 532, 533, 550
intellectual disabilities, 573, 575

intellectual functioning, 579
intelligence, 17, 199, 200, 203, 204, 210, 407, 409, 411, 412, 413, 414, 427, 430, 435, 437, 442, 503, 511, 512, 516, 589, 596, 597
intelligence quotient, 17, 411, 511
intelligence tests, 409, 411
intensity, xiii, 12, 18, 82, 83, 102, 129, 130, 153, 157, 340, 367, 368, 524
intensive care unit, 237
intentions, 751, 753, 755, 768, 777
interaction, 54, 62, 66, 77, 78, 79, 80, 83, 84, 88, 93, 97, 102, 107, 125, 143, 145, 147, 259, 262, 321, 351, 415, 543, 580, 592, 613, 618, 642, 643, 647, 657, 658, 661, 662, 666, 670, 671, 674, 683, 686, 688, 699, 729, 752, 774, 842
interactions, 29, 56, 57, 63, 66, 67, 69, 70, 71, 83, 138, 155, 169, 175, 185, 192, 263, 285, 299, 397, 477, 510, 524, 531, 580, 597, 613, 617, 618, 620, 640, 641, 642, 644, 646, 648, 655, 657, 658, 661, 662, 664, 666, 670, 674, 684, 688, 736, 759, 760, 811, 854
intercourse, 103
interdisciplinary, 514, 791
interest, 341
interface, 87, 91, 110, 287, 351, 516, 710, 801, 832
interference, 50, 61, 178, 179, 187, 189, 196, 237, 238, 240, 369, 372, 379, 380, 519, 643
interferon, 490, 804, 834
interleukin, 491, 804, 834
Interleukin-1, 479
interleukin-6, 491
internal consistency, 456, 594
internal environment, 77
internalizing, 456, 471
internet, 264, 300, 708
Internet, 292
interneurons, 3, 4, 5, 6, 29, 30, 31, 43, 186, 264, 290, 632, 636, 680, 800, 814
internists, 561
interpersonal factors, 759
interpersonal relations, 791
interpretation, 12, 15, 23, 34, 104, 361, 446, 482, 564, 580, 588, 589, 592, 594, 783, 826, 831
interrelationships, 468
interval, 84, 98, 119, 129, 132, 235, 285, 307, 424, 459, 627, 782
intervention, xiv, xv, xvii, 146, 147, 228, 251, 271, 285, 289, 343, 344, 345, 395, 396, 397, 400, 402, 499, 500, 501, 502, 503, 504, 506, 507, 510, 511, 513, 514, 515, 516, 548, 573, 574, 596, 605, 694, 735, 742, 785, 819, 833, 846

interview, 411, 461, 572, 695, 724, 735, 741, 754, 763, 764
interviews, 590, 764, 782, 783, 784
intestine, 658
intoxication, 383, 460
intracerebral, 104, 250, 252, 263, 272, 273, 287, 289, 314, 331, 361, 483, 540, 546, 549
intracerebral hemorrhage, 263, 272, 273, 287, 289, 314, 546, 549
intracranial, 9, 77, 249, 326, 375, 543
intraoperative, 231, 239, 240, 250, 252, 273, 286, 367, 373
intravenous, 250, 324, 350, 383
intravenous (IV), 195
intravenously, 237
intrinsic, 4, 5, 6, 7, 37, 38, 39, 45, 164, 166, 205, 212, 233, 265, 280, 290, 291, 313, 372
intuition, 750
invasive, 85, 234, 263, 271, 273, 278, 282, 283, 284, 286, 347, 395, 402
Investigations, 518, 614
investigative, 612
investment, 776
investments, 603
ion channels, 54, 408, 421, 650, 817
ion transport, 103
ionization, 523, 526
ions, 481
Iowa, xi, 141, 143, 145, 146, 149, 150, 152, 550, 554
Iowa Gambling Task, 149, 150, 152
IP, 479, 480, 491
IP-10, 479, 480, 491
ipsilateral, 157, 213, 274, 309, 310, 382, 387, 388, 413
IQ, 17, 411, 412, 414, 416, 418, 419, 420, 427, 431, 447, 461, 467, 579, 585, 610, 612, 615
IQ scores, 418, 612
Iraq, 611
Ireland, 578, 605, 731
iridium, 367, 373
iron, 167, 440, 476, 525
iron deficiency, 440
irradiation, 473
irrigation, 707
irritability, 345, 346, 375, 407, 421, 697, 732
IS, 169, 239, 266, 295, 330
ischaemia, 801, 804
ischemia, 439, 556, 559, 843, 852
ischemic, 297, 305, 314, 482, 542, 546, 547, 555, 556
ischemic stroke, 297, 547, 556
ISO, 239

isoelectric point, 523
isoenzymes, 422
isoforms, 478
isolation, 11, 639, 693, 723, 753, 815, 848
isomerization, 539
isotopes, 172
Israel, 42, 541, 571, 600, 602, 606
issues, xv, xvi, 13, 56, 59, 164, 259, 261, 263, 280, 282, 322, 347, 349, 396, 401, 414, 456, 499, 562, 588, 592, 594, 595, 596, 644, 707, 709, 734, 736, 757, 763, 764, 772, 776, 777, 780, 785, 786, 789, 795, 797
Italy, 153, 303, 365, 395, 475, 599, 601, 602, 603, 604, 606, 677, 691
IVF, 786

J

JAMA, 435, 472, 494, 495, 555, 605, 606
January, 36
Japan, xviii, 239, 499, 502, 511, 550, 602, 611, 656
Japanese, 502, 511, 517, 555
jaw, 710
Jaynes, 103, 113
JNK, 38
job satisfaction, 600, 603, 606, 607
jobs, 602, 733, 756
joints, 26, 564
Jordan, 77, 80, 84, 103, 166
joystick, 305, 366
judge, 27, 582
judgment, 19, 341, 579, 736, 755
jumping, 721
Jung, 526, 535
justification, 204, 446, 759

K

kainic acid, 322, 334, 335, 408, 412, 425, 441, 834, 839, 848
kappa, 491
kappa B, 491
Kentucky, 521
Kenya, 587, 596, 597
kernel, 13
ketamine, 383
kidney, 658, 660, 743
kidneys, 525, 581
kill, 727, 728, 729, 813
killing, 620, 727

kinase, 32, 180, 185, 190, 353, 355, 359, 360, 417, 494, 528
Kinase, 442, 487
kinases, 417, 427, 487, 531
kindergarten, 511, 515
kinetics, 172, 537, 839
King, 252, 298, 390, 436, 585, 607
kinship, 790
knockout, 53, 183
knowledge, 179, 188, 340
Korean, 45

L

labor, 11
lack of confidence, 776
lack of control, 706
lactate dehydrogenase, 528
lactating, 81, 84, 98, 99, 107
lactation, 74, 82, 97, 98, 100, 111
lactoferrin, 713
lakes, 235
lamina, 39, 212, 213
laminar, 39, 212, 213
laminin, 477, 544
Laminin, 554
landscape, 106, 212, 573
Langerhans cells (LC), 177, 180, 181, 184, 185, 186
language, xv, xvi, 8, 17, 31, 326, 368, 371, 372, 373, 374, 380, 407, 409, 410, 411, 412, 413, 414, 420, 421, 422, 427, 428, 476, 500, 502, 510, 541, 547, 579, 588, 589, 590, 591, 592, 594, 595, 603, 627, 694, 698, 764
language impairment, xv, 476
language processing, 371
language skills, 8, 17, 547, 603
languages, 588
laptop, 239
large-scale, 211, 274, 282, 526
larynx, 658
laser, 523, 526
latency, 8, 79, 130, 132, 243, 246, 260, 414, 429, 433
late-onset, xix, 536, 745, 746, 747, 758, 760, 765, 787
later life, 442, 758
lateral sclerosis, 313, 317, 522
laterality, 104, 256, 260, 273, 285
law, 519, 580, 581, 767, 790, 797
LDH, 528
LDL, 489, 544

lead, 6, 11, 15, 79, 82, 124, 143, 180, 238, 243, 245, 246, 247, 248, 249, 250, 252, 263, 283, 291, 323, 365, 366, 372, 376, 388, 422, 457, 463, 477, 478, 487, 502, 521, 522, 528, 529, 530, 531, 542, 544, 545, 547, 564, 591, 603, 610, 613, 615, 618, 636, 643, 686, 693, 697, 722, 724, 725, 730, 749, 756, 763, 767, 785, 809, 810, 814, 832
leakage, 287, 375, 528, 554
learned helplessness, 855
learners, 582
learning behavior, 120
learning difficulties, 419, 788, 789
learning disabilities, 191, 472
learning environment, 582
learning process, x, 115, 116, 117, 124, 128, 129, 133, 134, 135, 378, 410, 425
learning task, 172, 416, 448, 458, 466, 468, 502
lectin, 524, 531
left hemisphere, 8, 306, 309
legislation, 612
legs, 80, 95, 376
Leibniz, 115
lesion, 179, 192
Lesion, 246, 826, 839
lesion-associated proteins, 480
lesioning, 255, 260, 281, 334, 335, 354, 390
lethargy, 345, 612
leukaemia, 473, 479
leukemia, 672
leukocytes, 535
leukocytosis, 250
level of education, 568
levodopa, xiv, 228, 256, 257, 274, 275, 276, 304, 308, 309, 317, 366, 367, 369, 375, 376, 380, 382
Lewy bodies, 376, 556
liberal, 757
LIF, 479, 491
life cycle, 770
life expectancy, 261, 281, 563, 631
life experiences, 729
life satisfaction, 727, 729, 732
life span, 549, 579, 638, 699, 818
lifespan, 427, 581, 818, 829, 830
lifestyle, 552, 732
life-threatening, 355
lifetime, xiii, 340, 399, 606, 736
ligand, 53, 54, 64, 70, 477, 490, 491
ligands, 53, 66, 477, 491, 535
light, 6, 14, 20, 23, 26, 43, 64, 102, 167, 340, 347, 383, 397, 611, 634, 641, 649, 659, 709, 735, 749, 759

likelihood, 9, 13, 32, 342, 427, 516, 561, 566, 569, 582, 612, 645, 717, 719, 732, 748, 787
limbic system, 110, 125, 258, 261, 280, 281, 326, 612, 815, 816, 832
limitation, 86, 260, 262, 286, 421, 469, 591, 747
limitations, 239, 240, 256, 259, 263, 273, 275, 277, 278, 280, 282, 284, 285, 286, 306, 307, 402, 411, 458, 563, 582, 747, 776, 779, 781
Lincoln, 86, 95, 109, 795
line, 643, 728, 729, 779, 801, 802, 807, 841, 850
linear, 156, 220, 223, 230, 243, 456, 465, 468, 614, 749, 782
lingual, 13, 15, 17, 18
linguistic, 367, 371, 372, 373, 409
linkage, 55, 68, 137, 180, 285, 624, 629, 682, 687, 718, 730, 731, 736, 763, 771, 775, 807
links, 1, 180, 457, 671, 753, 767, 780
lipid, 355, 480, 481, 492, 521, 522, 528, 529, 532, 534, 535, 538, 551
Lipid, 492, 529, 535
lipid metabolism, 492
lipid oxidation, 481
lipid peroxidation, 481, 521, 522, 529, 534, 535, 538
lipids, 425, 543
lipopolysaccharide, 804, 834, 838
lipoprotein, 489
liquids, 699, 706, 707
liquor, 11
listening, 94, 170
literacy, 509, 510, 515
literature, 371, 372
lithium, 25, 29, 432, 442, 487
Lithium, 403
liver, 355, 525, 575, 658, 660
living conditions, 600, 603, 604
LOAD, 480
lobectomy, 428, 442
local anesthesia, 287
local government, 603
localization, xii, 2, 44, 64, 67, 70, 108, 164, 188, 212, 227, 233, 234, 235, 236, 237, 242, 248, 249, 252, 253, 254, 260, 273, 363, 364, 375, 413, 414, 539, 544, 618, 624, 632, 638, 639, 647, 648, 650, 653, 666, 668, 842, 847, 851, 855
location, 105, 106, 155, 201, 368, 433, 439, 578
loci, 191, 687
locomotion, 55, 95, 391
locomotor, 78, 92, 94, 106, 179, 183, 184, 187, 419, 427, 490, 612, 851
locomotor activity, 78, 179, 187, 419, 427, 490, 612

locus, 68, 136, 172, 177, 180, 185, 188, 189, 190, 191, 192, 193, 194, 195, 196, 376, 398, 646, 651, 773, 785, 797
locus coeruleus, 136, 177, 180, 185, 188, 189, 190, 191, 192, 193, 194, 195, 196, 376, 398
Locus of Control, 797
logical reasoning, 60
London, 30, 106, 135, 149, 169, 173, 212, 213, 254, 380, 435, 440, 459, 474, 583, 584, 595, 596, 606, 786, 788, 789, 790, 791, 792, 793, 794, 795, 796, 797
loneliness, 723
long period, 251, 314
longevity, 800, 821
longitudinal studies, 414
longitudinal study, 22, 58, 413, 420, 432, 492, 548, 553, 842, 853
long-term, 13, 19, 31, 40, 43, 52, 116, 124, 139, 143, 145, 146, 192
long-term memory, 31, 124, 410, 508
long-term potentiation, 139, 192, 408, 412, 417, 426, 438, 439, 441, 483, 834, 839
Long-term potentiation, 417, 435, 817
Long-term potentiation (LTP), 417, 817
lordosis, 80, 105, 111
loss of control, 758
losses, 145, 148, 507, 613, 756
love, 722, 782, 789
low molecular weight, 479, 530
low risk, 769, 776
low-density, 489
low-density lipoprotein, 489
LPS, 804, 823, 834
LSI, 292
LTD, 191, 417, 422, 427, 435
LTP, 139, 185, 191, 408, 417, 422, 423, 424, 425, 427, 435, 439, 441, 442, 817, 822, 825, 834
lubricants, 713
luminal, 476
lung, 563, 566, 572
lung function, 572
Luo, 15, 20, 35, 38, 334, 391, 838
lying, 500
lymph, 658
lymph node, 658
lymphoblast, 643
lymphocyte, 484
lymphocytes, 376, 684
lysis, 611
lysozyme, 713

M

machinery, 632, 681, 817
machines, 288
macrophages, 542
magnesium, 422
magnet, 397
magnetic, xvi, 28, 30, 34, 35, 38, 41, 44, 49, 59, 170, 190, 194, 196, 197, 225, 228, 230, 235, 236, 237, 238, 247, 251, 252, 253, 254, 263, 264, 288, 298, 299, 329, 332, 335, 364, 378, 396, 398, 411, 489, 522, 553, 556
magnetic field, 230, 263
magnetic properties, 230
magnetic resonance, xvi, 30, 34, 35, 38, 44, 49, 59, 170, 190, 194, 196, 197, 228, 235, 236, 237, 247, 251, 252, 253, 254, 264, 285, 364, 378, 398, 411, 489, 522, 553, 556, 632
magnetic resonance imaging, xvi, 34, 35, 38, 44, 49, 59, 170, 190, 194, 228, 237, 251, 252, 253, 254, 264, 285, 364, 378, 411, 489, 522, 553, 556, 633
Magnetic Resonance Imaging, 408, 548
magnetic resonance imaging (MRI), 190, 191, 193, 194, 195
magnetic resonance spectroscopy, 197
magnetic resonance spectroscopy (MRS), 197
magnetization, 230, 235
magnitude, 3, 19, 24, 412, 419, 446, 449, 450, 451, 452, 456, 458, 459, 463, 464, 465, 466, 467, 468, 469, 503, 748, 769, 821
main line, 220
maintenance, xvi, 10, 129, 167, 190, 289, 344, 528, 529, 532, 550, 562, 709, 842, 846
Maintenance, 849
major depression, xiii, 150, 267, 269, 294, 297, 298, 299, 300, 340, 346, 348, 395, 398, 697, 732, 733, 738, 843, 846, 855
major histocompatibility complex, 526
majority, 49, 158, 182, 277, 397, 399, 452, 459, 477, 510, 551, 564, 581, 587, 588, 590, 630, 639, 667, 719, 725, 730, 731, 754
maladaptive, 278
malaria, 596
malate dehydrogenase, 528
MALDI, 524
male animals, 116
males, 36, 73, 80, 83, 85, 89, 90, 93, 94, 111, 112, 447, 461, 512, 574, 721, 730, 733, 787
malignant, 350
malnutrition, 8, 11, 30
malondialdehyde, 522
mammalian brain, 800, 832, 839, 849

mammalian cells, 856
mammals, xx, 81, 82, 84, 103, 106, 218, 799, 800, 850
man, 107, 200, 204, 205, 210, 212, 253, 266, 295, 301, 307, 311, 330, 335, 373, 376, 444, 511, 573, 574
management, xvi, 259, 260, 263, 266, 268, 272, 273, 277, 289, 294, 300, 302, 331, 340, 341, 342, 346, 347, 348, 350, 354, 391, 428, 455, 458, 460, 467, 507, 562, 570, 571, 644, 691, 692, 693, 696, 700, 709, 712, 714, 715, 732, 741
mania, 274, 400
manic, 400, 733
manic episode, 400
manipulation, 167, 639
man-made, 573
mantle, 180, 200, 206
mapping, 44, 59, 199, 228, 238, 239, 240, 241, 242, 243, 244, 262, 283, 286, 287, 289, 310, 388, 624, 629
Marani, 300
marketing, 486
marketplace, 605
marriage, 729, 753, 767, 789, 793, 795, 797
married women, 603
marrow, 658
Maryland, 571
mask, 588
mass, 95, 523, 535, 578, 627, 658, 662
mass spectrometry, 523, 535, 658, 662
masseter, 242
matching, 192
materials, 292, 411, 447, 502, 589, 591, 710, 715
maternal, x, xvii, 73, 74, 77, 78, 79, 81, 82, 83, 84, 85, 86, 94, 95, 97, 99, 101, 102, 105, 106, 108, 109, 110, 111, 112, 418, 436, 511, 574, 575, 578, 579, 581, 612
maternal care, 81, 97, 112
mathematics, 515, 579
Matrices, 304, 368, 369, 380
matrix, 235, 248, 477, 523, 542, 545, 549, 554, 825, 856
matrix protein, 477, 545, 549
matter, iv, 1, 5, 10, 11, 12, 13, 14, 15, 16, 17, 18, 19, 20, 22, 23, 24, 25, 26, 27, 28, 29, 30, 31, 32, 33, 34, 35, 36, 37, 38, 39, 40, 41, 42, 43, 44, 45, 50, 143, 204, 213, 257, 261, 262, 276, 281, 283, 284, 297, 305, 314, 412, 413, 425, 500, 546, 547, 549, 551, 553, 556, 559, 579, 628, 754, 774, 815, 825, 848, 853
maturation, xi, xx, 5, 6, 8, 9, 129, 141, 187, 419, 424, 425, 518, 519, 799, 801, 832

maturation process, 9
maze learning, 113
MB, 30, 296, 470, 471
MCAO, 551
MCI, xv, 479, 480, 481, 487, 521, 522, 523, 525, 526, 527, 528, 529, 530, 531, 532, 544, 548
MCP, 375, 479, 480, 491
MCP-1, 479, 480, 491
MCS, 256
MDA, 175
MDH, 528
meanings, 83, 765, 791
measles, 611
measurement, 12, 109, 122, 235, 241, 270, 383, 445, 459, 503, 508, 515, 589, 592, 596, 615, 738
measurements, 74, 86, 89, 91, 94, 96, 98, 100, 233, 235, 239, 240, 503, 592, 736
measures, xv, 30, 37, 49, 50, 80, 91, 122, 123, 146, 147, 231, 253, 305, 309, 399, 411, 412, 414, 420, 429, 444, 446, 447, 448, 449, 450, 451, 452, 453, 455, 456, 458, 459, 461, 462, 463, 464, 465, 466, 467, 468, 469, 470, 471, 472, 485, 486, 488, 493, 496, 502, 503, 507, 508, 509, 510, 512, 513, 588, 590, 593, 594, 595, 596, 628, 694, 748, 757, 787
media, 136, 549, 581
medial prefrontal cortex, 21, 29, 30, 36, 41, 43, 78, 89, 94, 96, 98, 100, 101, 103, 108, 112, 115, 119, 120, 121, 123, 125, 128, 129, 136, 139, 150, 194, 518
median, 104, 108, 241, 242, 310, 335, 400, 603
mediation, 75, 103, 108, 113, 190, 194, 333, 514
mediators, 478
Medicaid, 601
medical, xiii, xvii, xix, 28, 234, 236, 238, 239, 260, 263, 287, 327, 353, 354, 355, 358, 389, 407, 411, 428, 470, 486, 563, 573, 574, 578, 581, 582, 590, 695, 699, 705, 709, 713, 714, 719, 723, 724, 732, 746, 764, 767, 769, 770, 775, 791, 797
medical care, xix, 705, 746
medicine, xii, xvii, 255, 315, 381, 562, 583, 590, 692, 707, 763, 792, 794, 797
medulla, 82
mellitus, 522, 553, 555
melt, 707
membranes, 71, 109, 296, 419, 421, 480, 648
memory capacity, 193, 194, 472
memory deficits, 62, 190, 419, 428, 430, 579
memory formation, 417, 425

memory function, 68, 116, 123, 128, 135, 136, 182, 185, 379, 410, 412, 414, 423, 460, 551, 584, 855
memory loss, xv, 522
memory performance, 60, 70, 71, 180, 182, 183, 187, 190, 192, 193, 197, 412, 417, 431, 439
memory processes, 124, 424
memory retrieval, 137, 139
MEMS, 292
men, 104, 367, 369, 418, 435, 555, 574, 721, 722, 730, 731, 756, 758, 773, 774, 793, 806
meninges, 386
menopause, 769
mental activity, 49
mental development, 407
mental disorder, 349, 547, 717, 718, 719, 723, 733
mental health, 511, 516, 581, 735
mental illness, ix, xiii, xix, 48, 144, 340, 348, 350, 718
mental impairment, 571
mental model, 764, 765, 789
mental processes, 211, 409
mental retardation, xiv, 408, 415, 576, 610, 611, 619, 622, 652
mental simulation, 749, 750, 780, 782
mental state, 502, 547, 568, 571, 632
Merck, 533
mercury, 610, 611, 614
Mercury, 611, 615
mesencephalon, 385, 386
mesocorticolimbic, 77
messages, 260, 619
messenger RNA, 44, 550, 834, 839
meta-analysis, 12, 14, 28, 32, 35, 37, 39, 49, 58, 60, 256, 267, 270, 273, 274, 293, 297, 298, 299, 302, 438, 515, 553, 752, 792
metabolic, 40, 59, 85, 99, 103, 172, 173, 273, 331, 333, 398, 411, 480
Metabolic, 403, 405, 435, 555
metabolism, 180, 312, 393, 398, 407, 417, 424, 431, 434, 435, 440, 475, 476, 477, 480, 492, 528, 530, 536, 538, 544, 550, 556, 636, 641, 648, 649, 651, 673, 812, 817, 828, 843
metabolite, 51, 118, 119, 488
metabolites, 54, 62, 69, 139, 480, 489
metabolizing, 421
metal ion, 481
metal ions, 481
metalloproteinases, 477
metals, 492, 497, 609, 610, 612, 613, 614
metaphor, xvi, 561, 562, 750
metaphors, 768

methamphetamine, 193
methodology, xii, 13, 21, 87, 204, 205, 227, 229, 231, 240, 241, 242, 456, 458, 460, 589, 698, 763, 789
methylene, 486
methylene blue, 486
methylphenidate, 177, 179, 183, 187, 193, 196, 469, 470, 471, 472, 473, 474
metric, 445, 458
metropolitan area, 503
metropolitan areas, 503
Mexican, 613
Mexico, 720
MFC, 157
MHC, 838
microaneurysms, 546
microarray, 39
microchip, 290
microdialysis, 116, 118, 122, 125, 137, 139, 140, 848
microelectrode, 231, 238, 239, 240, 242, 243, 246, 247, 248, 253, 254, 264, 291, 353, 354
microelectrodes, 239, 292
microelectronics, 265, 292, 293
microenvironment, xx, 799, 800, 804, 805, 820, 822, 823, 830, 832, 850
Microenvironment, 799, 804, 832
microfabrication, 292
microglia, 41, 478, 479, 490, 542, 804, 823, 837, 838, 840, 846, 851
microglial, 479, 490, 542, 545, 558
microglial cells, 479, 490, 542, 545
microorganisms, 707
microscopy, 21, 634
microstructure, 210
microtubule, 10, 21, 478, 539
microtubules, 21, 476, 478, 809, 817, 841
microvascular, 551, 554, 557, 558
midbrain, 67, 82, 104, 110, 123, 164, 186, 382, 389, 822, 823, 854
middle-aged, 603
midlife, 428
migration, 274, 386, 388, 426, 478, 523, 536, 552, 579, 611, 802, 821, 825, 829, 831, 837, 846, 847, 851, 853, 855
mild cognitive impairment, 491, 492, 500, 518, 533, 534, 535, 536, 538, 539, 544, 548, 572
mild cognitive impairment (MCI), 544, 548
Mild Cognitive Impairment (MCI), 480
milk, 106, 109
mimicking, 177, 188
miniature, 91
Mini-Cog, 547, 555

Mini-Mental State Examination, 376, 489, 569
Mini-Mental State Examination (MMSE), 376
Ministry of Education, 28
Minneapolis, 243, 259, 280, 356, 377
minority, xiii, 25, 319, 588, 724
minority groups, 588
MIP, 479
mirror, 17, 29, 161, 592
miscarriage, 613, 776
misfolded, 530
misinterpretation, 515
misunderstanding, 203, 204
misuse, 574, 581
MIT, 108, 166
mitochondria, 476, 480, 531, 643, 644, 681, 811, 812
mitochondrial, 10, 15, 29, 34, 353, 355, 358, 361, 425, 480, 491, 522, 528, 532, 535, 539, 544, 611
mitochondrial DNA, 522, 535
mitochondrial toxins, 673, 812, 818
mitosis, 539, 801
mitotic, 490
MK-80, 56, 71, 427, 441
ML, 36, 39, 45, 212, 213, 269, 301, 349, 350, 351, 554, 555, 557, 558, 570
MMA, 351
MMS, 472
MMSE, 304, 309, 376, 479, 483, 485, 488, 489, 496, 502, 547, 565, 568, 572
mobility, 257, 307, 362, 698
modalities, 75, 142, 298, 393
modality, 170, 179, 256, 289, 428
mode, 181
model, xix, 626, 628, 636, 639, 640, 642, 645, 646, 647, 649, 650, 652, 653, 658, 662, 663, 666, 669, 672, 674, 679, 685, 699, 745, 746, 747, 748, 751, 752, 759, 761, 764, 765, 766, 768, 774, 777, 778, 779, 783, 784, 786, 804, 805, 810, 816, 818, 819, 826, 827, 832, 837, 838, 839, 840, 842, 843, 844, 845, 846, 848, 849, 850, 851, 853, 854, 855, 856
modeling, 103, 551, 708, 791, 842
modelling, 293, 298, 764, 765, 793
modifications, 307, 366, 480, 481, 521, 522, 524, 532, 534, 587, 589, 593, 680, 698, 745
modulation, xii, xiii, 12, 51, 64, 68, 71, 74, 82, 83, 136, 138, 139, 174, 175, 185, 188, 190, 191, 192, 193, 196, 256, 264, 290, 314, 315, 320, 322, 323, 334, 335, 354, 367, 389, 421, 422, 442, 529, 554
modules, x, 141, 143, 507
molecular changes, 525

molecular mechanisms, 54, 482
molecular oxygen, 480
molecular weight, 479, 523, 530
molecule, 180
molecules, 65, 528, 530, 542, 544, 545, 549, 802, 842
Møller, 391, 393
money, 146
monitoring, 341
monkeys, 20, 29, 30, 32, 36, 56, 110, 124, 135, 139, 153, 154, 156, 165, 167, 168, 169, 170, 174, 175, 179, 181, 182, 183, 184, 185, 188, 189, 190, 192, 193, 194, 195, 196, 197, 203, 210, 211, 320, 351, 390, 430, 549, 844
monoamine, 6, 192
monoaminergic, 59, 75
monoclonal, 484
monoclonal antibodies, 855
monocytes, 479
monotherapy, 259, 280, 401, 418, 420, 426, 435, 436
monozygotic twins, 44, 152, 627, 652
Montenegro, 432
mood, 34, 261, 282, 283, 304, 306, 309, 343, 376, 395, 396, 401, 415, 420, 423, 435, 436, 439, 515, 625, 628, 694, 695, 696, 697, 734, 815, 816, 820, 831
mood change, 306, 376, 420
mood disorder, 34, 304, 401, 515, 697
moral imperative, 796
morals, 751
morbidity, 228, 234, 243, 361, 365, 432, 561, 563, 603, 740
morning, 231
morphemes, 374
morphological, 2, 6, 9, 10, 11, 15, 16, 25, 32, 40, 210, 211, 442, 551
morphological abnormalities, 2, 11, 25, 40, 822, 828
morphology, ix, 2, 10, 11, 12, 14, 16, 18, 19, 20, 21, 22, 23, 25, 27, 28, 32, 34, 36, 37, 42, 206, 212, 213, 517, 529, 802, 821, 822, 823, 832, 843
morphometric, 17, 29, 33, 34, 36, 38, 39, 41, 42, 45, 60, 633, 650, 651, 851
mortality, xvi, 228, 345, 378, 542, 543, 553, 561, 563, 570, 584, 645, 740, 743, 787, 838
Moses, 107
motherhood, 790, 791, 794, 795
mothers, 8, 84, 85, 94, 95, 414, 418, 419, 426, 427, 430, 433, 437, 510, 511, 574, 581, 613, 627, 756, 758, 788, 789, 791
motif, 477, 652

motion, xviii, 287, 656
motivation, ix, x, 48, 74, 75, 78, 79, 80, 82, 83, 87, 89, 91, 93, 99, 101, 102, 104, 105, 107, 109, 110, 111, 136, 137, 138, 146, 341, 507, 695, 747, 748
motives, 142, 725
motor activity, 374, 388, 612
motor area, 14, 17, 19, 169, 170, 207, 275, 278, 281, 284, 298, 304, 310, 341, 347, 351, 366, 388
motor control, xiv, 156, 157, 164, 165, 195, 351, 365, 370, 436, 590, 628, 815
motor coordination, 419
motor function, 276, 366, 372, 375, 384, 410, 568, 579
motor neuron disease, 378
motor neurons, 153, 164, 165
motor skills, 379, 500, 547, 613
motor system, 110, 372
motor task, 289, 305, 389, 420, 459
mouse, 43, 110, 129, 193, 323, 426, 439, 483, 493, 537, 550, 553, 554, 557, 558, 559, 613
mouse model, 323, 550, 553, 554, 558, 559, 613, 804, 805, 809, 810, 816, 817, 818, 819, 826, 827, 829, 832, 838, 842, 843, 845, 846, 848, 849, 850, 853, 854, 856
mouth, 81, 95, 564, 570, 610
movement, xii, xiii, xiv, 87, 98, 120, 157, 159, 162, 163, 166, 167, 169, 180, 239, 248, 255, 257, 258, 261, 262, 263, 265, 266, 267, 270, 276, 277, 278, 282, 283, 284, 286, 291, 292, 293, 294, 295, 297, 302, 303, 304, 305, 311, 312, 313, 314, 315, 316, 317, 320, 346, 347, 351, 354, 355, 365, 366, 372, 377, 378, 392, 393, 410, 420, 532, 567, 581, 625, 691, 693, 696, 706, 806, 808, 818
movement disorders, vi, xiv, 255, 257, 258, 261, 262, 266, 267, 270, 276, 277, 278, 282, 283, 284, 286, 294, 295, 297, 302, 303, 312, 313, 314, 315, 316, 320, 346, 347, 354, 355, 365, 366, 377, 378, 392, 393, 691, 693, 696, 808
mPFC, 73, 76, 77, 85, 87, 88, 89, 90, 92, 93, 94, 95, 96, 97, 99, 101
MPP, 41
MPTP, 172, 174, 381, 382, 383, 385, 386, 388, 389, 390, 391, 392, 393
MRI, 11, 39, 40, 41, 42, 43, 45, 49, 60, 191, 193, 195, 225, 228, 229, 231, 233, 234, 235, 236, 237, 247, 248, 249, 251, 252, 253, 254, 256, 268, 275, 278, 285, 288, 291, 299, 305, 306, 314, 326, 355, 356, 361, 367, 373, 375, 380, 382, 383, 385, 386, 387, 388, 391, 398, 408, 411, 413, 416, 419, 420, 423, 432, 433, 434,

483, 493, 517, 518, 548, 555, 556, 578, 628, 633, 651, 851
mRNA, 5, 8, 39, 44, 52, 65, 85, 175, 479, 550, 634, 635, 803, 817, 824, 834, 839, 850
mRNAs, 67, 633
MTC, 486, 496
MUA, 86, 91, 92
mucosa, 707, 708
multidisciplinary, xv, 263, 286, 428, 499, 514, 763
multidrug resistance, 530
multiple regression, 209, 445
multiple sclerosis, 257, 268, 270, 276, 299, 302, 491
multiplicity, 747, 754, 779, 784
multipotent, 800, 836
multivariate, 258
mumps, 611
murine model, 549, 558
murine models, 549
muscarinic receptor, 47, 52, 54, 62, 63, 64, 65, 66, 71, 173, 176
muscle, 144, 156, 167, 354, 355, 360, 361, 542, 544, 545, 547, 549, 550, 557
muscle cells, 542, 544, 545, 547, 549, 557
muscle contraction, 354
muscle spasms, 360, 361
muscles, 142, 166, 239, 242, 265, 291, 313, 625, 698
muscular dystrophy, 619, 645, 776, 794
mutagenesis, 678, 685
mutant, xviii, 183, 617, 618, 632, 636, 637, 638, 639, 640, 641, 642, 643, 644, 645, 646, 647, 648, 649, 651, 652, 653, 656, 658, 659, 660, 661, 662, 663, 664, 666, 667, 671, 674, 678, 683, 684, 685, 807, 809, 810, 811, 812, 813, 817, 819, 826, 827, 836, 838, 839, 841, 842, 845, 846, 852, 855
mutants, 477
mutation, 353, 355, 358, 359, 360, 361, 550, 554, 618, 619, 622, 624, 629, 630, 636, 644, 646, 649, 657, 660, 661, 670, 671, 679, 686, 688, 689, 691, 718, 730, 731, 732, 736, 758, 769, 775, 776, 787, 837, 839, 842, 844, 846, 848, 850, 853, 855
mutations, xv, xix, 312, 476, 478, 540, 546, 550, 618, 619, 625, 658, 679, 684, 688, 746, 811
myelin, 425, 440, 579
myelination, 500
myocardial infarction, 228, 487
myoclonus, xiv, 280, 312, 353, 354, 355, 358, 362, 363, 364, 818
myopia, 581

N

NAA, 8
N-acety, 30
naming, 222, 225, 301, 371, 411, 412, 414, 511
nanofibers, 265, 270, 292, 302
nanomaterials, 265, 292
nanotechnologies, 265, 293
nanotechnology, 265
narratives, 745, 751, 753, 765, 770, 771, 783, 790
NASA, 265
National Academy of Sciences, 210
National Health Service, 789
National University of Singapore, 171
natural, 53, 116, 153, 156, 163, 164, 165, 166, 210, 265, 284, 290, 292, 313, 425, 442, 460, 478, 487, 751, 755, 779, 785, 791
nausea, 273, 486, 612
Nazi Germany, 757
NC, 34, 334, 492, 554, 556
NCES, 258
neck, 155, 157, 166, 268, 278, 300, 312, 381, 383, 395, 397, 400
neck cancer, 801
necrosis, 424, 491, 543, 546, 611, 804
needs, 357
negative consequences, 718, 725
negative effects, 420
negative outcomes, 718
negativity, 68, 138
neglect, 285, 736
negotiating, 745, 748, 758, 778
negotiation, 753, 754, 755, 775, 781, 784
Nelson, 190
neocortex, 8, 16, 30, 40, 84, 170, 211, 212, 387, 388, 478, 632, 633, 801, 840, 841, 844, 846, 853
neonatal, 8, 29, 33, 37, 38, 412, 420, 429, 438, 441, 442, 612, 613
nerve, xiii, 40, 105, 111, 241, 242, 260, 284, 310, 320, 327, 331, 376, 395, 396, 397, 544, 578, 581, 641, 668, 707, 802, 821, 834
nerve growth factor, 641, 802, 834
nerves, 144
nervous system, xii, 41, 47, 67, 77, 81, 256, 264, 271, 275, 290, 292, 421, 457, 478, 479, 481, 483, 491, 551, 552, 555, 576, 581, 610, 611, 614, 617, 659, 707, 847, 848
nervousness, 144
Netherlands, 391, 587, 601, 604, 743, 793, 852
network, x, 7, 30, 43, 44, 82, 102, 110, 116, 153, 165, 217, 222, 223, 265, 275, 290, 291, 292,

298, 321, 435, 511, 514, 516, 655, 661, 662, 670, 671, 729, 751, 836, 849
NeuN, 821, 826, 828, 834
neural development, 519
neural function, 79, 291
neural network, xii, 256, 264, 268, 271, 272, 300, 421
neural networks, xii, 256, 264, 268, 271, 272, 300, 421
neural regeneration, 285
neural stem/progenitor cell, 839
neural systems, 60, 74, 83, 106, 191, 213, 264
neural tissue, 287
neuralgia, 272
neuritic plaques, 544
neuritis, 544
neuroanatomy, 36, 225, 517, 519
neurobehavioral, 615
neurobiological, 15, 17, 18, 19, 20, 24, 151, 395, 396, 397, 468, 514, 517, 556
neurobiology, ix, 2, 14, 22, 24, 26, 110, 293, 348, 350, 351
neuroblasts, 477, 801, 802, 825, 829, 849
neurochemistry, 47, 330
neurodegeneration, xviii, 33, 353, 355, 359, 360, 426, 432, 441, 442, 475, 478, 479, 484, 490, 491, 538, 546, 548, 551, 605, 617, 620, 632, 636, 638, 640, 644, 646, 647, 649, 651, 652, 653, 656, 672, 673, 678, 687, 806, 811, 813, 818, 829, 832, 833, 839, 841, 842, 849, 852, 854, 855
neurodegenerative, xi, xiv, 8, 171, 365, 391, 412, 480, 481, 489, 493, 522, 538, 543, 544, 547, 551, 552, 787
neurodegenerative disease, xiv, 365, 480, 481, 489, 552
neurodegenerative diseases, xiv, xviii, 365, 480, 481, 489, 552, 617, 620, 635, 643, 651, 678, 686, 841, 846, 849, 850
neurodegenerative disorders, 391, 493, 522, 538, 552, 619, 787, 806, 851
neurodegenerative processes, 8, 543
neuroendocrine, 108, 192
neurofibrillary tangles, 475, 476, 526, 536, 558
neurogenesis, xx, 20, 34, 38, 39, 44, 140, 425, 644, 672, 799, 800, 801, 802, 803, 804, 805, 806, 814, 816, 820, 821, 822, 823, 824, 825, 826, 828, 829, 830, 831, 832, 833, 836, 837, 838, 839, 840, 841, 842, 843, 844, 845, 846, 847, 848, 849, 850, 851, 852, 853, 854, 855, 856
neurogenic, 272

neuroimaging, 1, 2, 13, 15, 16, 20, 21, 27, 28, 36, 47, 49, 50, 60, 142, 145, 148, 150, 217, 222, 237, 238, 241, 249, 259, 261, 273, 275, 280, 285, 286, 299, 347, 351, 355, 372, 382, 398, 410, 500, 501, 511, 513, 518, 519, 815, 851
neuroimaging techniques, 1, 238, 249
neuroinflammation, 491, 525, 532, 550, 837
neuroleptic, 22, 36, 38, 41, 56, 58, 59, 61, 62, 63
neuroleptics, 20, 21, 22, 38, 44, 69, 399, 691, 694, 695, 696
neurologic disorders, 267, 296
neurological condition, 256, 259, 289, 355, 423, 564
neurological deficit, xiii, 322, 353, 354, 360, 573
neurological disease, 320, 710, 804, 844
neurological disorder, xii, xiv, 256, 263, 265, 289, 292, 408, 461, 541
neurologist, 286, 306
neuromodulation, xii, 255, 256, 264, 268, 271, 278, 290, 293, 295, 298, 300, 301
neuron, 191
neuron response, 191
neuronal apoptosis, 426, 427, 429, 613
neuronal cells, 480, 482
neuronal circuits, 6, 854
neuronal death, 9, 423, 426, 441, 529, 545, 812, 836, 844
neuronal degeneration, 539
neuronal density, 3, 4, 42
neuronal excitability, 407, 421, 422, 427
neuronal loss, xvi, 7, 522, 613
neuronal migration, 426, 478, 579, 855
neuronal plasticity, 407, 408
neuronal survival, 427, 823
neuropathic pain, 266, 267, 268, 272, 273, 293, 295, 299
neuropathological, 1, 2, 3, 6, 7, 9, 10, 11, 14, 15, 23, 27, 28, 32, 34, 41, 363, 376, 483, 484
neuropathologies, 479, 558
neuropathology, 4, 6, 21, 27, 30, 34, 41, 44, 363, 483
neuropathy, 272
neuropeptide, 813, 835
Neuropeptide Y, 803, 843
neuropeptides, 64
neuropharmacology, 47
neurophysiology, xii, 256, 330, 519
neuroplasticity, 44, 266, 284, 293, 295
neuroprotection, 29, 266, 393, 672, 828, 843
neuroprotective, xiv, 257, 276, 293, 313, 382, 420, 423, 424, 427, 429, 439, 479, 491, 822, 828, 847, 850

neuropsychiatric disorders, 29, 262, 281, 282, 284, 289, 302, 350
neuropsychiatry, 2, 854
neuropsychological assessment, xiv, 304, 377, 409, 412, 428, 443, 472, 597
neuropsychological tests, 1, 9, 37, 408, 427, 444, 445, 509, 628
neuropsychology, 36, 105, 217, 348, 379, 473
neuroscience, ix, xii, 58, 102, 144, 146, 199, 256, 260, 266, 300, 469, 502, 514, 516, 517, 582
neuroscientists, xi, 199, 204, 210
neuroses, 144
neurosurgeons, 263, 286
neurosurgery, xii, 227, 233, 234, 235, 236, 237, 239, 242, 251, 252, 253, 255, 256, 294, 346, 348, 367
neurotic, 144
neurotoxic, xvii, 9, 20, 270, 336, 380, 429, 534, 542, 545, 609, 612, 613, 614
neurotoxic effect, 429
neurotoxicity, 32, 44, 174, 423, 479, 483, 542, 610, 632, 639, 643, 849
neurotoxins, 818
neurotransmission, 48, 70, 139, 176, 325, 421, 422, 427, 529, 736
neurotransmitter, xx, 48, 57, 64, 342, 398, 407, 417, 421, 422, 423, 427, 526, 532, 576, 610, 612, 799, 800, 802, 804, 816, 820, 829, 838, 848, 850
neurotransmitters, 47, 74, 79, 86, 191, 260, 398, 408, 421, 422, 816, 825, 829
neurotrophic, 6, 10, 20, 34, 175, 408, 441, 477
neurotrophic factors, 20, 813
neutral, 78, 429, 590, 760
neutrophil, 536
neutrophils, 479
Nevada, 405
New England, 626, 721, 740, 741, 743, 791
New Orleans, 297
New World, 194
New Zealand, 606, 740
Newton, xvii, 587, 594, 595
next generation, 773, 780, 785
NFT, 526, 530, 531, 532, 544
NH2, 653, 856
nicotinamide, 813, 834
nicotine, 54, 55, 56, 67, 68, 70, 71
Nicotine, 47, 55, 67, 68, 71
Nielsen, 105, 381, 438, 540
nightmares, 192, 195
nigrostriatal, 173, 294, 836
NIH, 268, 299, 533
nimodipine, 437

NINDS, 547, 555
nitric oxide, 424, 542, 545, 554, 680, 827, 835, 838
nitric oxide synthase, 542, 545, 554, 680, 827, 835, 838
nitrogen, 530
nitrous oxide, 426
NMDA, 32, 56, 70, 71, 173, 175, 408, 417, 421, 422, 423, 425, 426, 427, 429, 438, 442, 579, 803, 812, 813, 818, 827, 828, 835, 838, 841, 842, 843, 848, 856
NMDA receptors, 56, 417, 421, 422, 423, 427, 429, 438, 442, 827
N-methyl-D-aspartate, 408, 435
NMR, 421, 843
Nobel Prize, 210
nociceptive, 272
nodes, xii, 256, 258, 261, 271, 278, 279, 280, 281, 283, 284
noise, 10, 184, 239, 708
non invasive, 236
non-enzymatic, 481
non-enzymatic antioxidants, 481
non-human, xi, 42, 50, 154, 155, 200, 203, 204, 205, 299, 549
non-human primates, 50, 155, 203, 205, 549
non-invasive, 234, 278, 283, 413
nonlinear, 147
non-pharmacological, 401
non-pharmacological treatments, 401
non-random, 14, 22
nonsmokers, 30, 67
non-smokers, 26
non-steroidal anti-inflammatory drugs, 478, 485
nonverbal, 416, 473, 512, 596
noradrenaline, 137, 805, 822, 823, 829, 831
Noradrenaline, 803, 816, 829, 830
norepinephrine, xi, 177, 178, 188, 190, 193, 195, 426, 844
norepinephrine (NE), xi, 177, 178, 180, 181, 182, 183, 184, 185, 186, 187, 188, 190, 193, 195
normal aging, xv, 522, 548, 556
normal children, 225, 442
normal conditions, xv, 522
normal development, 425, 429, 575
normal distribution, 448
normalization, 180, 343, 389
norms, 755, 789
North America, 268, 300, 322, 587
Northern Ireland, 578, 605
Norway, 112, 720, 742
NOS3, 492
novel stimuli, 820, 831, 833

novelty, 27, 119, 136, 509
novelty seeking, 27
NSAIDs, 485, 487, 494, 495
N-terminal, 477
NTS, 397
nuclear, 478, 522, 531, 535
nuclear family, 692
Nuclear Magnetic Resonance, 421
nuclei, xii, xiv, 31, 82, 137, 155, 156, 188, 233, 237, 239, 240, 241, 242, 254, 256, 258, 260, 261, 268, 271, 272, 274, 279, 281, 283, 284, 300, 301, 320, 321, 322, 323, 333, 335, 336, 350, 365, 366, 373, 376, 398, 636, 639, 681, 734, 814, 815, 856
nucleic acid, 680
nucleotides, 681, 685
nucleus accumbens, 32, 40, 63, 77, 84, 106, 119, 135, 138, 261, 268, 269, 281, 299, 301, 330, 345, 346, 347, 351, 370
nuisance, 755
null, 461, 636, 851, 854
null hypothesis, 461
nulliparous, 105
nurse, 81
nurses, 263, 286, 590, 700, 710, 715, 793
nursing, xvii, xix, 81, 84, 85, 94, 97, 99, 104, 502, 599, 600, 604, 605, 607, 692, 713
nursing home, xvii, xix, 502, 599, 600, 604, 605, 607, 692
nutrient, 542
nutrients, 476
nutrition, 596, 692, 699, 809
nutritional status, 576
nystagmus, 322, 335

O

Obese, 551
obesity, 401, 541
objectives, 711, 793
observational learning, 138
observations, x, xi, 40, 116, 199, 200, 262, 283, 308, 336, 397, 459, 480, 483, 493, 512, 530, 536, 546, 592, 628, 630, 636, 640, 664, 683, 755, 756, 764, 776, 782, 806, 821
observed behavior, 613
obsessive-compulsive, 43, 45, 60, 261, 267, 268, 269, 296, 298, 299, 300, 301, 340, 347, 348, 349, 350, 351
Obsessive-compulsive, xiii, 26, 268, 340, 347, 348, 349, 351

obsessive-compulsive disorder, 43, 45, 60, 261, 267, 268, 269, 296, 298, 299, 300, 347, 348, 349, 350, 351
obstacles, 411, 549, 726
obstetricians, 793
occipital cortex, 8, 13, 19, 50
occipital lobe, 14
occipito-temporal, 217
occlusion, 423, 551, 556
occupational therapy, 714
oculomotor, xix, 139, 153, 169, 195, 705, 836
odor memory, 851
odorants, 84, 110
odors, 83, 85, 90, 93, 99, 101, 105
OECD, 602, 606
offenders, 107
olanzapine, 20, 21, 22, 32, 36, 58, 692, 695, 697
old age, xvi, 561, 562, 563, 564, 565, 566, 571, 572
older adults, xv, xvii, 414, 487, 499, 500, 501, 503, 504, 507, 514, 516, 517, 519, 563, 565, 570, 600, 601, 602, 603, 604, 605, 606
older people, 563, 570, 571, 572, 600, 601, 602
olfaction, 101, 829
olfactory, 74, 76, 77, 80, 82, 85, 97, 98, 99, 101, 103, 106, 107, 108, 113, 800, 815, 829, 835, 837, 838, 841, 847, 849, 851
olfactory bulb, 82, 106
oligodendrocytes, 34
oligomeric, 483, 493
oligomerization, 851
oligomers, 483, 488, 493, 544, 553, 656
omega-3, 38, 487
omission, 134, 183, 783
oncogene, 111
online, 584
on-line, 179, 237, 238, 239, 240
on-line, 779
ontogenesis, 142
open-field, 81
operant conditioning, 116, 164
operations, 6, 76, 502
opioid, 272
Opioid, 112, 236
opportunities, xvii, 263, 293, 494, 549, 574, 588, 713, 772
opposition, 757, 767
optic nerve, 581
optical, 40
optimism, 442, 726, 773
optimization, 124, 134, 261, 274, 305
oral, 187, 193, 228, 309, 373, 436, 486, 797
oral cavity, xix, 705, 708, 709

oral health, xix, 705, 706, 707, 708, 710, 711, 713, 714
oral health problems, 713
oral lesion, 707
orbit, 231
orbitofrontal cortex, 10, 11, 14, 18, 19, 24, 36, 55, 76, 84, 111, 141, 142, 143, 145, 146, 148, 149, 150, 151, 152, 261, 283, 340, 341, 345, 351, 352, 398, 410
orchestration, 186
order, 632, 684, 693, 698, 699, 700, 731, 736, 749, 751, 752, 761, 763, 764, 770, 771, 774, 775, 776, 779, 782, 784, 785, 786, 807, 815
order statistic, 166
ores, 539
organ, 481, 556, 612
organelles, 636
organic, 611
organic disease, 723
organism, 77, 78
organization, xi, xiv, 78, 84, 96, 154, 169, 170, 178, 188, 210, 212, 213, 341, 365, 371, 377, 379, 412, 424, 431, 435, 515, 531, 552, 791
organizational culture, 788
organize, 290, 411, 752, 770
organs, 410, 557, 581, 660
orgasm, 104, 106, 110
orientation, xv, 79, 80, 89, 90, 91, 93, 96, 107, 136, 157, 165, 228, 369, 376, 391, 428, 475
originality, 353, 354
orthopaedic, 270, 302
oscillation, 190
oscillations, 6, 28, 30, 137
osteoarthritis, 272
osteopontin, 526, 536
outpatients, 35, 69, 398, 571
output, 194
ovarian cancer, 787
ovarian cancers, 787
ovariectomized, 84, 104
ovariectomy, 82, 111
ovary, 62
overlap, 2, 13, 14, 82, 101, 467, 564, 802, 823
overload, 69, 643, 751
overproduction, 557, 558
over-the-counter, 487
ownership, 740
ox, 746
oxidation, 38, 480, 481, 521, 522, 528, 529, 534, 537, 679, 680, 684, 685
oxidative, xvi, 8, 10, 15, 34, 425, 441, 475, 476, 480, 481, 482, 492, 497, 522, 524, 525, 528, 530, 531, 532, 534, 535, 538, 542, 546, 550, 551, 553, 557
oxidative damage, 475, 476, 481, 497, 530, 534, 535, 546, 680, 682, 687
oxidative stress, xvi, 8, 10, 15, 34, 425, 441, 480, 481, 482, 492, 522, 528, 534, 538, 542, 550, 551, 553, 557, 656, 677, 679, 680, 681, 684, 699
oxide, 424, 426, 542, 545, 554
oxygen, 381, 382, 383, 387, 388, 389, 393, 480, 528, 542, 545, 546
oxygen consumption, 381, 382, 387, 388, 389, 393

P

p21WAF1/CIP1, 667
P300, 238, 404, 413, 430, 433
p53, 534, 613, 662, 663, 667, 669, 671, 674, 853
pacemaker, 228, 249, 250, 321, 366
pacemakers, 228
pacing, 104, 113
pain, 144, 236, 250, 255, 256, 259, 260, 263, 266, 267, 268, 272, 273, 277, 282, 283, 286, 287, 293, 294, 295, 297, 298, 299, 300, 302, 312, 313, 315, 317, 361, 362, 366, 400, 484, 564, 706, 713, 723
pain management, 259, 273, 294
pain therapy, 273
pairing, 119, 550
palliative, 328, 707, 712, 713, 716
palpebral, 576
pancreas, 658, 660
panic disorder, 27, 38, 45, 146, 148, 149, 152, 282, 298, 730
paper, 154, 184, 484, 609
Paper, 596
paradigm, 765, 789, 796
paradox, 372
paradoxical, 325
parallel, xii, xiv, 5, 6, 75, 116, 127, 208, 230, 231, 232, 235, 246, 287, 304, 306, 307, 312, 313, 341, 347, 365, 390, 481, 489, 561, 593, 649, 784, 805, 815, 820, 836
parallel processing, 390
paralysis, 612
parameter, 266, 294, 480
parameters, 628, 629, 687, 697, 736, 748
paranoid schizophrenia, 34
paraoxonase, 614
paraphilia, 111
parasympathetic activity, 707
paraventricular, 82

paraventricular nucleus, 82
Parellada, 41, 59
parenchyma, 545, 549, 849
parenchymal, 288, 545, 549, 550
parenthood, 756, 789, 790, 795
parenting, 745, 758, 764, 765, 768, 771, 772, 773, 774, 778, 779, 780, 784, 788
parents, 408, 460, 461, 469, 510, 511, 604, 619, 620, 751, 753, 758, 768, 769, 770, 771, 777, 785, 786, 788
paresthesias, 400
parietal cortex, 20, 21, 26, 36, 76, 119, 165, 167, 170, 185, 194, 278, 291, 341, 389, 413, 417
parietal lobe, 14, 155, 175, 194, 212, 469, 501, 567, 579
parietal lobes, 501
Paris, 66, 107, 431, 606
parity, 105
Parkinson's Disease, 251, 256, 303, 365, 366, 367, 369, 377, 378, 380, 390, 391
Parkinson's disease (PD), xiv, 256, 382, 806
parkinsonism, xiii, 137, 257, 274, 276, 304, 315, 355, 363, 376, 383, 391, 695, 696
Parkinsonism, 269, 301, 303, 304, 305, 316, 378, 379, 391, 393
paroxetine, 697
partial seizure, 258, 279, 280, 322, 414, 424, 437, 440
participants, 50, 83, 146, 401, 447, 448, 449, 452, 453, 454, 455, 456, 458, 460, 461, 462, 488, 489, 502, 503, 507, 508, 510, 511, 513, 591, 592, 627, 649, 718, 724, 725, 726, 728, 729, 730, 731, 734, 751, 754, 759, 762, 764, 765, 767, 768, 769, 770, 772, 773, 774, 775, 776, 782, 783, 784, 785, 788, 797
particles, 610
partnership, 781
parvalbumin, 3, 4, 5, 27, 35, 37, 637, 814
parvicellular, 342
passive, 147, 419, 484, 496, 513, 736, 754
pathogenesis, ix, 10, 44, 277, 341, 342, 347, 475, 476, 478, 480, 481, 490, 521, 534, 535, 536, 538, 544, 553, 622, 629, 632, 639, 640, 641, 642, 643, 644, 646, 648, 655, 657, 658, 661, 662, 664, 666, 667, 669, 670, 671, 674, 685, 844, 851
pathogenic, 475, 476, 478, 544, 809
pathological gambling, 143, 149
pathology, x, 6, 12, 20, 26, 32, 33, 45, 47, 48, 52, 53, 55, 57, 60, 65, 113, 145, 235, 417, 476, 483, 490, 492, 493, 534, 542, 543, 544, 545, 549, 550, 551, 554, 555, 618, 620, 632, 635, 637, 638, 639, 641, 649, 651, 658, 672, 674, 688, 816, 817, 818, 820, 825, 829, 832, 836, 851, 853
pathophysiological, 20, 24, 26, 34, 47, 59, 258, 351, 494, 817
Pathophysiological, 330, 351
pathophysiological mechanisms, 817
pathophysiology, 8, 18, 35, 37, 40, 54, 176, 266, 269, 278, 281, 282, 301, 340, 347, 397, 416, 556, 713, 839, 853
pathways, xviii, 26, 68, 108, 168, 192, 207, 212, 237, 260, 274, 284, 285, 290, 321, 370, 374, 389, 415, 490, 522, 528, 532, 544, 552, 609, 618, 641, 642, 644, 655, 657, 664, 667, 670, 678, 686, 749, 814, 847
patient care, 700
patterning, 109, 292
Pavlovian, 138
Pavlovian conditioning, 138
PC12 cells, 41
PCO, 522, 524, 527, 528, 530
PCR, 630
PDG, 786
pedagogies, 582
pediatric, 186, 431, 433, 440
pedigree, 796
peer, 510
peers, 144, 146, 510, 613
pelvic, 91, 92, 107, 111
pelvis, 91
penetrance, 618, 619, 620, 625, 629, 631, 649, 787, 807
penicillin, 320
Pennsylvania, 510, 511
pension, 601
pentylenetetrazol, 321, 333
pentylenetetrazole, 327, 329, 331, 332
penumbra, 804
peptide, 477, 478, 483, 490, 491, 493, 534, 540, 544, 545, 550, 554, 669
peptides, 526, 639, 673, 674, 856
perceived control, 747
perception, x, 30, 73, 74, 77, 79, 97, 99, 119, 123, 128, 135, 138, 147, 155, 164, 184, 185, 272, 341, 409, 415, 420, 442, 747, 748, 755, 765, 767, 786, 788, 795, 796
perceptions, 605, 736, 753, 756, 765, 766, 770, 772, 774, 786, 787, 790, 793
perfusion, 307, 310, 315, 383, 542, 544, 546, 558
pericytes, 542
perinatal, 8, 9, 361, 420, 438
periodic, 486
periodontal, xix, 705, 706, 707, 711, 714
periodontal disease, xix, 705, 707, 711, 714

Peripheral, 405, 492
peripheral blood, 685, 687
peripheral nerve, 260, 284
peripheral nervous system, 67, 81, 275, 617, 658
peristalsis, 397
periventricular, 272
permeability, 542, 546, 555, 559
permit, 734, 768
peroxidation, 480, 481, 492, 521, 522, 529, 534, 535, 538
peroxide, 491
peroxiredoxins, 480
peroxynitrite, 530, 682
persecutory delusion, 515
personal, 146, 410, 578, 599, 601, 602, 603, 604, 605, 606
personal accounts, 761
personal communication, 146, 285, 293, 315, 632, 750
personal hygiene, 698
personality, xix, 27, 36, 41, 146, 149, 156, 236, 344, 346, 350, 500, 622, 700, 705, 717, 718, 722, 729
personality characteristics, 350, 700
personality disorder, 27, 36, 41, 146, 622
Personality disorders, 149
personality factors, 729
personality traits, 27, 149, 346, 350
persons with disabilities, 603
perspective, 347, 351, 363
persuasion, 792
perturbations, 230, 551
pessimism, 729, 735, 738
pesticide, 612
pesticides, 610, 612, 613, 614, 615
PET, v, 40, 49, 50, 62, 171, 172, 173, 174, 175, 176, 180, 191, 224, 286, 305, 311, 312, 329, 334, 355, 370, 374, 380, 381, 382, 383, 384, 385, 386, 388, 389, 398, 410, 411, 417, 434, 513, 548, 736, 737, 815, 835, 849
PET scan, 381, 383
Petroleum, 708
PFC, xi, xv, 43, 48, 49, 50, 51, 52, 55, 56, 57, 73, 74, 75, 77, 79, 85, 87, 91, 92, 94, 95, 97, 101, 102, 177, 178, 179, 180, 181, 182, 183, 184, 186, 187, 188, 209, 217, 499, 500, 501, 502, 511, 513, 514
phantom limb pain, 272
pharmaceutical, 355
pharmaceuticals, xii, 255
pharmacokinetic, 436, 485
pharmacokinetics, 69, 438, 494

pharmacological, xiv, 51, 52, 62, 64, 66, 70, 94, 228, 266, 282, 293, 323, 324, 325, 346, 347, 373, 395, 396, 398, 399, 402, 416, 440, 459, 604
pharmacological treatment, 266, 282, 293, 347, 399, 401, 402, 459, 604
pharmacology, 47, 51, 438
pharmacotherapy, 342
phencyclidine, 70
phenol, 70
phenomenology, 348, 631, 750
phenothiazines, 494
phenotype, xviii, 8, 23, 35, 208, 617, 625, 630, 636, 639, 646, 647, 652, 805, 807, 808, 809, 810, 811, 817, 818, 819, 820, 831, 840, 847
phenotypes, 258, 278, 478, 619, 635, 638, 639, 644, 645, 651, 653, 852
phenotypic, 206, 211, 414, 477
phenytoin, 250, 408, 418, 432, 436, 437, 439, 441
pheochromocytoma, 651
pheromone, 84
Philadelphia, 69, 151, 167, 212, 254, 380, 404, 430, 672
philosophy, 234
philtrum, 576
Phoenix, 791
phone, 164, 410, 475
phonological, 309, 368, 369, 370, 372
phosphatases, 531
phosphate, 383, 423, 711, 813, 834
phosphatidylethanolamine, 529
phospholipids, 530
phosphorylation, 21, 422, 487, 530, 531, 532, 539, 544, 628, 643, 645
Phosphorylation, 531, 539
photon, 59
phthalates, 613
physical abuse, 604
physical activity, 838
physical characteristics, 774
physical exercise, 830
physical interaction, 658
physical resemblance, 760
physical treatments, 350
physicians, xvii, 262, 263, 282, 397, 429, 562, 731
physiological, x, xi, 9, 78, 116, 129, 133, 134, 143, 154, 237, 239, 240, 250, 261, 269, 292, 301, 336, 351, 363, 366, 411, 415, 435, 477, 516, 612, 613
Physiological, 59, 60, 129, 167, 402
physiological factors, 78
physiological regulation, 435

physiology, x, 69, 74, 103, 105, 140, 243, 340, 348, 379, 836
physiopathology, 321
picture processing, 224
pig, 104, 382, 383, 393
pigmentation, 376
pigs, 321, 333, 391
pilot studies, 328, 763
pilot study, 269, 296, 300, 302, 317, 321, 326, 328, 331, 332, 336, 439, 495, 519
pitch, 230, 231, 233, 289
PKC, 427
placebo, 61, 64, 65, 67, 103, 195, 262, 273, 283, 313, 314, 322, 323, 329, 418, 424, 458, 459, 460, 469, 482, 483, 484, 485, 486, 487, 488, 489, 493, 494, 495, 697, 699
placenta, 401, 575, 658
planar, 254, 287, 292
planning, 75, 116, 117, 124, 125, 126, 129, 130, 132, 134, 135, 140, 172, 173, 174, 178, 187, 188, 192, 237, 238, 239, 241, 263, 341, 356, 371, 410, 428, 471, 474, 500, 580, 628, 694, 717, 719, 726, 751, 755, 788, 789, 808
planning decisions, 789
plant growth, 539
planum temporale, 35
plaque, 479, 482, 483, 484, 485, 490, 536, 549, 557, 558, 708, 710, 711, 712, 713
plaques, 377, 475, 476, 478, 543, 544, 550, 554, 557, 558
plasma, 51, 140, 187, 383, 426, 479, 480, 481, 484, 485, 490, 522, 523, 525, 535, 557, 611, 687, 818
plasma levels, 140, 187
plasma membrane, 611, 818
plastic, 143, 612, 613
plasticity, 9, 10, 11, 15, 20, 33, 36, 37, 108, 185, 211, 278, 296, 323, 407, 408, 412, 417, 423, 425, 427, 432, 435, 438, 439, 513, 516, 517, 644, 688, 811, 816, 817, 822, 831, 832, 833, 848, 850, 852, 853, 854
plastics, 610, 613
platelet, 525
Platelet, 847
platinum, 367, 373
play, 2, 10, 31, 47, 53, 56, 57, 82, 87, 101, 124, 186, 258, 291, 293, 347, 367, 371, 389, 462, 475, 477, 478, 479, 480, 481, 508, 509, 511, 529, 532, 610, 611, 746, 763, 784
playing, 82, 145, 410, 513
pleasure, 151
pleiotropic factor, 479
plexus, 272

plug-in, 242
plumbism, 614
pneumonia, 706, 707, 714, 721, 767
poisoning, 392, 610, 611
polar, 20, 229, 810, 849
polarity, 241, 242, 246
polarization, 793
policy, 606
polyacrylamide, 523, 535
polychlorinated biphenyl, 612
polygraph, 87, 91, 239
polymerase, 681
polymerization, 478, 525
polymorphism, 10, 34, 40, 53, 492, 513, 536, 558, 651, 677, 681, 684
polymorphisms, 10, 55, 65, 550, 629
polypeptide, 483, 493
polypeptides, 493
polyphenols, 487
polysomnography, 242
polyunsaturated fat, 480
polyunsaturated fatty acid, 480
polyunsaturated fatty acids, 480
polyurethane, 88
PON1, 614
pons, 155
pools, 680, 681
poor, xvii, 17, 18, 39, 53, 86, 178, 180, 182, 184, 256, 260, 275, 312, 340, 345, 355, 358, 409, 412, 414, 421, 422, 425, 428, 456, 459, 529, 543, 561, 563, 564, 565, 567, 569, 574, 580, 595, 602, 603, 612, 683, 698, 706, 707, 710, 808
poor performance, 18, 412, 414, 561
population, ix, xiii, xiv, xvi, xviii, 48, 55, 86, 162, 163, 186, 259, 261, 276, 280, 295, 300, 314, 340, 354, 355, 361, 362, 369, 377, 408, 412, 427, 518, 542, 544, 555, 563, 569, 570, 574, 579, 587, 588, 589, 590, 596, 599, 601, 602, 625, 629, 677, 678, 679, 682, 697, 714, 715, 717, 719, 720, 721, 723, 724, 725, 726, 727, 728, 731, 733, 736, 743, 751, 760, 787, 800, 807, 820, 824, 841, 846
population group, 588
portability, 288
positive correlation, 479, 546, 550, 567
positive reinforcement, 708, 709
positron, 49, 58, 59, 60, 86, 172, 175, 191, 377, 379, 381, 382, 383, 391, 411, 417, 434, 513, 528, 736, 815, 835
positron emission tomography, 49, 59, 60, 86, 172, 175, 379, 381, 382, 411, 417, 434, 513, 528, 736, 815, 835

posterior cortex, 185, 217, 417
postmortem, 20, 31, 34, 41, 42, 63, 157, 383, 386, 611, 613
postnatal exposure, 426
postoperative, 250, 252, 260, 261, 263, 274, 276, 282, 287, 309, 313, 356, 375, 383, 386, 387, 388, 473
postpartum, 81, 84, 98, 105, 109, 113, 428, 613
Postpartum, 112
postpartum period, 109, 428
poststroke, 313, 316
post-stroke, 258, 267, 277, 284, 285, 297, 298, 299, 303, 312
post-stroke, 316
post-stroke, 316
postsynaptic, 5, 44, 310, 415, 417, 421, 422, 427
post-translational, 522, 524
post-translational modifications, 522, 524
posttraumatic stress, 45
post-traumatic stress, 423
post-traumatic stress, 439
posttraumatic stress disorder, 45
post-traumatic stress disorder, 423
post-traumatic stress disorder, 439
postural instability, 275, 307
posture, 80, 275, 305, 306, 311, 314
potassium, 237, 424, 438
potato, 146
potential benefits, 726
potential energy, 117
powder, 564, 571
power, 12, 14, 87, 89, 91, 94, 95, 96, 98, 100, 106, 130, 131, 132, 133, 134, 264, 291, 327, 447, 458, 469, 470, 568, 749, 750, 792
practical intelligence, 597
praxis, xvi, 561, 562, 564, 567, 568, 571
precipitation, 481
preclinical, 47, 173, 326, 347, 438, 475, 483, 487, 533
precursor cells, 800, 841, 843, 848
predictability, 132
prediction, 123, 139, 151, 283, 291, 293, 328, 473, 533, 535, 738
predictor variables, 456
predictors, 57, 58, 395, 567, 572
pre-existing, 588
preference, xvi, 77, 80, 81, 83, 104, 105, 106, 109, 110, 111, 113, 285, 351, 377, 562, 570, 661, 711
prefrontal cortex (PFC), xv, 43, 73, 75, 177, 499, 500

pregnancy, 8, 82, 111, 401, 418, 419, 428, 574, 575, 576, 577, 581, 582, 584, 758, 767, 771, 773, 776, 777, 785, 786, 795
pregnant, 418, 426, 573, 576, 581, 584, 771, 777, 791
pregnant women, 581, 791
premature death, 724, 782
premenstrual syndrome, 515
premotor cortex, xi, 16, 24, 26, 154, 165, 207, 278, 298, 305, 307, 366, 389
prenatal care, 760
preoperative, 367, 373, 375
preparation, iv, 157, 311, 323, 333, 342, 594, 644, 747, 763, 764
preprocessing, 12
preschool, xv, 424, 474, 499, 501, 509, 510, 511, 513, 515, 519
preschool children, xv, 499, 511, 513, 519
preschoolers, 509, 512, 516
prescription drug, 796
prescription drugs, 796
preservation, 204, 237, 361, 425
pressure, 236, 248, 384, 397, 492, 543, 546, 550, 698, 736, 789, 795, 796
pressure sore, 698
presynaptic, 3, 64, 182, 342, 417, 422, 529
pretraining, 507
prevention, xvi, 15, 326, 342, 409, 438, 494, 495, 500, 514, 519, 533, 542, 552, 555, 639, 694, 711, 715, 716, 791
preventive, 500, 509, 516, 605
primacy, 57
primary care, 81, 510, 570
primary caregivers, 510
primary school, 448, 461, 462
primary visual cortex, 21
primate, xi, 5, 33, 37, 42, 82, 139, 140, 149, 167, 168, 169, 170, 174, 200, 203, 204, 205, 210, 211, 212, 213, 299, 331, 348, 349, 351, 352, 381, 382, 391, 392, 557, 801, 837
primates, xi, 6, 50, 70, 76, 82, 112, 154, 155, 164, 168, 169, 170, 182, 185, 189, 203, 204, 205, 208, 210, 211, 212, 213, 382, 390, 438, 549
priming, 283, 564
principles, 123, 124, 125, 135, 300, 377, 430, 576, 592, 606, 646, 751, 752, 761, 784
private, 509, 601, 602, 603, 604, 748, 753, 754, 757, 766, 779, 785
proactive, 178, 573, 582, 774, 785, 786
proactive interference, 178
probability, xiii, 22, 81, 108, 137, 320, 421, 445, 450, 601, 760, 797
probe, 125, 210, 243, 246, 300

problem behavior, 442, 510
problem behaviors, 442
problem solving, 1, 124, 129, 134, 510, 579, 627, 628, 754
problem-solving, ix, 48, 115, 133, 142, 224, 509, 510, 571
problem-solving behavior, 115, 133
problem-solving skills, 509, 510
procedures, 366, 372
processing stages, 140
procrastination, 749
procreation, 780
producers, 478
production, 217, 220, 263, 371, 372, 373, 375, 385, 422, 425, 476, 479, 480, 481, 485, 491, 493, 494, 526, 528, 529, 531, 532, 539, 544, 546, 550, 588, 590, 618, 643, 841, 848, 852
profession, 769
professionals, 573, 582, 585, 590, 699, 712, 713, 724, 728, 736, 758, 769, 781
profit, 145
progenitor cells, 800, 837, 838, 839, 844, 845, 847, 848, 851, 854
progesterone, 103, 105, 111
prognosis, 340, 355, 361, 433, 767
prognostic factors, 315
program, 57, 106, 146, 147, 252, 292, 501, 502, 503, 507, 509, 510, 511, 512, 513, 514, 515, 517, 698, 699, 709, 710, 718, 724, 725, 726, 728, 732, 734, 740, 796, 797, 849
programmability, 274
programming, 257, 259, 264, 268, 277, 288, 289, 291, 299, 306, 366
progressive, 375
progressive neurodegenerative disorder, xviii, 617, 622, 656, 677, 678
progressive supranuclear palsy, 257, 276, 478
proinflammatory, 478, 526, 542
project, 5, 22, 23, 28, 75, 155, 180, 184, 235, 397, 423, 507, 510, 518, 632, 718, 739
prolactin, 852
proliferation, 20, 36, 41, 426, 484, 526, 531, 801, 802, 820, 822, 823, 825, 826, 828, 829, 830, 831, 832, 833, 836, 837, 838, 839, 840, 841, 842, 843, 844, 845, 846, 847, 848, 850, 854, 855, 856
proline, 483, 493, 640
promote, 97, 105, 147, 423, 428, 460, 510, 516, 546, 591, 802
promoter, 55, 638, 667, 811, 817, 818, 819, 831
promoter region, 55
propagation, xiii, 258, 279, 301, 320, 323, 326, 421

properties, 641, 674, 801, 844, 854, 855
property, 216, 373, 375, 409
prophylaxis, 568, 709
propofol, 237, 240
proposition, 79, 101
prosocial behavior, 510
prospective memory, 50
prostheses, 296
prosthesis, 285, 290, 710
prosthetic device, 296
protection, 331, 423, 479, 685, 709, 748, 830
protective factors, 490
protective role, 618
protein, 3, 5, 11, 21, 30, 32, 52, 63, 65, 66, 67, 108, 180, 185, 191, 363, 417, 422, 424, 425, 426, 427, 428, 441, 476, 477, 480, 481, 482, 484, 490, 495, 497, 521, 522, 523, 524, 525, 526, 528, 529, 530, 531, 532, 534, 535, 536, 537, 538, 539, 540, 544, 546, 547, 550, 551, 557, 801, 802, 805, 806, 807, 809, 810, 811, 812, 817, 818, 819, 820, 824, 828, 833, 834, 835, 837, 838, 840, 842, 848, 852, 853
protein family, 532, 666
protein folding, 531
protein function, 528
protein kinase C, 180, 185, 417, 427
protein kinase C (PKC), 427
protein kinases, 427
protein misfolding, 637
protein oxidation, 480, 521, 522, 537
protein structure, 67, 620
protein synthesis, 528, 531, 532, 547
protein-protein interactions, 642, 644, 670, 811
proteins, 10, 21, 35, 37, 64, 427, 440, 477, 478, 480, 490, 521, 522, 523, 524, 525, 526, 527, 528, 529, 530, 531, 532, 534, 535, 536, 537, 538, 539, 540, 543, 545, 549, 554, 617, 618, 620, 634, 635, 636, 637, 639, 640, 641, 642, 647, 649, 655, 656, 657, 658, 659, 660, 661, 662, 664, 665, 666, 667, 668, 669, 670, 671, 672, 673, 681, 683, 684, 685, 687, 688, 692, 694, 707, 802, 810, 811, 813, 814, 817, 827, 843
Proteins, 526, 527, 530
proteolipid protein, 425
proteolysis, 476, 638, 639, 640, 648, 813, 840, 847
proteome, 535
proteomes, 522
proteomics, 521, 522, 523, 524, 525, 532, 534, 535, 537, 538
protocol, xiii, 71, 228, 236, 238, 248, 310, 320, 327, 366, 383, 518, 595, 712, 718, 738, 786

protocols, 238, 260, 273, 277, 280, 286, 287, 289
provocation, 342
proxy, 204
PRP, 493
pruning, 9, 10, 34, 36
PSD, 812, 827, 835
PSEN1, xv, 476
PSEN2, xv, 476
pseudogene, 539
PSG, 242
PSI, 511
PSP, 417
psychiatric diagnosis, 282
psychiatric disorder, 67, 69, 156, 256, 260, 261, 262, 263, 268, 271, 282, 283, 286, 293, 296, 299, 300, 348, 351, 457, 500
psychiatric disorders, vi, 67, 69, 156, 256, 260, 261, 262, 263, 268, 271, 282, 283, 293, 296, 299, 300, 348, 351, 457, 500, 623, 688, 723, 730, 829, 832, 833
psychiatric illness, 55, 67, 262, 283, 348, 731
psychiatric patients, 262, 281, 282, 518
psychiatrist, 143, 447, 448
psychiatrists, 262, 263
psychiatry, 59, 260, 261, 262, 272, 282, 283, 396, 606, 742
psychobehavioural, xv, 476
psychobiology, 137
psychological distress, 725, 730, 788
psychological problems, 514, 730
psychological processes, 27
psychological stress, 186, 192
psychologist, 729, 740
psychology, 102, 112, 379, 594, 596, 745, 747, 763, 784, 788, 791, 792, 793, 794, 795, 797
psychometric properties, 587, 588, 592
psychopathic, 151
psychopathology, 149, 473, 517
psychopathy, 143, 148, 151, 517, 518
psychopharmacological, 25
psychopharmacology, 472
psychoses, 23, 50
psychosis, ix, 1, 9, 15, 18, 22, 26, 29, 30, 31, 32, 33, 37, 39, 40, 41, 53, 70, 228, 274, 697
psychosocial functioning, xiii, 31, 57, 340, 346, 416
psychosocial stress, 841
psychosocial variables, 414, 428
Psychosomatic, 795
psychosurgery, 260, 281, 282, 347, 348
psychotherapy, 142, 146, 344, 472
psychotic, 18, 22, 27, 40, 50, 156
psychotic symptoms, 50, 695

psychotropic drug, 342
psychotropic drugs, 342
PT, 37, 187, 213, 331
PTSD, 182, 192, 195, 423
puberty, 362
public, 573, 575, 581
public awareness, 573
public concern, 766, 785, 788
public health, 575
publishers, 167
publishing, 700
pulse, 87, 236, 242, 280, 306, 325, 332, 333, 353, 356, 357, 361, 366, 368, 375, 381, 383, 397, 400, 412
pulses, 311
pumps, 272, 528
punishment, 17, 107, 116, 119, 124, 125, 126, 127, 128, 130, 132, 135, 143, 148, 150, 509, 548
purification, 524
purines, 687
Purkinje, 194, 419, 809
Purkinje cells, 419, 809
putrescine, 539
pyramidal, 1, 3, 4, 5, 6, 7, 11, 25, 27, 29, 30, 31, 32, 33, 37, 39, 40, 42, 43, 44, 45, 60, 86, 154, 182, 199, 206, 207, 208, 209, 210, 211, 212, 361, 362, 439, 440
pyramidal cells, 5, 30, 39, 86, 206, 208, 209, 440, 809
pyrrole, 55
pyruvate, 528

Q

QOL, 424
qualitative differences, 824
qualitative research, 789, 791, 793, 795, 796
quality assurance, 237
quality control, 632
quality of life, xvii, 256, 257, 267, 274, 275, 276, 284, 298, 304, 336, 340, 415, 418, 423, 424, 440, 442, 489, 574, 596, 600, 603, 604, 606, 692, 695, 700, 706, 713, 714, 715, 726, 727, 758, 759, 767, 772
Quality of life, 350
quantification, 205, 239, 240, 734, 821, 824
Queensland, 211, 717, 743
questioning, 767
questionnaire, 228, 305, 512, 547, 590, 719, 725, 764, 782, 794
questionnaires, 503, 594
quetiapine, 20, 58, 697

quinolinic acid, 835, 837, 842, 846, 849, 850, 851, 853
Quinones, 851

R

rabies virus, 156, 165, 168
radar, 581, 585
radiation, 234
radicals, 422, 424, 480, 482, 530, 813
radio, 570
radiofrequency, 83, 84, 343, 344
radiological, 231, 233, 234, 243, 247, 249, 250, 280
radiopaque, 251
radon, 789
rail, 420, 606, 786
rain, 18, 28, 440, 800, 820, 845
Raman, 494
random, 22, 508
range, xiii, xvi, xvii, 19, 50, 51, 52, 57, 129, 215, 218, 219, 220, 221, 222, 223, 234, 290, 304, 306, 341, 354, 367, 369, 412, 427, 445, 448, 449, 457, 462, 467, 501, 507, 561, 562, 565, 568, 569, 573, 574, 577, 579, 581, 617, 619, 626, 627, 628, 630, 631, 666, 735, 748, 750, 762, 772, 776, 777, 785, 808, 818
rapamycin, 811, 834
raphe, 188, 342, 376, 398
rapid-cycling, 401
rating scale, 261, 282, 362, 378, 396, 401, 423, 450, 455, 456
ratings, 179, 423, 444, 448, 449, 450, 452, 456, 460, 462, 463, 464, 467, 756
rational, 180
rationality, 751
RCI, 446
reaction time, 50, 120, 121, 122, 123, 130, 131, 132, 134, 194, 448, 451, 459, 462, 463, 468, 512
reactions, 74, 83, 86, 130, 143, 163, 341, 380, 420, 592, 718, 726, 730, 731, 732, 736, 742, 767, 780, 782, 795, 807
reactive gliosis, 490
reactive nitrogen, 530
reactive oxygen, 528, 542, 545, 546
reactive oxygen species, 528, 542, 545, 546
Reactive Oxygen Species, 480
reactive oxygen species (ROS), 528, 546
reactivity, 194, 416, 436, 509, 551, 613
reading, 135, 145, 163, 373, 419, 462, 472, 502, 513, 517, 612, 763
reading comprehension, 373, 612

reading disorder, 472
real time, 234, 235
reality, 262, 281, 388, 500, 514, 591, 604, 786
reason, 719, 725, 751, 759, 763, 773, 778, 779, 829
reasoning, 60, 117, 125, 142, 224, 294, 309, 503, 508, 512, 589
reasoning skills, 512
recall, 120, 356, 370, 376, 412, 414, 420, 547, 627, 709, 710, 783
receptive field, 119, 196, 253
receptor, 180, 181, 182, 183, 185, 186, 188, 189, 190, 194, 196, 197
receptor agonist, 62, 66, 69, 70, 190, 818
receptors, ix, 5, 21, 23, 26, 32, 40, 43, 44, 47, 48, 50, 51, 52, 54, 55, 56, 61, 62, 63, 64, 65, 66, 67, 69, 70, 71, 82, 107, 139, 172, 173, 175, 176, 180, 181, 182, 183, 184, 185, 186, 188, 190, 193, 195, 197, 310, 417, 421, 422, 423, 425, 426, 427, 429, 435, 438, 441, 442, 477, 491, 542, 579, 641, 652, 672, 673, 802, 812, 817, 828, 838, 853
recognition, 61, 68, 85, 103, 108, 125, 137, 153, 164, 167, 225, 260, 272, 341, 378, 411, 412, 417, 507, 510, 556, 585, 592, 600, 618, 627, 685, 714, 724, 748, 749, 755, 784, 847
recollection, 580
recombination, 126, 127, 624, 629, 688
recommendations, iv, 248, 298, 565, 607, 711
reconstruction, 237, 633
recovery, 151, 284, 285, 296, 313, 346, 375, 416, 709, 845
recovery process, 284
recovery processes, 284
recruiting, 141, 322, 327
rectification, 241
recurrence, 312, 361, 408, 437, 552
redox, 488, 522, 524, 532
Redox, 527, 534, 535, 537, 538
reduction, xi, 1, 3, 4, 5, 6, 7, 8, 9, 10, 11, 13, 14, 15, 16, 17, 18, 19, 20, 21, 22, 23, 24, 25, 26, 27, 40, 41, 45, 50, 54, 81, 112, 117, 128, 135, 171, 172, 173, 342, 343, 344, 345, 346, 359, 360, 366, 371, 372, 373, 375, 376, 413, 420, 422, 423, 425, 432, 444, 465, 467, 480, 488, 497, 513, 525, 531, 536, 538, 546, 564, 579, 805, 810, 811, 816, 817, 820, 822, 824, 828
refining, 629
reflection, 123, 135, 747, 755
reflexes, 83, 623
refractory, 28, 51, 62, 255, 256, 257, 258, 261, 266, 268, 269, 270, 272, 274, 275, 277, 281, 282, 286, 293, 296, 297, 300, 302, 325, 327,

328, 331, 332, 336, 343, 344, 345, 347, 348, 349, 350, 366, 413, 424, 436, 440
refuge, 755
regeneration, 284, 285, 490
regenerative capacity, 831
region, 624, 629, 639, 640, 649, 651, 656, 660, 677, 679, 680, 809, 811, 825, 830
regional, 5, 10, 12, 16, 17, 18, 19, 22, 25, 26, 27, 30, 32, 38, 39, 43, 44, 45, 50, 58, 59, 60, 61, 136, 149, 172, 173, 183, 189, 190, 192, 211, 424, 433, 534, 537, 550, 555, 763, 837
region-of-interest, 11
region-of-interest (ROI), 11
regression, 131, 209, 416, 445, 832
regression line, 131, 209
regressions, 203
regular, 462, 509, 756
regulation, xi, xviii, 4, 5, 9, 10, 28, 29, 32, 37, 41, 56, 63, 64, 65, 67, 71, 81, 82, 108, 113, 119, 138, 177, 178, 179, 180, 184, 186, 188, 189, 190, 196, 341, 395, 401, 426, 435, 441, 478, 508, 509, 511, 514, 515, 518, 531, 536, 538, 539, 550, 553, 644, 650, 670, 671, 678, 747, 824, 831
regulators, 544
rehabilitation, 57, 284, 285, 286, 343, 502, 507, 515, 518, 691, 694, 698, 699, 700, 709
rehabilitation program, 57, 284, 507, 698
reinforcement, 101, 103, 105, 109, 110, 111, 112, 117, 124, 128, 133, 135, 136, 137, 138, 143, 146, 568, 569, 708, 709, 710, 795
reinforcement learning, 136, 137, 143
reinforcers, 81, 87, 97, 105, 146
rejection, 759
relapse, 58, 399
relationship, xvi, 16, 17, 18, 28, 29, 30, 35, 37, 38, 48, 69, 75, 140, 155, 162, 179, 203, 204, 219, 220, 221, 222, 237, 239, 242, 250, 333, 401, 426, 438, 460, 468, 470, 538, 542, 553, 562, 564, 566, 570, 581, 593, 594, 597, 606, 611, 626, 627, 628, 643, 645, 647, 707, 730, 739, 752, 755, 762, 774, 783, 805, 811, 836, 842
relationships, 66, 70, 75, 125, 127, 194, 204, 345, 438, 441, 456, 460, 468, 472, 510, 567, 614, 752, 755, 765, 768, 775, 781, 783, 794
relative size, 204
relatives, 68, 262, 375, 376, 630, 693, 700, 719, 722, 723, 724, 736, 742, 770, 782, 853
relaxation, 104, 698
relevance, xi, 117, 176, 178, 181, 185, 187, 211, 224, 323, 327, 485, 514, 537, 548, 549, 587, 603, 630, 640, 850

reliability, 2, 11, 130, 362, 457, 458, 470, 473, 588, 589, 593, 776
Reliability, 445, 594
relief, 272, 273, 341, 389, 706, 713, 725
rem, 180
REM, 180, 242
remission, 346, 398, 399, 400, 401, 622
remitters, 399
remodeling, 9, 15, 536, 549, 553
remorse, 500
renal, 485
renal failure, 723
renin, 138
repair, 94, 266, 293, 477, 480, 492, 677, 678, 679, 680, 682, 683, 684, 685, 686, 687, 688, 689, 801, 806, 807, 813, 820, 824, 826, 829, 832, 840, 847, 850
replication, 14, 281, 458, 614, 678, 679, 680, 683, 685, 686, 688
repression, 770
repressor, 675
reproduction, 77, 746, 755, 756, 757, 758, 760, 764, 766, 767, 768, 769, 770, 772, 774, 776, 777, 780, 783, 785, 786, 792, 794
reproductive state, 74, 97, 98, 99, 100
requirements, 77, 122, 128, 238, 308, 445, 460, 480, 486, 761, 813
RES, 661, 672, 675
research, xi, xv, xvi, xvii, 2, 11, 28, 47, 55, 58, 79, 102, 146, 147, 168, 178, 181, 182, 185, 187, 188, 211, 218, 372, 379, 417, 428, 444, 457, 458, 459, 460, 463, 468, 472, 476, 499, 500, 502, 522, 544, 555, 561, 562, 564, 568, 578, 581, 582, 587, 590, 594, 595, 597, 607, 609, 611
research design, 790
research reviews, 468
researchers, xix, 86, 183, 217, 274, 395, 446, 508, 510, 511, 514, 566, 578, 592, 595, 610, 612, 613, 614, 618, 641, 644, 679, 682, 685, 692, 726, 786
resection, 327, 328
Reserpine, 190
reserve capacity, 428
residues, 478, 479, 635
resilience, xv, 499, 513, 516
resin, 87
resistance, 260, 399, 401, 402, 425, 530, 538, 541, 551, 563, 622, 826, 827, 828, 831, 842
resolution, 11, 59, 74, 86, 102, 120, 129, 234, 240, 262, 283, 307, 415
resources, 11, 178, 185, 188, 223, 263, 486, 700, 726, 729, 736, 779

respect, 644, 682
respiration, 292, 706
respiratory, xvi, 77, 236, 250, 543, 553, 562, 565, 569, 570, 698, 706, 838
respiratory problems, 250
response time, 172
responsiveness, x, 48, 84, 108, 167, 194, 280, 444, 456, 459, 570, 612
restless legs syndrome, 376
restoration, 502, 707, 710, 715, 805
restorative material, 710
restorative materials, 710
restriction fragment length polymorphis, 629
restrictions, 698
restructuring, 748, 770, 778, 785
retardation, xiv, 408, 415, 441, 576, 578, 610, 611, 619, 622, 652
retention, 228, 563, 567, 569, 572, 580, 604, 605, 706
reticulum, 477, 532
retinol, 525, 526, 535
retinol-binding protein, 525, 526, 535
retinopathy, 623
retirees, 572
returns, 145, 274
reversal learning, 149, 172
reversal performance, 195
rewards, 142, 150, 151, 349
Reynolds, 5, 29, 41, 65, 71, 461, 473, 588, 597, 816, 823, 828, 829, 831, 850
rheumatoid arthritis, 564
Rho, 34
rhythm, 137, 139, 140, 400, 435
rhythmicity, 138
rhythms, 53, 65, 129, 135, 138, 139, 242, 314
ribonucleic acid, 835
ribosome, 531
ribosomes, 534
rice, 37
right hemisphere, 180, 309, 579
rigidity, 249, 274, 275, 277, 305, 306, 308, 313, 314, 366, 373, 375, 383, 389, 390, 806
ring chromosome, 324, 329
rings, 262, 436
risk aversion, 151
risk factors, xvi, 428, 429, 480, 481, 522, 541, 543, 544, 546, 552, 553, 714
risk perception, 147, 747, 786
risks, 288, 328, 347, 428, 429, 430, 697, 698, 706, 709, 731, 745, 746, 747, 748, 750, 756, 757, 758, 760, 763, 765, 766, 767, 768, 769, 770, 771, 772, 773, 774, 776, 777, 778, 779, 780, 784, 786, 787, 794, 795, 797

risperidone, 20, 22, 38, 50, 56, 58, 61, 64, 70, 695, 697
Rita, 365
RNA, 44, 426, 478, 481, 534, 550, 619, 622, 623, 643, 659, 661, 662, 834, 835, 839
RNA processing, 619
RNAi, 671
robotic, 292
rodent, 55, 81, 103, 174, 197, 417, 420, 551, 556
rodents, 51, 55, 70, 76, 81, 116, 136, 412, 438, 548, 829
rofecoxib, 494
ROI, 11, 12, 21, 24, 26
rolls, 727
Rome, 605
root, 372, 449, 710
ROS, 480, 481, 528, 546
Roses, 554
rowing, 476
Royal Society, 212, 213, 584
rubber, 708
rubella, 611
rules, 75, 115, 124, 196, 284, 410, 446, 450, 456, 462, 692
rural, xvii, 587, 590, 596
rural areas, 726
Russian, 797
Rwanda, 778

S

saccades, 67, 157, 158, 160, 166, 170
saccadic eye movement, 169, 808
sacrifice, 821, 824
sadness, 376
safeguard, 262, 284
safety, 229, 239, 288, 289, 299, 300, 315, 320, 328, 347, 482, 484, 485, 489, 494, 693, 698, 796
SAHA, 640
saline, 183, 384, 420
saliva, 707
salivary gland, 707
salivary glands, 707
salts, 611
sample, 12, 14, 23, 25, 50, 86, 124, 192, 215, 218, 309, 344, 355, 369, 371, 398, 399, 400, 401, 451, 456, 458, 468, 473, 488, 525, 555, 570, 588, 591, 600, 601, 602, 603, 763, 830
sampling, 87, 91, 136, 151, 205, 762
SAP, 478, 490
SAR, 248
SARA, 572

Index

Sarin, 68
satisfaction, 599, 602, 603, 606, 607, 727, 728, 729, 732
saturation, 412
savannah, 595
sawdust, 98
scaffolding, 530
scaling, 210
scalp, 239, 243, 271, 322, 512
scar tissue, 287
scavengers, 629
scheduling, 711
schema, 752
schemas, 768
Schiff, 524
Schiff base, 524
schizoaffective disorder, 39, 42, 58
schizophrenia, ix, xix, 1, 2, 3, 4, 5, 6, 7, 8, 9, 10, 11, 12, 13, 14, 15, 16, 17, 18, 19, 20, 21, 22, 23, 24, 25, 26, 27, 28, 29, 30, 31, 32, 33, 34, 35, 36, 37, 38, 39, 40, 41, 42, 43, 44, 45, 47, 48, 49, 50, 51, 52, 53, 54, 55, 56, 57, 58, 59, 60, 61, 62, 63, 64, 65, 66, 67, 68, 69, 70, 156, 191, 293, 299, 472, 500, 516, 517, 519, 650, 692
Schizophrenia, ix, 1, 2, 4, 11, 29, 30, 31, 32, 33, 35, 38, 41, 44, 47, 48, 49, 50, 52, 54, 57, 67
schizophrenic patients, 31, 35, 38, 40, 41, 42, 48, 54, 58, 62, 64, 65, 68, 69, 282, 697
schizotypal personality disorder, 36
Schmid, xvii, 609, 610, 611, 612
school, 148, 218, 367, 409, 415, 424, 427, 428, 448, 461, 462, 509, 510, 514, 516, 581, 768
school activities, 462
school failure, 511
school performance, 218
school success, 511
science, xi, 169, 178, 268, 300, 715, 768, 795
scientific community, 611, 832
scientific knowledge, 284
scientists, 47
sclerosis, 12, 257, 268, 270, 276, 299, 302, 326, 417, 432, 491
scope, 501, 561, 736
scores, 90, 100, 275, 277, 304, 306, 307, 308, 322, 344, 346, 353, 359, 360, 369, 373, 375, 383, 398, 399, 400, 409, 411, 419, 420, 423, 424, 436, 445, 446, 447, 448, 450, 451, 452, 454, 455, 456, 459, 463, 479, 483, 489, 508, 510, 566, 567, 568, 571, 587, 589, 592, 593, 698, 726, 732, 734

search, ix, x, 2, 59, 74, 79, 83, 99, 115, 131, 327, 328, 373, 378, 470, 503, 522, 523, 624, 641, 659, 662, 684, 749, 751, 765, 794, 796
search engine, 523
searches, 662
searching, 78, 524, 535, 728
second generation, 21, 22, 50, 51
secrete, 478, 542, 801, 802
secretion, 424, 490, 532, 540, 542
secrets, 518
security, 508
sedation, 236, 400, 419
sedative, 189, 713
sedatives, 237, 426
sediment, 251
seeds, 344, 806
segmentation, 11
segregation, 54, 636
seizure, xiii, 65, 259, 265, 279, 280, 291, 301, 320, 321, 322, 324, 325, 326, 327, 328, 330, 331, 332, 333, 334, 335, 336, 337, 344, 395, 408, 412, 413, 414, 416, 417, 420, 421, 422, 423, 424, 425, 426, 428, 429, 430, 433, 434, 437, 438, 441, 824
seizure-free, 320
selecting, 75, 273, 282, 354
selective attention, 470
selective serotonin reuptake inhibitor, 351, 804, 805
selectivity, 53, 66, 160, 628, 812
self, 179, 189, 342
Self, 151, 561, 568, 796
self injurious behavior, 732
self-awareness, 411, 579
self-care, xvi, 562
self-concept, 510
self-control, 146, 189, 510
self-efficacy, 747
self-esteem, 756
self-identity, 756
self-recognition, 592
self-regulation, 508, 509, 514, 515, 518
self-renewal, 852
self-repair, 801, 820, 829
self-report, 179, 467, 503
semantic, xii, 215, 216, 217, 218, 219, 221, 223, 224, 225, 309, 368, 369, 370, 371, 372, 409, 410, 548, 589, 591
semantic association, 225
semantic memory, 225
semantic processing, 225, 548
semantics, 224
seminars, 507

senescence, 531, 539
senile, 377, 476, 479, 501, 526, 536, 545, 557
senile dementia, 501
senile plaques, 377, 476, 479, 526, 545, 557
sensation, xii, 304, 409
sensations, 248, 409
sensitivity, x, 93, 97, 138, 141, 143, 144, 146, 147, 148, 187, 238, 449, 455, 457, 458, 459, 587, 588, 589, 590, 613, 653, 712, 813, 826, 828, 831, 845, 848, 849
sensitization, 9, 37, 42, 56, 193, 652, 853
sensorimotor cortex, 211, 305, 389
sensorimotor gating, 69
sensors, 292, 681
sensory cortices, xi, 178, 184, 185
sensory memory, 68
sensory modalities, 75, 142
sentences, 373, 374, 502, 517
separation, 95, 99, 123, 129, 521
septum, 82, 113, 229, 849
sequelae, 250, 270, 275, 279, 291, 324, 468
sequencing, 36
sequential behavior, 75, 96
series, 49, 55, 69, 77, 79, 90, 94, 146, 210, 343, 344, 345, 361, 377, 420, 449, 457, 460, 528
serine, 554, 643
serotonergic, 118, 119, 139, 194
serotonin, 118, 119, 173, 340, 342, 351, 426, 804, 805, 822, 823, 829, 833, 844
Serotonin, 398, 803, 816, 823, 829, 830, 836, 855
sertraline, 697
serum, 419, 428, 429, 438, 478, 479, 480, 491, 545
severity, 19, 37, 44, 53, 54, 272, 285, 321, 344, 345, 346, 400, 412, 414, 444, 456, 458, 459, 465, 600, 625, 628, 632, 651, 657, 733, 786, 812, 828, 837
sex, 35, 54, 79, 83, 104, 105, 107, 109, 110, 111, 517, 717, 731, 776, 854
sex offenders, 107
sex steroid, 105, 111
sexual behavior, 79, 80, 81, 82, 83, 84, 87, 89, 91, 92, 93, 97, 101, 102, 103, 104, 105, 106, 107, 108, 111, 112, 113
sexual behaviour, 104, 579, 581
sexual contact, x, 74, 80
sexual intercourse, 103
sexual motivation, x, 74, 79, 80, 83, 87, 91, 93, 104, 111
shame, 144, 736
shape, xii, 11, 216, 218, 224, 234, 421, 532, 747, 752, 760, 774
shaping, 215, 221, 749, 771, 773, 784

sharing, 579, 619, 700, 753
sheep, 111, 483
shock, 134, 140, 423, 480, 530, 537, 538, 673, 732, 818
shocks, 81
short period, 273
short term memory, 132, 309
shortage, 588, 602, 713
shortness of breath, 400
short-term, xv, 38, 62, 76, 102, 116, 123, 124, 136, 194, 398, 399, 401, 420, 422, 424, 437, 444, 467, 494
Short-term, 63
short-term memory, 116, 123, 124, 136, 194, 420, 422, 627
shoulder, 312, 564
showing, x, 48, 51, 52, 53, 54, 56, 134, 173, 220, 221, 222, 241, 246, 310, 311, 312, 322, 325, 330, 344, 345, 376, 387, 400, 415, 449, 452, 487, 488, 558, 591, 633, 679, 686, 718, 749
shy, 757
shyness, 144
SI, 34, 35, 45, 273, 289, 312
SIB, 70
sibling, 24, 774, 780, 783, 796
sibling support, 780
siblings, 14, 23, 35, 625, 629, 721, 763, 769, 774, 781, 782, 783
sickle cell, 791
side effects, ix, 48, 51, 53, 186, 189, 238, 257, 260, 276, 278, 325, 343, 366, 419, 420, 421, 428, 429, 432, 695, 697, 706
Siemens, 229, 383
sign, xiii, 132, 304
signal transduction, 10, 525, 528, 530, 817
signaling, 6, 9, 38, 123, 125, 137, 180, 185, 484, 490, 491, 530, 532, 552, 845, 846, 852, 853
signaling pathway, 490, 641, 667, 669
signaling pathways, 490
signalling, xviii, 167, 477, 540, 678, 802, 811, 817, 823, 845
signals, 77, 78, 87, 91, 93, 106, 110, 125, 126, 127, 129, 143, 144, 147, 165, 265, 291, 292, 313, 351, 426, 529, 542, 826, 847
signal-to-noise ratio, 10
signs, xv, 1, 27, 104, 259, 280, 376, 377, 379, 383, 386, 522, 544, 667, 694, 697, 722, 728, 732, 733, 812, 818
similarity, 87, 99, 175, 210, 216, 224, 751
simulation, 7, 749, 750, 780, 782
simulations, 445, 749
sine, 326
sine wave, 326

Singapore, 171
single test, 445, 446
singular, 130
singularities, 134
sinoatrial node, 397
sinus, 311
sites, 56, 64, 82, 87, 109, 153, 157, 182, 188,
    206, 208, 243, 249, 250, 278, 285, 289, 327,
    351, 429, 477, 482, 484, 487, 489, 490, 578,
    589, 595, 825, 832, 837, 855
Sjogren, 553
skeletal muscle, 628, 649, 652, 850
skilled personnel, 238
skilled workers, 604
skills, 8, 17, 239, 373, 379, 428, 500, 502, 508,
    509, 510, 512, 547, 551, 573, 579, 588, 589,
    603, 604, 613, 729, 733, 734, 768
skills training, 428
skin, 143, 145, 148, 265, 292, 322, 326, 355, 612,
    658, 757
skin conductance, 143, 145, 148
skin conductance responses, 145
Skinner box, 80
sleep, xiii, 65, 99, 180, 240, 242, 304, 335, 415,
    416, 434, 435
sleep-wake cycle, 808, 848
slow-wave, 434
Sm, 404
SMA, 305, 311, 312, 366, 388
small intestine, 658
small vessel disease, 551
smoke, 55
smokers, 26, 30, 55, 67
smoking, 35, 55, 67, 68, 69, 541, 543
smooth muscle, 542, 544, 545, 549, 550
smooth muscle cells, 542, 544, 545, 549
smoothing, 13
SNR, 323, 324, 325
social, 373
social activities, 764
social and emotional learning, 510, 516
social attitudes, 517, 736
social behavior, 41, 82, 612
social behaviour, 474
social care, 601, 605
social cognition, 17, 28
social competence, 510
social consequences, 718
social context, 693, 700, 757, 782, 785
social costs, 736
social development, 511
social evaluation, 785
social events, 717, 719

social factors, 615
social influence, 596, 748
social influences, 596, 748
social interactions, 580, 613, 736
social isolation, 11
social learning, 797
social learning theory, 797
social life, xv, 499, 500, 514, 723
social network, 729, 748
social psychology, 793
social relations, 345, 700
social relationships, 345, 700
social sciences, 790
Social Services, 606
social skills, 8, 428, 729
social skills training, 428
social support, 726, 729, 747, 765, 766, 772, 780,
    783
social withdrawal, ix, 48
social workers, 700
socialization, 518
society, xvii, 48, 574, 600, 606, 700, 726, 790
socioeconomic, 510
sociology, 763, 789
sodium, 421, 422, 438, 523, 535, 646, 711, 712,
    840
sodium dodecyl sulfate (SDS), 523
software, 234, 235, 238, 239, 240, 259, 260, 263,
    264, 265, 266, 267, 280, 291, 293, 294, 298,
    356, 523, 525, 764
soil, 612
solubility, 534
solution, 117, 127, 128, 135, 250, 258, 314, 699,
    712
somata, 6
somatic cell, 687
somatic marker, 145, 148, 150, 151
somatomotor, 77, 112, 388
somatosensory, 25, 30, 75, 76, 77, 78, 96, 97,
    113, 185, 196, 197, 207, 228, 241, 258, 277,
    278, 360, 557
somatosensory cortical neurons, 185
somatostatin, 6, 39, 813, 835
Sonic hedgehog, 845, 846
sorting, 17, 224, 448, 452, 458, 508, 571
sounds, 90, 93, 166, 169, 511
South Africa, 719, 731, 740
space, 706
Spain, 227, 238, 601
spasticity, 293, 298, 316, 360, 623
spatial, 17, 34, 39, 63, 68, 71, 75, 78, 86, 90, 96,
    108, 138, 139, 155, 157, 164, 167, 169, 172,
    173, 175, 179, 180, 183, 187, 189, 190, 191,

193, 195, 196, 197, 230, 234, 235, 240, 241, 253, 263, 286, 291, 299, 309, 376, 409, 410, 412, 420, 423, 424, 432, 437, 439, 469, 470, 471, 513, 548, 579
spatial learning, 39, 63, 412, 423, 424
spatial location, 155, 191
spatial memory, 76, 138, 139, 439, 579
spatial representations, 579
special education, 419, 583, 596
specialisation, 210, 445
specialists, xvii, 263, 562, 700
specialization, 60, 211, 213, 239
species, xi, 76, 77, 78, 81, 109, 200, 201, 203, 204, 205, 207, 208, 210, 528, 530, 542, 545, 546, 549, 635, 804, 810, 813, 824, 829
specificity, 5, 22, 33, 39, 42, 59, 238, 291, 455, 477, 548, 568
SPECT, 59, 183, 307, 310, 313, 315, 388, 398, 555
spectroscopy, 197, 843
spectrum, xvi, xvii, 24, 57, 146, 278, 416, 434, 555, 562, 573, 576, 580, 582, 583, 585, 610, 613, 615, 619, 680, 696
speculation, 374, 446
speech, 167, 280, 289, 373, 374, 415, 427, 428, 500, 509, 579, 595, 623, 625, 698, 699, 706, 714
speed, ix, 48, 56, 414, 420, 462, 464, 467, 468, 502, 503, 507, 508, 516, 698, 768
spending, 601, 764
sperm, 77, 788, 854
spermatogenesis, 807, 811
spermiogenesis, 537
spin, 537
spinal cord, 106, 155, 260, 272, 292, 426, 855
spinal cord stimulation, 272
spine, 3, 5, 6, 8, 21, 30, 31, 32, 33, 36, 37, 38, 39, 41, 43, 45, 80, 206, 207, 212, 268, 300, 518, 581, 821, 828, 853
spines, 3, 4, 5, 6, 8, 9, 11, 21, 27, 32, 43, 182, 206, 207, 208, 209, 529
spirometry, xvi, 561, 562, 568, 572
spleen, 658, 660
sponge, 712
sponsor, 486
spontaneity, 579
sporadic, 134, 240, 355, 490, 546, 555
sports, 357
spouse, 601, 791
Sprague-Dawley rats, 337
Spring, 558
sprouting, 407, 821, 837
SPSS, 218, 369

square wave, 367
SSA, 595
stability, 30, 172, 304, 306, 372, 373, 446, 447, 451, 457, 682, 698
stabilization, 708, 709, 715
stabilize, 21, 476, 477, 478, 483
stabilizers, 401, 694, 697
stages, xii, xv, xx, 78, 115, 120, 121, 122, 124, 129, 130, 131, 133, 134, 135, 140, 173, 188, 216, 222, 223, 262, 275, 282, 285, 389, 425, 448, 475, 478, 490, 528, 531, 532, 544, 551, 589, 613, 614, 748, 749, 752, 762, 768, 770, 777, 780, 781, 782, 799, 808, 812, 814, 824, 830, 836
stakeholders, 753
standard deviation, 369, 445, 449, 451, 461
standard error, 89, 90, 91, 94, 96, 98, 100, 503, 515
standardization, 588, 589, 592
standards, 262, 547, 692, 751
Standards, 442
starvation, 542, 644
state, vi, xv, xvii, 7, 30, 69, 73, 78, 79, 87, 89, 90, 93, 94, 97, 99, 101, 134, 135, 143, 145, 172, 177, 180, 181, 186, 190, 204, 210, 249, 256, 258, 271, 272, 274, 282, 284, 304, 308, 341, 358, 367, 371, 375, 389, 415, 416, 417, 428, 463, 500, 502, 509, 522, 547, 563, 568, 571, 602, 632, 640, 663, 670, 674, 712, 733, 734, 735, 753, 764, 783
statin, 489
statins, 488
Statins, 488, 495, 497
statistical analysis, 305, 309, 460
Statistical Package for the Social Sciences, 369
statistics, xiv, 14, 166, 443, 446, 466, 468, 472, 563, 606, 738, 760
status epilepticus, 322, 324, 331, 408, 440, 441, 442, 696
steady state, 358
steel, 87
Stem cell, 843
stem cells, 800, 802, 820, 829, 830, 832, 836, 837, 839, 840, 841, 842, 845, 846, 847, 848, 849, 851, 852, 853, 854, 855, 856
stenosis, 228, 343, 547
stereotypical, 417
steric, 66
sterilization, 757, 776
steroid, 105, 109
steroid hormone, 105
steroid hormones, 105
steroids, 111, 838

stigma, 428, 724
stigmatized, 260, 792
stimulant, xv, 179, 180, 187, 188, 189, 443, 444, 448, 449, 450, 452, 455, 456, 457, 458, 459, 460, 462, 467, 468, 470, 472, 473, 474
stimulants, 187
stimuli, 178, 179, 181, 191
Stimuli, 83, 84, 85, 367
stimulus, x, 49, 61, 73, 74, 78, 79, 80, 85, 98, 101, 108, 116, 119, 120, 122, 124, 125, 129, 132, 134, 138, 139, 143, 152, 153, 157, 159, 160, 163, 164, 165, 179, 196, 242, 290, 314, 326, 372, 409, 445, 457, 501, 628, 829
stock, 796
stomach, 659
storage, x, 115, 116, 190, 191, 372, 508, 827
storytelling, 750
strain, 91, 294, 551, 804, 821, 824, 826
strains, 110, 558
strategies, xiv, xvi, 75, 115, 116, 117, 129, 134, 135, 173, 340, 346, 361, 372, 391, 395, 396, 411, 428, 495, 500, 507, 542, 552, 573, 581, 582, 588, 605, 627, 629, 644, 645, 656, 694, 698, 699, 712, 715, 724, 748, 749, 750, 756, 766, 769, 773, 777, 780, 782, 785, 846
strategy, 634, 643, 726, 735, 765, 770, 774, 781, 792
streams, 155, 168
strength, 117, 128, 129, 185, 265, 290, 417, 450, 544, 698
stress, xvi, 8, 10, 11, 15, 27, 29, 34, 37, 38, 41, 45, 95, 119, 125, 130, 132, 135, 137, 139, 151, 177, 181, 182, 186, 187, 189, 190, 191, 192, 351, 411, 423, 425, 439, 441, 480, 481, 482, 491, 492, 509, 518, 521, 522, 528, 534, 536, 538, 540, 542, 545, 550, 551, 553, 557, 599, 610, 656, 677, 679, 680, 681, 684, 687, 699, 707, 724, 726, 727, 734, 756, 781, 784, 796, 813, 822, 828, 841, 846, 847
stress level, 119
stressors, 185, 500, 833
stretching, 22
striatum, xi, xviii, 20, 40, 63, 66, 135, 143, 171, 174, 175, 224, 240, 261, 281, 283, 323, 324, 341, 342, 347, 349, 352, 367, 371, 617, 628, 629, 632, 633, 634, 636, 637, 640, 667, 670, 673, 674, 677, 679, 683, 685, 686, 687, 801, 804, 806, 809, 812, 813, 814, 815, 816, 818, 822, 823, 825, 826, 828, 829, 830, 831, 839, 840, 844, 847, 848, 849, 850, 853, 854
stroke, 258, 267, 277, 282, 284, 285, 286, 293, 295, 296, 297, 298, 299, 301, 303, 312, 316, 368, 482, 487, 513, 519, 543, 546, 547, 550,
551, 552, 553, 554, 556, 558, 566, 710, 711, 714, 836, 838, 849, 854
strokes, 355, 547, 550
structural changes, 8, 628
structural protein, 529, 531
structure, xii, 10, 11, 13, 16, 17, 23, 27, 28, 32, 38, 53, 67, 69, 83, 84, 86, 87, 95, 97, 99, 113, 134, 136, 181, 199, 204, 205, 206, 208, 210, 211, 213, 216, 217, 224, 248, 258, 279, 321, 323, 371, 372, 409, 411, 417, 423, 430, 439, 478, 484, 517, 539, 551, 553, 579, 593, 597, 614, 635, 644, 652, 666, 668, 681, 682, 710, 749, 750, 763, 765, 783, 794, 810, 811, 815, 817, 825
structuring, 783
students, 179, 187, 219, 510, 581
style, 13, 144, 148, 711, 727, 732, 738, 749
subacute, 330
subarachnoid hemorrhage, 346
subcortical nuclei, 233
subcortical structures, xi, 13, 20, 86, 87, 154, 178, 184, 185, 187, 320, 346
subdural haematoma, 228
subdural hematoma, 324
subgranular zone, xx, 799, 800, 835, 843
subgroups, ix, 2, 16, 19, 25, 353, 355, 402, 487
subjective, 11, 117, 239, 304, 306, 414, 461, 464, 559, 755, 767
sub-Saharan Africa, 596
Sub-Saharan Africa, 595
substance abuse, xiii, 13, 143, 340, 730
substances, 25, 112, 480, 522, 609, 611
substantia nigra, xi, xiv, 156, 168, 171, 279, 329, 331, 334, 336, 342, 365, 370, 374, 376, 390, 394, 398
Substantia nigra, 296, 331
substantia nigra pars compacta, xi, 156, 171, 376, 835
substitutes, 591
substitution, 61, 502
substrate, 18, 20, 24, 133, 139, 150, 151, 187, 204, 277, 292, 351, 377, 477, 484, 490, 649, 680, 682
substrates, 42, 82, 101, 136, 390, 398, 484, 485, 683, 836
subventricular zone, xx, 294, 799, 800, 835, 836, 838, 841, 843, 846, 849, 854
success rate, 272, 279, 568, 721
succession, xiv, 396, 462
suffering, xiii, 261, 280, 285, 293, 303, 304, 311, 312, 320, 321, 322, 324, 346, 361, 376, 580, 601, 602, 603, 604
sugar, 680, 707

suicidal, 275, 344
suicidal behavior, 737, 740
suicidal ideation, 344, 693, 695, 697, 717, 724, 727, 728, 729, 730, 732, 733, 734, 735, 736, 738, 740, 742
suicide, xix, 54, 262, 283, 288, 343, 344, 345, 346, 500, 518, 697, 717, 718, 719, 720, 721, 722, 723, 724, 725, 726, 727, 728, 729, 730, 731, 732, 733, 734, 735, 736, 737, 738, 739, 740, 741, 742, 743, 806, 808
suicide attempts, 262, 283, 288, 344, 717, 719, 722, 723, 728, 729, 730, 734, 736, 737
suicide rate, 717, 719, 720, 721, 728, 729, 731, 733
sulfate, 523, 535
summaries, 504, 506
Sun, 32, 34, 36, 171, 269, 277, 301, 536, 539, 640, 652, 670, 672, 812, 825, 826, 843, 853
superimposition, 244
superior parietal cortex, 389
superior temporal gyrus, 2, 12, 13, 14, 15, 17, 19, 22, 24, 26, 63, 155, 398, 479
superiority, 272, 275, 281
superoxide, 480
superoxide dismutase, 480
supervision, xvi, 562, 601
supervisors, 508, 788
supplementation, 38, 488, 495, 694
supplements, 488
supply, 542, 602, 606
support services, 599, 601
suppression, 251, 276, 279, 321, 323, 324, 325, 332, 333, 334, 371, 372, 619
suprachiasmatic, 137
suprachiasmatic nuclei, 137
surface area, 200, 203, 209
surgeons, 289
Surgeons, 296
surgery, xii, xiii, 228, 229, 231, 233, 236, 237, 238, 239, 247, 248, 249, 250, 251, 252, 253, 254, 255, 256, 257, 263, 268, 271, 272, 273, 275, 276, 277, 278, 286, 287, 300, 304, 306, 307, 309, 312, 313, 319, 326, 345, 353, 354, 358, 359, 360, 361, 362, 363, 365, 367, 368, 369, 370, 371, 372, 373, 374, 375, 376, 379, 392, 395, 428, 455, 470, 473
Surgery, 227, 229, 249, 298, 315, 316, 375
surgical, xii, 228, 229, 234, 235, 236, 237, 238, 239, 242, 248, 250, 256, 268, 271, 273, 274, 275, 278, 280, 289, 299, 305, 324, 328, 334, 340, 343, 344, 345, 347, 354, 356, 361, 384, 396, 397, 400
surgical intervention, xii, 256, 344, 396, 397, 400

surgical removal, 250
surgical resection, 328
surgical technique, 343, 347
surveillance, 600, 786, 787
survival, x, 31, 32, 77, 116, 426, 427, 484, 536, 646, 672, 804, 811, 819, 821, 823, 832, 838, 841, 843, 844, 846, 850, 855
survival rate, 821
surviving, 285, 563
survivors, 473
susceptibility, 10, 34, 40, 191, 230, 329, 412, 425, 644, 706, 717, 718, 748, 826, 849, 850, 852
susceptibility genes, 40
swallowing, 360, 362, 808
Sweden, 229, 235, 343, 601, 602, 605, 607, 717, 724, 728, 731, 732, 741, 796
swelling, 827, 828
switching, 307, 309, 372, 375, 376, 515, 810
Switzerland, 199, 796
symbols, 782
sympathetic, 568
symptom, xiii, 36, 44, 48, 53, 343, 344, 345, 346, 354, 355, 374, 447, 458, 461, 465, 467, 628, 693, 696, 706, 722, 723, 732, 733
symptomatic treatment, 547
symptomology, 250
synapse, xvi, 6, 10, 42, 51, 187, 285, 478, 522, 529, 531, 854
synapses, 4, 21, 39, 43, 61, 185, 192, 206, 208, 314, 412, 542
synaptic plasticity, 33, 36, 108, 417, 423, 435, 438, 439, 644, 688, 811, 848, 852
synaptic strength, 185, 265, 290, 417
synaptic transmission, 421, 648, 817
synaptic vesicles, 3, 641, 667
synaptogenesis, 10, 424, 426, 477, 516, 825
synaptophysin, 3, 21, 33
synchronization, 10, 43, 129, 138, 139, 239, 240, 366, 421
synchronize, 43
syndrome, x, xvi, xvii, 28, 33, 48, 94, 168, 191, 193, 259, 261, 266, 267, 269, 270, 272, 273, 280, 281, 293, 295, 296, 298, 300, 302, 312, 322, 324, 336, 354, 355, 358, 362, 363, 364, 376, 416, 434, 435, 441, 442, 446, 476, 515, 541, 543, 544, 555, 573, 582, 583, 584, 610, 618, 619, 652, 733, 787, 807, 808, 852, 853
synergistic, 56, 71, 281, 544
synergistic effect, 544
syntactic, 370

synthesis, 5, 15, 51, 69, 102, 210, 348, 421, 425, 427, 430, 491, 526, 528, 531, 532, 547, 549, 628, 682
systems, 19, 38, 47, 54, 56, 57, 60, 67, 74, 75, 83, 97, 99, 106, 112, 117, 118, 119, 124, 128, 129, 133, 135, 137, 139, 144, 166, 169, 170, 174, 180, 187, 189, 191, 212, 213, 341, 348, 379, 409, 411, 414, 423, 432, 530, 532, 571, 580, 582, 610, 611, 612, 614, 757

T

T cell, 484
T cells, 484
T lymphocyte, 376
T lymphocytes, 376
tangles, 475, 476, 479, 526, 536, 539, 554, 558
tardive dyskinesia, xiv, 51, 62, 354, 695
target population, 569, 587, 588, 590
task demands, 174
task performance, 50, 146, 157, 173, 286, 352, 371, 469
taste, 76
tau, 476, 478, 482, 483, 486, 487, 490, 493, 494, 496, 530, 531, 532, 538, 539, 541, 544
tax credit, 601
tax credits, 601
taxonomy, 747, 754, 760
TBP, 623, 640
T-cell, 484
teachers, 509, 510, 581
teaching, 511, 570, 710, 779
teams, 263, 286, 342
technetium, 59
technicians, 263, 286, 568, 588
techniques, xiv, xv, 1, 13, 74, 84, 85, 86, 104, 110, 199, 204, 205, 238, 242, 249, 252, 286, 287, 292, 296, 314, 327, 328, 340, 343, 346, 379, 395, 396, 443, 445, 457, 523, 567, 571, 572, 578, 581, 587, 590, 592, 594, 629, 708, 709, 710, 776, 796, 807
technological advancement, 172
technological advances, 204, 256, 346
technological change, 776
technological developments, 776, 779
technologies, 203, 291, 771
technology, 172, 204, 263, 265, 266, 267, 286, 288, 291, 292, 293, 297, 300, 327, 564, 569, 713, 765, 773, 776, 792, 797
Technology Assessment, 789, 797
teenagers, 574
teeth, 581, 706, 707, 708, 710, 711, 712
Tel Aviv, 541

telencephalon, 848
telephone, 264, 292
television, 591
television viewing, 591
temperament, 149, 512, 515, 518, 806
temperature, 81, 292, 383, 384
temporal lobe, xvi, 2, 12, 13, 14, 18, 23, 24, 33, 36, 41, 75, 212, 222, 258, 269, 270, 279, 280, 301, 302, 321, 322, 323, 326, 327, 328, 329, 330, 332, 335, 336, 337, 408, 410, 411, 412, 413, 417, 430, 431, 432, 433, 434, 437, 442, 522, 579, 816, 840
temporal lobe epilepsy, 12, 36, 269, 280, 301, 322, 323, 327, 329, 330, 332, 335, 336, 337, 408, 413, 430, 431, 432, 433, 434
tenants, 620
tendinitis, 294
tension, 546
tensions, 341, 784
teratogenesis, 429
teratogenic, 575, 576, 584
terminally ill, 713
terminals, 3, 8, 34, 39, 42, 43, 45, 182, 342, 422, 529, 544, 641, 810, 825
territory, 785
terrorism, 45
Tesla, 288
test data, 446, 587
test scores, 511, 565, 569, 589, 593
test statistic, 449
testing program, 631, 718, 724, 725, 726, 727, 728, 731, 732, 734, 740, 764
testis, 646, 659, 839
testosterone, 105, 111
Testosterone, 362
Texas, 287, 288, 474, 795
textbook, 430, 742, 767
TGF, 479, 549, 550, 558
thalamocortical pathways, 274, 415
thalamocortical system, 258, 277
thalamus, xiv, 4, 6, 7, 8, 10, 12, 13, 14, 15, 17, 18, 19, 22, 25, 26, 32, 37, 40, 44, 75, 76, 77, 103, 155, 165, 180, 181, 193, 239, 242, 249, 257, 258, 272, 277, 279, 294, 296, 297, 307, 316, 320, 321, 322, 323, 330, 331, 332, 333, 334, 337, 340, 341, 342, 343, 345, 347, 365, 366, 388, 390, 398, 410, 412, 413, 414, 417, 434, 579, 734, 814, 849
theatre, 237, 238, 239, 264, 291
theory, 8, 17, 136, 144, 165, 212, 509, 597, 611, 745, 747, 751, 752, 761, 764, 765, 772, 776, 784, 788, 789, 790, 792, 794, 796, 811
therapeutic agents, 47, 51, 70

therapeutic approaches, 28, 340, 342, 552
therapeutic effects, xii, 180, 188, 256, 271, 340, 344
therapeutic interventions, 541, 552, 627, 644, 818
therapeutic targets, xx, 52, 521, 533, 618, 680, 799
therapeutic use, 263
therapeutics, 48, 55, 57, 67, 522, 532, 644, 645
therapist, 698, 708
therapists, 263, 273, 286, 699, 706, 708
therapy, xvi, 70, 147, 148, 175, 228, 256, 259, 260, 261, 262, 266, 273, 274, 276, 280, 281, 282, 284, 285, 295, 306, 314, 315, 327, 328, 342, 348, 349, 351, 354, 365, 373, 391, 396, 398, 399, 401, 408, 420, 435, 436, 437, 438, 440, 488, 489, 494, 495, 552, 554, 556, 561, 562, 563, 565, 568, 645, 694, 695, 698, 706, 708, 709, 710, 714, 729, 752, 763, 790, 794, 804, 834, 848
thermoregulation, 77
theta, x, 6, 28, 30, 115, 116, 129, 130, 131, 132, 133, 134, 135, 136, 137, 138, 139, 140, 283, 313
theta band, 6
thimerosal, 611
thin film, 292
thinking, xii, 142, 144, 145, 187, 216, 411, 419, 422, 427, 471, 512, 732, 734, 756, 763, 790, 808
thinking styles, 145
thinning, 60, 851
thiobarbituric acid, 480
thioridazine, 62
Thomson, 253
thoracic, 581
thoughts, ix, xiii, 1, 178, 340, 419, 708, 727, 728, 732, 734, 774, 788
threat, 142, 144, 509, 542, 726, 748, 774, 775, 794
threatening, 354, 360, 551, 756, 759, 764, 770
threatening behavior, 739
threats, 756, 757, 759, 793
three-dimensional, 235, 241, 247, 253, 264, 291
threshold, 7, 15, 83, 84, 157, 158, 238, 239, 242, 307, 309, 321, 325, 333, 412, 417, 421, 566, 567, 620, 647, 810, 828
thresholds, 567, 571
thrombin, 525
thrombosis, 250
thrombus, 551
thymidine, 801
thymus, 659, 660
thyroid, 612, 845

tibialis anterior, 239, 242
tic disorder, 195
tics, 261, 281
time bomb, 615
time constraints, 759
time consuming, 366, 595
time resolution, 120
timing, 138, 289, 350, 576, 580, 769, 779, 823, 848
tissue, xix, 32, 63, 64, 237, 240, 242, 259, 261, 263, 264, 265, 268, 283, 287, 288, 292, 300, 326, 331, 383, 392, 393, 426, 428, 480, 483, 486, 536, 544, 546, 548, 554, 558, 612, 613, 614, 622, 647, 658, 667, 670, 673, 679, 683, 746, 809, 813, 832
tissue perfusion, 558
titanium, 391
titration, 260, 263, 274, 287, 420, 424
TJ, 31, 32, 35, 44, 267, 270, 297, 298, 302, 331, 554, 556
TLE, 408, 412, 413, 414, 417
TM, 28, 38, 211, 213, 252, 364, 469, 489, 553
T-maze, 73, 81, 87, 88, 112, 556
TNF, 479, 487, 497
toddlers, 595, 597
tolerance, 274, 361, 420, 528, 774, 776
tones, 117, 118, 119, 126, 127, 130, 140
tonic, 136, 160, 162, 181, 182, 186, 208, 257, 258, 277, 279, 280, 321, 323, 416, 422, 433, 814
tonic-clonic seizures, 279, 321, 323, 433
tooth, 707, 709, 710, 711, 712, 713
Topiramate, 419, 436, 438
topographic, 412, 435
torsion, 278, 362, 363
torticollis, 363
Toshiba, 235
total-trapping antioxidant potential, 480
Tourette's syndrome, 852
toxic, xvii, 55, 408, 419, 441, 480, 481, 485, 488, 490, 530, 542, 544, 609, 610, 612, 614
toxic effect, 634
toxic metals, 610, 612
toxic products, 811
toxic radicals, 480
toxicity, 29, 419, 423, 424, 428, 438, 483, 485, 487, 488, 491, 492, 493, 495, 497, 538, 544, 552, 553, 614, 617, 632, 636, 637, 638, 639, 647, 648, 650, 656, 658, 662, 664, 666, 672, 674, 679, 685, 811, 812, 841, 842, 845, 848, 850
toxin, 29, 312, 355, 363, 611, 694, 696, 706, 714
toxins, xvii, 530, 609

toys, 821
TP53, 668, 669
trace elements, xvii, 610, 615
tracking, 783
tracks, 241, 243, 244, 248
trade, 11
trade-off, 11
traditions, 790
traffic, 664, 665, 666, 721
trafficking, xviii, 529, 530, 618, 634, 636, 641, 642, 644, 648, 649, 650, 656, 666, 667, 671, 674, 678, 817
training, 118, 119, 120, 121, 122, 125, 126, 127, 129, 130, 142, 239, 428, 501, 502, 503, 507, 511, 512, 513, 514, 516, 517, 518, 519, 548, 563, 564, 565, 566, 567, 568, 581, 587, 590, 594, 600, 603, 604, 698, 708, 710, 712
training programs, 603
traits, 27, 149, 346, 350
trajectory, 237, 239, 241, 243, 250, 326, 356, 375, 752, 770
trans, 292, 531, 539
transcranial direct current stimulation, 262, 396
transcranial magnetic stimulation, xiii, 28, 41, 255, 270, 293, 294, 295, 296, 297, 298, 299, 300, 301, 302, 320, 330, 331, 332, 333, 334, 335, 336, 349, 396
transcription, 527, 531, 532, 636, 640, 641, 642, 648, 650, 651, 652, 656, 660, 661, 662, 670, 671, 674, 675, 681, 802, 810, 811, 817, 849, 853, 856
transcription factors, 640, 641, 652, 660, 661
transcriptional, 538
transducer, 91
transduction, 10, 525, 528, 530, 817
transfer, xv, 77, 165, 168, 181, 382, 481, 499, 501, 502, 503, 513, 514, 519
transference, 514
transformation, xii, 216, 448, 463
transformations, 448, 463
transforming growth factor, 557, 558, 837
Transforming Growth Factor beta, 479
transgene, 644, 685
transgenic, 478, 482, 483, 484, 492, 494, 495, 544, 549, 557, 558, 804, 809, 810, 811, 816, 818, 820, 821, 822, 826, 828, 831, 832, 838, 839, 841, 842, 843, 844, 845, 846, 847, 848, 849, 850, 852, 853, 854, 856
Transgenic, 549, 550
transgenic mice, 478, 482, 483, 484, 492, 495, 544, 549, 557, 558
transgenic mouse, 483, 557

transition, 116, 130, 132, 223, 239, 241, 262, 266, 386, 387, 526, 533, 628
transition metal, 526
transitions, 241, 320
translation, 201, 210, 527, 531, 532, 539, 540, 589, 591, 592, 596, 597, 811
translocation, 639
transmembrane, 65, 476, 477, 547
transmembrane region, 476
transmission, xii, 68, 97, 99, 118, 139, 172, 173, 177, 180, 187, 191, 193, 256, 271, 290, 292, 383, 417, 421, 529, 618, 620, 627, 648, 692, 707, 805, 806, 817
transparency, 284
transparent, 146, 262
transplantation, 743, 807, 820, 833
transport, xviii, 10, 34, 81, 103, 168, 172, 476, 490, 528, 531, 545, 548, 611, 634, 636, 641, 642, 643, 645, 646, 647, 653, 666, 667, 671, 678, 809, 811, 817, 837, 841
transportation, 530
trauma, 266, 295, 312, 804, 855
travel, 768
treatment-resistant, 22, 47, 62, 267, 281, 295, 296, 298, 346, 395, 396, 398, 399, 401
treatments, 180, 187
trees, 206, 208, 209, 764, 765
tremor, xiii, 227, 228, 241, 251, 252, 254, 256, 257, 258, 266, 267, 268, 269, 270, 274, 275, 276, 277, 284, 285, 295, 297, 298, 299, 300, 301, 302, 305, 307, 308, 311, 312, 313, 314, 316, 353, 354, 362, 366, 373, 389, 392, 622, 623, 696, 806, 808, 818
trend, xvi, 24, 129, 130, 206, 207, 208, 468, 562, 601, 603
trends, 734
trial, 55, 64, 70, 117, 120, 122, 123, 124, 126, 128, 129, 130, 131, 132, 133, 134, 135, 267, 279, 280, 281, 294, 295, 296, 298, 299, 311, 313, 322, 327, 328, 329, 331, 336, 350, 355, 398, 399, 400, 401, 431, 448, 458, 460, 462, 463, 471, 474, 482, 483, 484, 485, 486, 487, 488, 489, 493, 494, 495, 496, 497, 502, 507, 510, 511, 514, 515, 516, 517, 518, 556, 571, 697, 707
trial and error, 431
trigeminal, 82, 272, 273, 293, 327, 331
trigeminal nerve, 327, 331
triggers, 124, 313, 417, 661
trisomy, 476
trisomy 21, 476
trophic support, 832
tropism, 836, 856

trust, 597
trypsin, 523, 524
tumor, 491, 671, 856
tumor necrosis factor, 491
tumors, 355
tumours, 832
turnover, 74, 86, 137, 600, 603, 605, 607
twins, 24, 35, 44, 152, 627, 652
two-dimensional, 535
two-way, 138, 593
type 1 diabetes, 559
Type I error, 445
typology, 748
tyrosine, 29, 42, 649
tyrosine hydroxylase, 29, 42

U

U.S. Department of Labor, 602, 603, 606
ubiquitin, 526, 536, 537, 545
Ubiquitin, 526, 536, 554
ubiquitous, 206, 539
uncertainty, 130, 137, 138, 259, 273, 284, 564, 719, 725, 726, 745, 750, 755, 758, 759, 760, 774, 776, 785, 792, 797
unconditioned, 109, 116, 120, 122, 124, 125, 128, 130, 131, 132, 133, 134, 137
unconditioned response, 130, 131, 132, 133, 134
underlying mechanisms, 260, 424
unfolded, 540
unfolded protein response, 540
uniform, 205, 625
United, xvii, 344, 345, 503, 561, 563, 568, 573, 574, 581, 585, 600, 605, 711, 712, 740, 741
United Kingdom (UK), xvii, 151, 253, 344, 345, 367, 369, 561, 563, 565, 568, 570, 573, 574, 578, 581, 584, 725, 739, 789, 791, 795
United Nations, xvii, 574, 585
United States (USA), 31, 33, 38, 39, 40, 43, 45, 61, 65, 68, 103, 106, 107, 138, 166, 168, 169, 171, 176, 177, 191, 192, 195, 196, 210, 229, 235, 239, 243, 268, 328, 344, 356, 391, 393, 432, 435, 490, 491, 494, 503, 517, 518, 521, 536, 538, 539, 554, 557, 563, 578, 581, 587, 588, 596, 597, 600, 605, 609, 672, 705, 711, 712, 718, 719, 720, 721, 728, 729, 731, 740, 741, 775
universality, 590
university students, 219
unmasking, 274
unplanned pregnancies, 574, 771, 777
unpredictability, 130, 289
updating, 129, 217, 508

uranium, 610
urban, xvii, 587
uric acid, 480
urinary, 236, 249, 251, 343
urinary tract, 249
urinary tract infection, 249
urine, 251, 346, 355, 481, 487, 549, 557, 844
uterus, 659

V

Vaccination, 483, 484, 496
vaccinations, 611
vaccine, 553, 611
vagus, xiii, 320, 376, 395, 396, 397
vagus nerve, xiii, 320, 395, 396, 397
valence, 18, 150
validation, 281, 518, 555, 556, 571, 596, 794
validity, 2, 11, 225, 458, 473, 589, 590, 593
valuation, 117, 176, 518, 710, 794
values, 87, 90, 99, 101, 118, 119, 121, 122, 123, 127, 130, 131, 146, 203, 219, 221, 229, 240, 250, 306, 341, 359, 360, 369, 390, 465, 588, 594, 595, 746, 748, 749, 751, 752, 753, 755, 756, 757, 760, 765, 766, 771, 772, 773, 774, 778, 780, 781, 783, 784, 789, 792
valve, 564
variability, 11, 12, 15, 16, 27, 242, 286, 289, 373, 446, 447, 448, 449, 457, 468, 482, 593, 594, 619, 664, 678, 689
variable, ix, 1, 105, 218, 361, 369, 448, 451, 456, 463, 517, 826
variables, 18, 53, 263, 357, 414, 415, 428, 442, 456, 462, 472, 603, 614, 751
variance, 305, 309, 414, 457, 463, 468, 523, 588, 592, 594, 626, 678, 807
variation, 38, 205, 206, 208, 211, 212, 213, 445, 448, 456, 468, 592, 781, 853
variations, xii, 10, 206, 216, 235, 580, 601, 632, 664, 679, 720, 765, 768, 777, 782
varieties, 75
vascular dementia, 491, 541, 543, 546, 547, 548, 550, 551, 552, 553, 555, 556, 558
Vascular dementia, 481, 541, 542, 547, 552, 555
vascular disease, 480, 552
Vascular disease, 542
vascular diseases, 480
vascular risk factors, 543, 552
vascular system, 387
vascular wall, 543, 549
vasculature, xx, 558, 799, 800, 824
vasculogenesis, 550
vasodilatation, 388

vector, 157, 166
VEGF, 802, 803, 805, 822, 830, 835, 840
Venezuela, 670, 687, 688, 689, 807
venlafaxine, 697
ventricle, 228, 234, 235, 236, 247, 248, 367, 801, 830, 834, 839, 849
ventricles, 12, 235
ventrolateral prefrontal cortex, 25, 26, 150, 155, 157, 212, 518
verbal fluency, 1, 16, 173, 225, 309, 311, 314, 369, 372, 375, 412, 420, 487, 495, 503
Vermont, 469
versatility, 771, 777
vesicle, xviii, 21, 417, 424, 618, 643, 667, 678, 811, 817
vessels, 543, 544, 545, 546, 549, 550, 551, 554, 557, 826
veterans, 195
victims, 284, 285, 500, 518, 722
Victoria, 47
video-recording, 239
videos, 512
Viking, 795
violence, 719, 720, 734
violent, 264, 376
virus, 33, 156, 165, 168, 613
viruses, 382
viscosity, 546, 551
visible, 11
vision, 156, 157, 224, 289, 292, 368, 501, 564, 569, 788
visual, 372, 376, 378
visual area, 212
visual attention, xv, 56, 194, 378, 444, 448, 467, 468, 474
visual memory, 16, 412, 414, 416
visual perception, 290
visual stimuli, 105, 161, 165, 166, 169
visual stimulus, 153, 157, 165, 179
visual system, 169, 170, 211, 212
visualization, 367, 388
visual-spatial attention, 195
visuospatial, 68, 70, 157, 285, 379, 428, 513, 548
vital, xi, 178, 188
vitamin E, 487
vitamin supplementation, 495
vitamins, 694
vocabulary, 511, 590, 591
vocalizations, 80, 93, 113, 164
voice, 103, 159, 164, 257, 269, 276, 301, 400
voiding, 680, 758, 777
vomeronasal, 103
vomiting, 612

voxel-based morphometry, 12, 17, 29, 31, 33, 35, 36, 39, 43, 45
vulnerability, xvii, 8, 9, 24, 29, 38, 351, 419, 425, 434, 610, 614, 632, 633, 688, 727, 758, 760, 806, 813, 837, 844, 847, 851
Vygotsky, 509

W

wage level, 599
wages, 603, 604
waking, 180, 181, 186, 190
Wales, 570, 578, 605
walking, 73, 89, 90, 91, 94, 95, 96, 164, 307, 308, 309, 310, 311, 698
war, 789
warfare, 612
warts, 296
Washington, 57, 404, 442, 469, 515, 585, 596, 597, 615
water, 81, 138, 381, 383, 393, 412, 418, 420, 426, 707, 711, 713, 722, 804, 805
water maze, 138, 412, 418, 420, 426
WCST, 17, 424, 459, 469, 508
weakness, 313, 564, 622, 623, 698
wealth, 656
Wechsler Intelligence Scale, 461, 474
weight changes, 808
weight gain, 343, 400
weight loss, 401, 627, 695, 699, 818
weight reduction, 692
welfare, xv, 499, 514, 601
welfare state, 601
wellbeing, 757, 758
well-being, 423, 439, 606, 726, 727, 732
western blot, 662
Western blot, 524
western countries, 757
Western countries, 578
white matter, 11, 12, 25, 34, 35, 39, 204, 213, 257, 261, 262, 276, 281, 283, 284, 297, 305, 314, 412, 413, 425, 547, 549, 551, 559, 579, 825, 848, 853
WHO, 589, 591, 597
wholesale, 754
wild type, 183, 655, 656, 657, 658, 659, 661, 662, 670, 684, 835
Wilson's disease, 355
wine, 487
winter, 8
wireless, 264, 292
wireless connectivity, 264
wires, 263, 291

Wisconsin, 17, 49, 175, 371, 412, 424, 459, 508
Wistar rats, 550
withdrawal, ix, x, 2, 10, 48, 79, 80, 141, 142, 143, 144, 420, 428, 489
wives, 722
women, 24, 36, 367, 369, 428, 429, 553, 574, 575, 577, 581, 582, 584, 585, 730, 731, 751, 756, 758, 759, 763, 773, 774, 775, 783, 788, 791, 792, 794, 795, 805, 808, 855
wood, 88
word processing, 371
word recognition, 417
words, 178, 341
work, 341, 349, 355
workers, 398, 399, 564, 590, 599, 603, 604, 605, 606, 683, 700, 711, 719
workforce, 599, 600, 606
working women, 601
workplace, 789
workstation, 235
worldwide, 256, 275, 276, 575, 678, 718, 725, 730, 736, 737, 743, 775
worms, 666
worry, 143, 145, 149, 773, 778
wound healing, 526
wrists, 721, 728
writing, 374, 411, 547, 579

X

X-axis, 206, 207, 208
xerostomia, 706, 707, 708, 710, 711, 712, 713
X-linked, 756, 787, 792
X-rays, 311

Y

YAC, 638, 647, 819, 820, 827, 828, 835, 842, 856
Yale University, 176
yeast, 641, 658, 662, 666, 835
yes/no, 708
yield, 278, 445, 448, 452, 459
young adults, 420, 430, 471, 563
young people, 580, 581, 582
yttrium, 344
Yugoslavia, 720, 742

Z

zinc, 477, 492, 495, 497
zippers, 849
ziprasidone, 22
Zn, 480, 481